Competing in World-class Manufacturing:

America's 21st Century Challenge

Competing in World-class Manufacturing:

America's 21st Century Challenge

NATIONAL CENTER FOR MANUFACTURING SCIENCES
Ted Olson, VP Technology Transfer

Craig Giffi — Deloitte & Touche
Aleda V. Roth — Duke University
Gregory M. Seal — Deloitte & Touche

BUSINESS ONE IRWIN
Homewood, IL 60430

Senior editor: Susan Glinert Stevens, Ph.D.
Production manager: Carma W. Fazio
Printer: Maple-Vail Book Manufacturing Group

Library of Congress Cataloging-in-Publication Data

Giffi, Craig

Competing in world-class manufacturing : America's 21st century challenge / National Center for Manufacturing Sciences : Craig Giffi, Aleda Roth, Greg Seal.

p. cm.

Includes index.

ISBN 1–55623–401–5

ISBN 1–55623–516–X (Deloitte Touche Edition)

1. United States—Manufactures. 2. Competition, International. I. Roth, Aleda V. II. Seal, Greg. III. National Center for Manufacturing Sciences (U.S.) IV. Title.

HD9725.G53 1990

338.0973—dc20 90–45315

Printed in the United States of America

2 3 4 5 6 7 8 9 0 MP 7 6 5 4 3 2 1 0

Table of Contents

Foreword vii
Preface ix
Acknowledgements xv

1 The World Surpasses American Manufacturing 1

Competitiveness and World-class Manufacturing. Establishing an Operating Framework. Organization of the Remaining Chapters.

2 The Quality Revolution 21

Quality as a Way of Managing Business. Shifting Strategies for Achieving Quality. Quality Gurus. Quality Toolkits. Practices in Japan. Design for Customer Satisfaction. Conclusion.

3 Manufacturing Strategies and Agendas 71

The Birth of Business Strategy. Manufacturing Strategy for the 1990s. Revamping Strategic Directions and Priorities: U.S. on the Defensive. Building Strategic Bills of Materials. Closing the Competitive Gaps. Future Directions and Key Action Programs. Strategy Development Process. Conclusion.

4 The Power of Manufacturing 113

Staking Out the Global Competitive Agendas. Winning through Manufacturing. Linking Perceived Quality and Performance. Time-based Competition and Performance. The Impact of a Lean Operation on Manufacturing Performance. The Impact of Engineering Change Orders on Performance. The Impact of Holistic Manufacturing Strategies on Performance. Conclusion.

5 Reconciling Accounting and Manufacturing 147

Defining Manufacturing Excellence. What are Good Performance Measurement Systems? How Do Current Financial Measurement Systems Measure Up? Toward New Measures of Financial Performance. Today's Trends, Tomorrow's Practices. Conclusion.

6 Value-added Performance Measurement 183

Productivity Measurement. Removing Time Barriers. Measuring Quality Progress. Conclusion.

7 Adopting Best Manufacturing Practices 203

"Best" Manufacturing Practices. Restructuring the Design Engineering Process. Just-in-Time Manufacturing Practices. Material Requirements/Resources Planning. Product and Process Simplification with Value Analysis/Value Engineering. Maintaining the Production Machine. Conclusion.

8 Restructuring the American Workplace 241

The Shrinking Role of Headquarters Staff. The Many Shapes of Restructuring. The Process of Organizational Change. Conclusion.

9 Developing Human Assets 263

The Changing Workforce. Manufacturers' Responses to the Changing Workforce. New Methods of Workforce Management. The Extinction of the "American Made" Supervisor. The Quandary Over Engineering Resources. Rewards and Recognition. Conclusion.

10 Technology Investment and Realization 301

The National Challenge. Manufacturing Technology. Current Progress. What Has Caused Our "Falling Out" with Advanced Manufacturing Technology? A Framework for Success. The Essence of Failure. The Implications of Success. Conclusion.

11 Attributes of World-class Manufacturing 327

Operating Principles of World-class Organizations. The Challenge for the Future.

Notes 341

Bibliography 353

Index 399

Foreword

Competing successfully in the global marketplace during the 1990s and into the twenty-first century will require many major changes in American manufacturing companies. These changes will involve enlightened manufacturing strategies employing world-class capabilities, continuous quality improvements, and the increased utilization of advanced technology.

Until now, few, if any, publications have addressed these issues in a format that is easy to read and of sufficient detail to adequately cover the multidimensional problem of such integrated change. This book describes the operating principles that world-class companies have followed to establish manufacturing preeminence in global markets. It presents a framework, centered on quality and the customer, for initiating these principles, and then describes the management approach that develops and promotes superior manufacturing strategies. It analyzes manufacuring capabilities, describes the best manufacturing practices to put in place, relates manufacturing strategies to both technology and human resource management, and discusses problems encountered in implementing these practices. A summary of operating principles of world-class manufacturers provides a global perspective of what will be required to carry companies into the twenty-first century.

At stake is not only the survival of countless companies—but also the survival of American leadership. Our companies, large and small, have to emerge as leaders to ensure our national economic and defense security. This book has the potential of serving as a launching pad for substantial improvement and increased competitiveness.

Thomas J. Murrin
Deputy Secretary
U.S. Department of Commerce

PREFACE

Forty-five years ago, America's industrial might was second to none. Buoyed by a strong manufacturing base, a vast reservoir of technical know-how, and the temporary weakness of war-torn Europe and Japan, the United States dominated the global economy. In fact, the U.S. accounted for roughly half of the world gross national product in the years immediately following World War II.

But now the U.S. is rapidly losing its leadership position in world markets. During the last decade, U.S. manufacturers have suffered devastating losses to the Japanese and other foreign competitors in such basic industries as steel, automobiles, machine tools, electronics, chemicals, and textiles. As a result, America's trade deficit in 1989 soared to $108.6 billion, while its staggering national debt jumped to $152 billion.

Theories abound concerning why the U.S. has failed to keep pace. The most likely culprit is the diversification of U.S. manufacturing companies and their focus on short-term financial results. Most of these problems date back to the 1950s. At that time, American companies were reaping huge profits from pent-up demand for a wide variety of consumer goods that had been unavailable during the war. But, management failed to invest those funds in modernizing its manufacturing operations.

There are several reasons why this occurred. Europe and Japan were still in a shambles, so U.S. companies had most markets to themselves. U.S. companies focused too much attention on short-term results and too little on improving their manufacturing operations. No chief executive officer wanted to explain to investors that quarterly earnings were down because of reinvestment into an already profitable operation.

American industry was also hurt by the concept of diversification and the growing popularity of portfolio management theory. Prior to World War II, U.S. companies focused almost exclusively on one industry and were committed to excelling in that field. But in the late 1950s, they began diversifying, buying and selling companies in much the same way that investors trade stocks and bonds. As a result, management rarely invested the time and money required to fix troubled companies. When one of its operations lost market share or saw its margin decline, management simply sold it and used the proceeds to buy another company.

Those problems were compounded by the emergence of a tough new competitor in the 1970s. By then, Japan had rebuilt its war-ravaged economy and was equipped with a more modern manufacturing base than the U.S. Armed with the capital and technology to compete across the industrial spectrum, the Japanese also had the advantage of a lower wage base. Because of its relatively small domestic market, the country set a goal of gaining success in international competition. Japan, however, understood the importance of manufacturing in creating added product value (their only viable export) and producing national wealth.

To achieve that objective, the Japanese invited an obscure American statistician, W. Edwards Deming, to help them improve productivity and quality in the 1950s. He taught them to eliminate waste, not just physical scrap, but capital waste and waste from not doing the job right the first time. Deming also stressed the manager's role in getting everyone involved in upgrading productivity and quality. Soon, the Japanese were producing goods that were less expensive and widely perceived as offering better quality than U.S. products.

But, the seriousness of the threat from Japan was not apparent in the 1970s. The U.S. managed to maintain its share of the world market by devaluing the dollar. Although U.S. steel, automobile, and textile companies lost a sizable share of their domestic markets to imports during these years, the weak dollar enabled them to offset those losses with a growing share of overseas trade.

When the value of the dollar dropped in the 1980s, the results were predictable. The United States saw its share of the world market shrink. Indeed, the U.S. now has only a 19.2 percent share of the global market compared to 21.5 percent in 1980 and 26 percent in 1960. In 1981, the U.S. was a high-technology exporter to the tune of $27 billion. By 1986, the high-technology deficit was $2 billion.

To meet the Japanese challenge, U.S. companies began looking for ways to become more competitive. They soon discovered an easy remedy. To take advantage of low wages in the developing world, American companies began building components outside the U.S. Eventually, that led to the construction of U.S. factories overseas. Those moves helped cut costs. But, they also strengthened our foreign rivals by giving them access to our manufacturing and product technology.

The results have been disastrous. The consumer electronics industry has fallen victim. The domestic automobile and machine tool manufacturers are also in serious trouble. Both continue to lose market share to offshore rivals. And soon the Japanese plan to launch assaults on two of our healthiest industries, computers and aerospace.

The U.S. must regain a globally competitive position in manufacturing if it expects to maintain its current standard of living. To make matters worse, many U.S. business leaders are still unaware of the problem, unsure of how to address it, or unwilling to step up to the challenge. If they do not act soon, America's long-term future will be at risk. The United States could follow in the footsteps of England, declining from a position of industrial and economic leadership to a second-rate status among industrialized nations.

Purpose of the Book

The purpose of this book is two-fold. First, it sounds an alarm to the North American business community about the magnitude of the threat posed by overseas competitors and the complacency to date of much of the North American manufacturing base in responding to that threat. Second, and more importantly, it describes various pathways North American manufacturers can follow to become world-class competitors and directly confront this foreign competition.

The material presented in this book is based on a research study commissioned by the National Center for Manufacturing Sciences (NCMS) in 1988. Conducted by Deloitte & Touche, the study defined world-class manufacturing and developed a vision of the manufacturing agenda required to make America's manufacturers successful in competing on a global basis into the twenty-first century.

The bulk of the research for the initial study was conducted over a six-month period stretching into early 1989. The study focused on the elements of world-class manufacturing from the manufacturers' perspectives. Extensive interviews were conducted with executives of both NCMS member companies and non-member companies. Field research was conducted to determine the critical components of manufacturing excellence and the impact of these components on business performance. Additionally, technology vendors and system integrators were interviewed to provide perspectives on state-of-the-art technology application.

The research study established a framework that underlies the operations and strategies of world-class manufacturers. The framework is presented in Chapter 1. The book expands upon the research study by updating the data to the 1990 timeframe with an analysis of the status quo and a prescription for success into the twenty-first century in each area of the framework. This book is not to be construed as an exhaustive discussion of the topics presented or the issues raised. The material presented is what leading manufacturing executives, academicians and business consultants currently believe to be the best practices which, if successfully implemented, will help a company become a world-class competitor.

While this book has attempted to pull all of the best manufacturing practices together for the first time, it is merely a snapshot in time. The global marketplace and the practices utilized are quite dynamic and will undoubtedly change as we progress through the 90s.

As you read this book, the NCMS is continuing to pursue additional research to provide a better vision of the world of tomorrow as it relates to new products, new technologies, and new challenges from foreign competition. Again, the NCMS goal continues to be to help North American manufacturers become once again be the pre-eminent manufacturers in the global marketplace of the twenty-first century.

About the National Center for Manufacturing Sciences

The National Center for Manufacturing Sciences is a not-for-profit cooperative research corporation organized under the National Cooperative Research Act of 1984. The Center currently has a membership of more than 100 corporations committed to making U.S. and Canadian manufacturing globally competitive through development and implementation of next-generation manufacturing technologies. The effort has marshalled the support of the leading manufacturers, government, and the education and business communities.

The organization traces its roots to the mid-1980s. It was then that a group of interested organizations representing government and industry held a series of meetings to address growing concerns about America's ability to develop and competitively produce advanced technology for defense and commercial needs. The Center was incorporated in November, 1986.

Since that time, a formal organization has been developed. A staff of full-time professionals has been recruited from the top ranks of industry. A national research agenda has been established and a host of short-term and strategic research projects started. Cooperative activities with government, academia, and private manufacturing research organizations have been undertaken. The process of transferring technology and know-how to member companies has begun. The Ann Arbor, Michigan-based consortium has become the largest North American consortium, with the resources and strength to lead North American manufacturers to a position of renewed global competitiveness.

ACKNOWLEDGEMENTS

Our deepest appreciation goes to the National Center for Manufacturing Sciences, Manufacturing Practices Strategic Initiative Group for its vision in initiating this project and its experience, direction, and input to the original study and this book. Thanks to Paul Rexford, General Motors Corporation and chairman of the Project Committee; Ronald Lang, NCMS group director and project manager; Larry Cundy, Gilbert/Commonwealth, Inc.; Peter J. Joseph and Bill Lawrence, Texas Instruments, Inc.; Lawrence F. Kmiec, The Cross Company; Raymond J. Lipa, Bresson Rupp Lipa & Co.; Art Mason, Midwest Brake Bond Co.; David R. Nalven, AT&T; J. Robert Roark, Pratt & Whitney; Edward Sherwin, United Technologies; Pete Warren, Kingsbury Corporation; Harrison Williams, Fabreeka International; Brian Wilson, U.S. Air Force; R. Jeffrey Wilson, GE Aerospace & Electronics.

This book would not have been complete without the information derived from extensive interviews with Eli Karter, Mead Corporation; Peter J. Joseph, Texas Instruments; Robert F. Bescher, Pratt & Whitney; Peter Borrows, Foxboro Company; Fran Duverneux, Campbell Soup Co.; Ernest O. Vahala, General Motors Corporation; Berkley Merchant, Mentor Graphics; R. Jeffrey Wilson and Herbert Schneider, GE; Frank Cassidy, Digital Equipment Corporation; Allen Rose, Johnson & Johnson; David Auld, Baxter Healthcare; Jaap Johanssen, Van Leer Containers, Inc.; and Robert Badelt, Northern Telecom. Their comments, experiences, and interest in this project were both vital and gratifying. A hearty thanks also to all those organizations who responded to our surveys and to those who were willing to talk about, in print, both their positive and negative experiences on the road to becoming world-class organizations. They are mentors to us all.

We would like to acknowledge the editorial efforts of Francine Hyman, Communitec, Inc. and Hillary Handwerger, NCMS, and the rest of the

Communitec and NCMS staffs for helping us convert a state-of-the-art survey into a finished book. Thanks also to James Throop, Learnstar, Inc.; Dr. John Ettlie, The University of Michigan, School of Business Administration; and Dr. Robert Howell, New York University, Stern School of Business; for their comments and critique of the material. And thanks, finally, to Bruce Beavis, Paul Alvey, Michael Fradette, Paul Baier, Matt Ferko, Bill Ross, Sharon Nelson, and Peter Marton for all their efforts during the research phase of the project.

The National Center for Manufacturing Sciences is appreciative for the funding of this book made possible by NCMS member companies with an additional grant from Eastman Kodak Company.

Ted Olson
Craig Giffi
Aleda V. Roth
Gregory M. Seal

CHAPTER 1

THE WORLD SURPASSES AMERICAN MANUFACTURING

Manufacturing is vital for the economic health of the nation and for North American world leadership. It is discussed at the highest levels of government, studied by some of the most respected scholars in academia, and written about in most of the prominent journals published today. Books dealing with manufacturing have become best-sellers and, in corporate circles, manufacturing has emerged as a major concern of those charged with guiding the course of their companies into the future. This sudden increase in manufacturing interest is due to a single factor—increased international competitiveness.

In 1989, Japan out-invested the United States for the first time in history, as America failed to keep pace with its major trading partners. In 1989, Japan spent $549 billion to modernize and expand its industries, compared with $513 billion for the United States, even though the U.S. economy is more than twice the size of Japan's. "In effect," says Kent Hughes, president of the U.S. Council on Competitiveness, "Japan is putting twice the tools in the hands of the Japanese worker."

> America is looking like an aging athlete—still on top, but trying to ignore all the younger talent that is breaking into the lineup.

These were the words Hughes used as he presented the Council's *1989 Report on Competitiveness*. The report tracks U.S. economic performance since 1970 compared with Japan, West Germany, Great Britain, France, Canada, and Italy.

To understand America's competitive woes of today, it is important to briefly review its manufacturing successes and failures of the past century.

Since the mid-nineteenth century, American manufacturing has pioneered new ideas and methods in industry. The concept of interchangeable standard parts was first developed to mass produce firearms for the Civil War. Prior to that, most industrial goods were custom-made. Companies such as DuPont and Carnegie Steel developed new forms of corporate organization. DuPont was the first company to establish a divisionalized structure. Carnegie was one of the first to vertically integrate an industry from raw materials to the finished product. In industry after industry, American firms were at the leading edge.

Around the turn of the century, American business practices became internationally synonymous with efficiency and effectiveness. Henry Ford revolutionized the production of automobiles through the initiation of a new production methodology. Frederick Taylor created the theory of scientific management and the time study technique. Frank Gilbreth developed the motion study technique. Henry Gantt created the Gantt Chart. The achievements of these men are what became *modern management.*

Until the 1970s, modern management methods were a roaring success, and American management took for granted the superiority of its technology and management practices. In 1971, however, America registered a negative merchandise trade balance for the first time in this century (see Figure 1-1). By 1987, the negative balance was over $150 billion.

The U.S. advantage in gross dollars paid (GDP) per capita narrowed significantly by the late 1980s (see Figure 1-2). For the first time, America's global competitors came within striking distance on this critical measure.

FIGURE 1-1 Merchandise and Manufacturing Trade Balances, 1960 - 1988

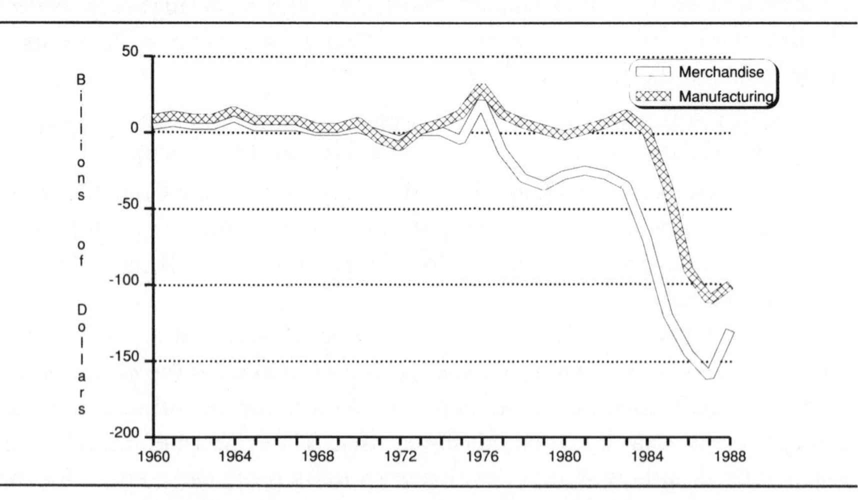

Source: U.S. Department of Commerce, Bureau of Economic Analysis, *Business Conditions Digest* (Washington DC: U.S. Government Printing Office), September 1989.

FIGURE 1-2 GDP per Capita in 1988 U.S. Dollars

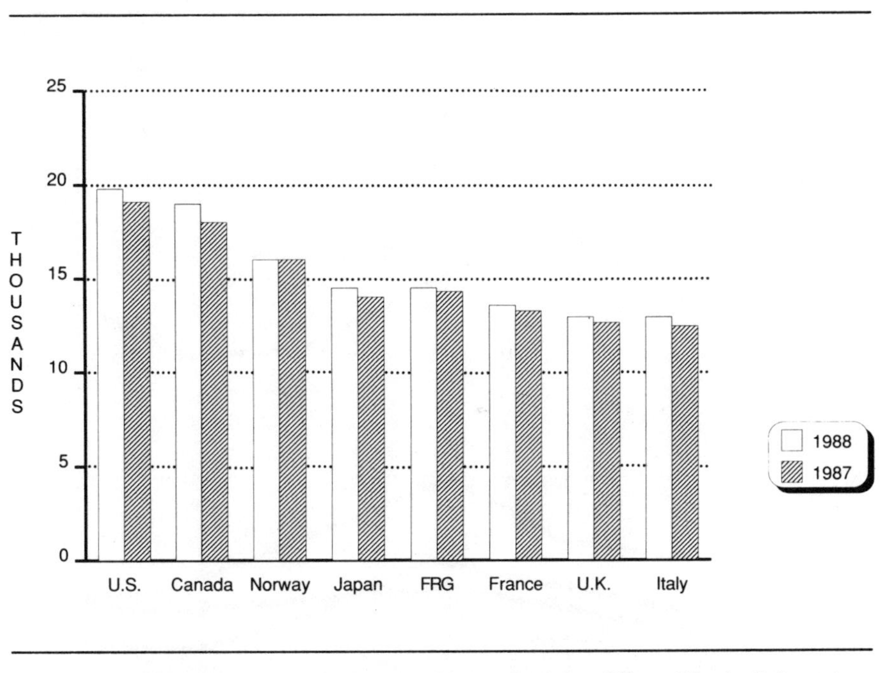

Source: U.S. Department of Labor, Bureau of Labor Statistics, Office of Productivity and Technology, unpublished data, August 1989.

The U.S. real GNP growth, however, is modest (see Figure 1-3). Relative to the rest of the world, the rate of U.S. productivity growth had been declining. By 1988, annual productivity growth in manufacturing was smaller in the U.S. than in any of its global rivals. Japan, once ridiculed for cheap products and poor quality, became America's worst competitive nightmare. Alone among advanced industrial countries, Japan managed in the 1980s to combine great productivity growth in manufacturing with rising manufacturing employment, wages, benefits, and output (see Figure 1-4).[1]

In several high-visibility industries, such as consumer electronics, automobiles, semi-conductor equipment, and machine tools, Japan has surpassed the U.S. to gain a global leadership position.

Semi-Conductor Equipment: In the late 1970s, the U.S. held a tremendous leadership position in the semi-conductor equipment market, owning more than 75 percent of the market. By 1988, the U.S. market share had

FIGURE 1-3 Real GNP Growth

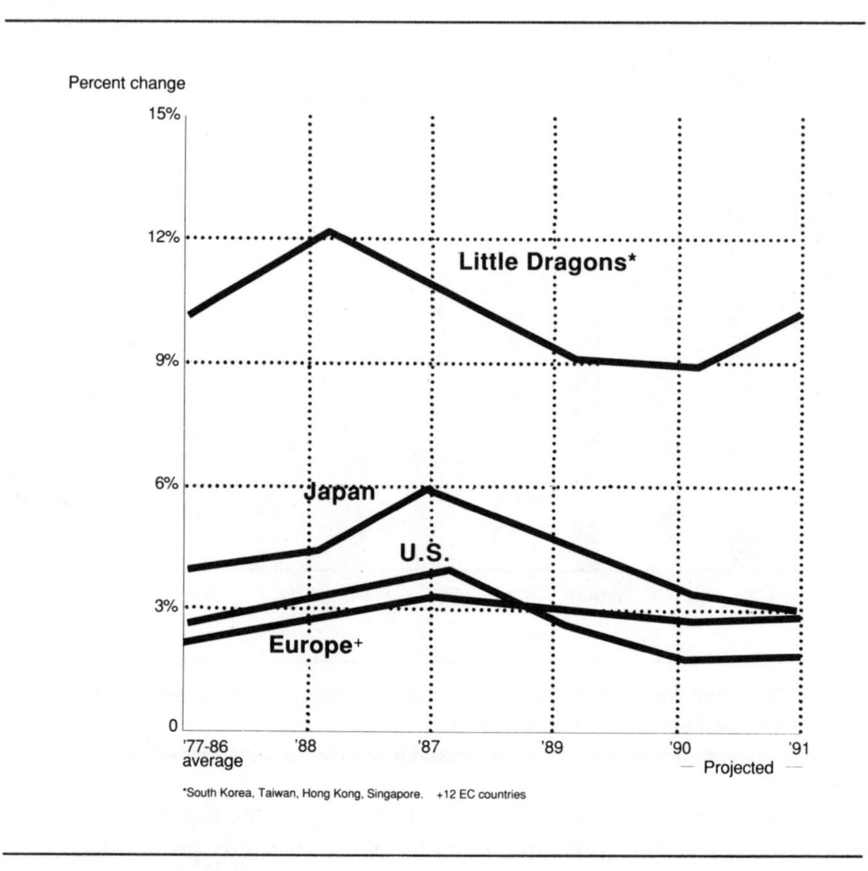

Source: S. Neumeier, "Markets of the World Unite," *Fortune,* July 30, 1990, p. 102. Reprinted with permission from FORTUNE Magazine, 1990.

dropped to 47 percent. By the end of 1990, it is estimated that Japan will hold the number one position.

Automobiles: In the early 1970s, Japanese cars were virtually nonexistent on America's highways. Some smaller Japanese companies, such as Honda, which had never even produced automobiles, were just beginning to enter the market. By the late 1980s, Honda's Acura Division had led the J.D. Powers customer satisfaction ratings three years running, and non-American makes accounted for 8 of the top 10 rated cars in 1989. As they enter the 1990s, the Japanese now command over 26 percent of the U.S. car

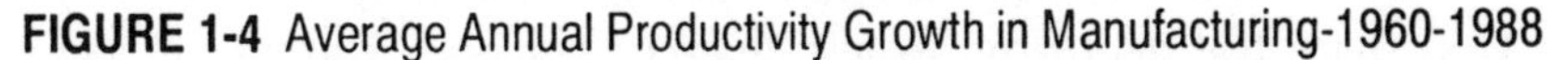

FIGURE 1-4 Average Annual Productivity Growth in Manufacturing-1960-1988

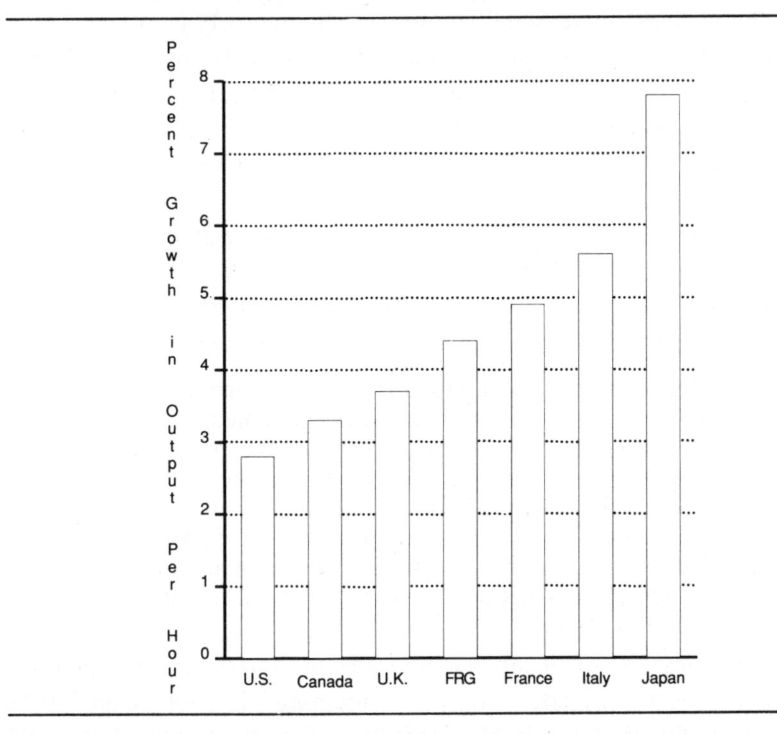

Source: U.S. Department of Labor, Bureau of Labor Statistics, "International Comparisons of Manufacturing Productivity and Labor Trends, 1988," June 1989, Table 1.

market, with many analysts questioning the ability of all Detroit's Big Three to survive the decade intact.

Consumer Electronics: In the consumer electronics industry, the once dominant position of the U.S. is a thing of the past. In 1990, the U.S. is no longer a competitor, much less a competitive threat. In 1955, 96 percent of all radios sold in the U.S. were made in the U.S. By 1965, the proportion was down to 30 percent. By 1975 it was near zero.[2] U.S. manufacturers likewise abandoned the $6.0 billion plus commercial television market to the Japanese, leaving only Zenith to compete. The U.S. entered, then backed out of, the booming VCR marketplace. The Japanese now hold such a competitive advantage in the consumer electronics market place that it is unlikely that the U.S. will ever again be a serious competitor.

Machine Tools: In the machine tool industry, the U.S. dominated the 1960s with over a 25 percent share of the global market. By the mid-1980s, the U.S. share had decreased to less than 10 percent. During the same period, the Japanese increased their global share to nearly 25 percent, after having a less than 10 percent share in the late 1960s. In the robotics industry, the Japanese are now the clear leaders with over 50 percent of the global market. By 1984, nearly 67 percent of the world's installed base of industrial robots were of Japanese manufacture, compared to less than 15 percent manufactured in the U.S.[3]

What happened? How could the U.S. have allowed Japan to take over these major industries? Robert Hayes and Steven Wheelwright, in their book *Restoring Our Competitive Edge: Competing Through Manufacturing*, argue that the U.S. grew complacent during the 1970s and 1980s.

> This complacence (one might go so far as to call it arrogance) tended to blind U.S. managers . . . to the rapid improvements taking place elsewhere. It kept them from studying other countries and/or from making fundamental changes in their own approaches. The shock of suddenly discovering that the United States might not be the repository for all the best and brightest approaches to managing either individual companies or the national economy undoubtedly contributed to American society's sense of malaise and its business community's loss of confidence.[4]

Complacency and arrogance stemmed from American manufacturers always being the world leaders. Why fix something if it is not broken? Why adopt new strategies and technologies if the old are still serving you well?

Competitiveness and World-class Manufacturing

The manufacturing game today has changed from one in which the field is defined by national boundaries to one in which there are no boundaries. Manufacturers from every industrialized nation in the world are expanding their markets in territories which the U.S. once considered safe and sheltered. The market today is the world. Companies who want to compete successfully are striving to become *world-class manufacturers*.

Defining "World-class Manufacturing"

Few references to or uses of the term *world-class* in the context of manufacturing can be found prior to 1986. Until then, *manufacturing excellence* was the most common way to refer to the goal of achieving superior manufacturing capabilities.

Each of the more recent books on manufacturing has its own definition of world-class manufacturing. Robert Hayes, Steven Wheelwright, and Kim Clark, in their book *Dynamic Manufacturing*, identify these attributes of world-class manufacturers:

- **Becoming the best competitor.** "Being better than almost every other company in your industry in at least one aspect of manufacturing."
- **Growing more rapidly and being more profitable than competitors.** "World-class companies can measure their superior performance by observing how their products do in the marketplace and by observing their cashbox."
- **Hiring and retaining the best people.** "Having workers and managers who are so skilled and effective that other companies are continually seeking to attract them away from your organization."
- **Developing a top-notch engineering staff.** "Being so expert in the design and manufacture of production equipment that equipment suppliers are continually seeking one's advice about possible modifications to their equipment, one's suggestions for new equipment, and one's agreement to be a test site for one of their pilot models."
- **Being able to respond quickly and decisively to changing market conditions.** "Being more nimble than one's competitors in responding to market shifts or pricing changes, and in getting new products out into the market faster than they can."
- **Adopting a product and process engineering approach which maximizes the performance of both.** "Intertwining the design of a new product so closely with the design of its manufacturing process that when competitors 'reverse engineer' the product they find that they cannot produce a comparable one in their own factories without major retooling and redesign expenses."
- **Continually improving facilities, support systems, and skills that were considered to be "optimal" or "state-of-the-art" when first introduced,** so that "they increasingly surpass their initial capabilities. This emphasis on continual improvement is the ultimate test of a world-class organization."[5]

in their book, *The Spirit of Manufacturing Excellence—An Executive Guide to the New Mind Set*. They also identify "lower cost, higher quality, better service, and more flexibility than competitors" as the factors behind the success of leading Japanese companies both in Japan and in the U.S.[6]

Richard Schonberger offers a third definition in his book, *World-class Manufacturing: The Lessons of Simplicity Applied*. "Today there is wide agreement . . . that continual improvement in quality, cost, lead time, and customer service is possible, realistic, and necessary," and that "one more primary goal, improved flexibility, is also part of the package." "With agreement on the goals, the management challenge is reduced to speeding up the pace of improvement."[7]

The concepts of total quality, Just-in-Time (JIT) manufacturing, and people involvement constitute the substance of manufacturing excellence, according to Robert Hall, author of *Attaining Manufacturing Excellence*. Hall believes these concepts represent a fundamentally different way of operating a company and are not techniques that can be grafted onto a current organization. He stresses the importance of continuous improvement and concludes by stating that "one interesting aspect of manufacturing excellence is that if such a thing is attainable at all, it will be by those who realize that no such condition exists."[8]

World-class manufacturers, then, are those that are able to create high-value products and earn a superior return over the long run through the application of competitive strategies. World-class manufacturers today are applying concepts designed to demolish the obsolete methods, systems, and cultures of the past that have impeded their competitive progress. In their place, these manufacturers are building a more effective competitive structure, using improved tools, materials, and techniques, on a solid foundation of human resources, organizational development, and cultural understanding. The result in each case is an organization capable of competing on a global basis.

Establishing an Operating Framework

World-class manufacturers, regardless of industry or size, operate within a common framework. This framework, shown in Figure 1-5, was derived from an exhaustive study of the strategies and operating practices of dozens of leading manufacturers in the U.S. and around the globe. At its center, the framework highlights quality and the customer. *A commitment to quality and the customer affects every aspect of the organization:*

- Management Approach
- Manufacturing Strategy
- Manufacturing Capabilities
- Performance Measurement
- Organization
- Human Resources
- Technology

Each of these elements is introduced in the following paragraphs and discussed in detail in subsequent chapters.

FIGURE 1-5 World-class Manufacturing Framework

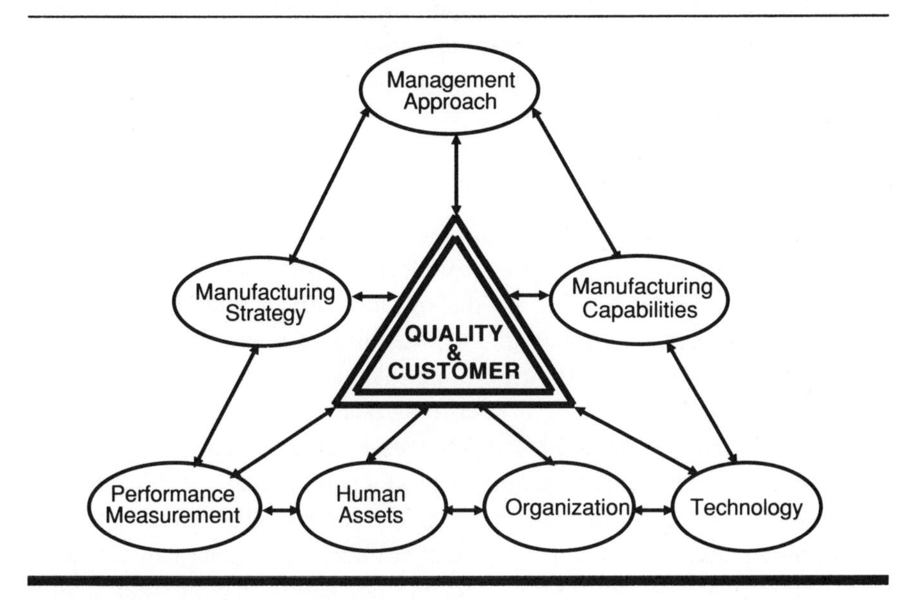

Quality and Customer

Manufacturers in every corner of the world have made quality a top priority. Quality expectations are on the rise, as are manufacturers' abilities to deliver quality products. The Japanese have demonstrated that superior quality can be attained with current technologies and known techniques. It is reasonable to expect that the 1990s will bring levels of quality never before achieved, and that quality will be required not just of world-class companies but of all companies competing in open markets.

World-class companies define quality from their customers' perspectives. These companies strive to remain close to their customers; closeness to customer is critical to developing and understanding quality requirements. While conformance to specifications is important, world-class companies understand that quality is what your customer says it is. The axiom: "conformance quality drives perceived quality and perceived quality drives market success" is well understood by world-class companies around the world. These companies have found that achieving superior quality, however, is like hitting a moving target. To be successful over the long run, a company must go beyond product and even process quality to build customer loyalty.

Matching marketing, engineering, and manufacturing capabilities is what differentiates world-class competitors. Determining what the customer needs and which attributes add value to the product, and creating operations capable of satisfying those requirements are characteristic of world-class companies.

Management Approach

A hallmark of world-class manufacturers is a closeness to both the product and the critical process technology. World-class manufacturers realize that an understanding of the product drives manufacturing and service requirements, and that an understanding of manufacturing process capabilities can be used to define competitive advantages.

Marketplace leaders have a keen understanding of the products they manufacture and the manufacturing processes critical to their current and future success. This understanding of product and process enables them to differentiate between effective and ineffective strategic moves regarding the implementation of new process technology. Those that can articulate the competitive elements of process technology carefully guard their unique technological advantage.

Manufacturing Strategy

Long held in contempt by manufacturing executives, strategic manufacturing planning is returning to its rightful place in world-class companies.

The importance of developing a realistic manufacturing strategy cannot be overstated. The vision of where you are going and how you are going to get there is fundamental to success.

Global trends that widen gateways into nontraditional markets, move toward competitively favorable trade agreements and partnerships, and diminish national boundaries are forcing an increasing number of industries into the global arena. The Japanese have capitalized upon these trends to a greater extent than either the Europeans or the Americans. Research has shown that Japan places significantly greater emphasis on the development of new products for new markets than other world regions, including the U.S. and Europe.[9]

The Europeans appear to be ahead of the U.S. in breaking into Japanese markets. From 1986 to 1987, Japanese imports of European products increased nearly 26 percent measured in dollars, or 9 percent measured in yen, while U.S. imports increased only 8 percent in dollars and declined when measured in yen.[10] The Europeans still have a long way to go, but they are rapidly becoming global competitors. The South Koreans, too, are now nudging their way into the global arena.

The need to meet or exceed the evolving standards set by world-class manufacturers is most apparent in highly visible markets such as electronics and automobiles. Here, the products of various competitors are in head-to-head competition and the customer chooses the victor. Although manufacturers who supply intermediary parts and components are less likely to perceive the need to change their practices, they may be more vulnerable to global pressures. Intermediary parts manufacturers have been forced to comply with increasingly stringent standards of performance. The importance of manufacturing strategy runs vertically through the entire supplier base.

Manufacturing Capabilities

Many companies are focusing on a first-level, time-based capability, such as dependable on-time delivery. This focus is apparent in the significant attention being given to shipping schedules and production floor schedule performance. While dependable delivery times are important, they are only a starting point. After dependable delivery comes delivery speed, or reduced cycle time. The capability to produce a high-quality product, on time and faster than the competition, represents a formidable competitive advantage. This has implications for reduced cycle time as a key manufacturing strategy.

The Japanese are also focusing on flexibility as a key ingredient of their manufacturing strategies. Manufacturers that can design new products rapidly and introduce them to the marketplace ahead of their competition will be the leaders. They will be able to shift quickly between product lines and product volumes, keeping pace with rapidly changing market demands.

Flexibility is not achievable, however, without a foundation of superior product quality and delivery performance. As more and more manufacturers achieve their quality and dependability goals, flexibility will increase dramatically—as will costs. Balancing these elements is the competitive challenge of the future.

Being good in only one area will not lead to success. World-class manufacturers understand the value of simultaneously designing product and process, improving both, and reducing the time required to pull product and process from concept to market. They are breaking down the traditional organizational barriers between product design engineers, production engineers, and manufacturing engineers. They are removing functional goals and substituting shared objectives. Simultaneous engineering will be a requirement for world-class manufacturers in the twenty-first century.

Performance Measurement

World-class manufacturers are changing their performance measurement systems to encourage manufacturing excellence. Performance measurement systems in the future will be tailored to a company's strategic action programs. Traditional cost accounting systems at leading manufacturers will be dismantled and replaced with performance management systems.

Accounting systems in the future will be reshaped to enable manufacturers to examine the activities that drive costs rather the than departments that collect costs. Cost drivers that have a direct influence on operational performance will be identified and managed. Overhead will not be equally spread over all manufacturing workcenters, based upon changes in direct labor input, but will be directly traceable and accountable to individual cost drivers. Costs will be segregated into value-added and nonvalue-added categories. Nonvalue-added costs are those that can be eliminated without detriment to the product or service. Target costs that are not based upon internally generated cost standards, but rather upon externally defined competitive costs, will be developed.

Performance systems will be expanded to include the critical nonfinancial or operational measures of performance. Time-based measures of performance such as manufacturing cycle time, setup time, on-time delivery, engineering cycle time, and order processing cycle time are important barometers and regular measures of performance. Quality measures will be expanded to include not only manufacturing quality but quality in all functional areas of the enterprise. Customer feedback and performance ratings are an integral part of the enterprisewide system of world-class competitors. This quality integration will have to be a common part of the performance measurement systems of all contenders.

Organization

World-class manufacturers in the twenty-first century will improve their operations on an ongoing basis. They will strive to eliminate waste and create lean organizations that bolster their competitive postures.

Excellent manufacturing organizations today are revising organizational charts, sequential approaches to work, and functional and hierarchical divisions of duties and responsibilities – traditional elements that have often hindered the ability of organizations to respond to competitive challenges. The organization of a world-class manufacturer in the 1990s will be one that capitalizes on the strengths of all employees. Clearly, the most important factor is the fostering of shared motivations and common goals, reducing segregation and isolation in the performance of duties.

Manufacturing cells typically focus on production parts or product lines, which can produce an organization resembling factories within factories, when multiple cells make up a system. These systems, built around specific product lines, are being implemented in larger, more diverse manufacturing organizations. Process manufacturers may have entire plants devoted to the manufacture of a small number of products that share not only these systems, but also common customer and product relationships. Efforts to focus the organizations within organizations are usually accompanied by the implementation of work teams and the reduction of job classifications to a manageable few.

World-class manufacturers have developed unique corporate cultures, which foster an attitude of teamwork. All employees feel as though they have a stake in the business, because the management style is participatory. Successful competitors in the future will be those able to effectively dissolve the line between management and worker and between functionally segregated staffs. They will merge individual functions into multifaceted, multitalented teams. Innovation will be expected and rewarded.

All manufacturers in the 1990s will exert ever-increasing pressure on and control over their suppliers. World-class manufacturers will exercise this control in the form of partnerships aimed at improving their competitive advantages and those of their suppliers.

Exercising effective vertical control over supplier networks and the channels of distribution will be a critical element of world-class manufacturing in the 1990s and beyond. World-class manufacturers have found that it is not enough to achieve excellence in only their in-house operations. The full effective length of their operations, including both suppliers and distributors of component parts and services, is of equal importance.

While the debate over the correct amount of vertical integration will continue for years to come, excellent manufacturers will continue to improve their management of the entire supply chain, from vendor through customer. World-class manufacturing organizations in the 1990s and beyond will develop unique ways to strengthen their customer and vendor relationships, resulting in improved manufacturing operations and competitive advantages.

Human Assets

Traditionally, workers have been viewed as a part of the production process, much as machines have. Repetitive work was accomplished, people could be moved in and out of a job without disruption, and little or no decision making was required. Since little knowledge and minimal skill were required, the pool from which to draw workers was very large; companies had little difficulty finding people with the necessary qualifications. Skilled workers were usually trained on the job and remained with the company for the duration of their work lives.

Today, however, advanced technology in the manufacturing environment requires workers who can act as managers of machines and processes—workers who have much higher levels of skill and knowledge than their predecessors. Companies now realize that qualified workers are a critical resource and must be considered a company asset. World-class manufacturing companies manage their human resources just as they manage other corporate assets. Human resources are incorporated into the strategic and operational plans, budgeting for acquisition, strategies for improvement, and organizing to take optimal advantage of the potential at the company's disposal.

To take full advantage of the potential of human resources, however, major problems will have to be addressed.

Organization: New organizational structures will have to emerge to give workers more opportunities to utilize their skills and knowledge. World-class manufacturers have drastically reduced their levels of management, placing more decision making and authority at the levels where work is accomplished. To take full advantage of existing expertise, the concept of team-driven efforts has emerged as a lean, efficient method of accomplishing work goals. Decisions are made by those who are directly involved with the process. As a result, companies that have implemented this type of organizational structure have found that a built-in benefit is constant improvement in the processes managed by the teams. Since process improvement will be a key element of competitiveness in the twenty-first century, world-class companies will be organized around the team concept.

Development: A major problem currently exists in acquiring and retaining qualified workers. The changing demographics of the workforce—more women and minorities entering the workforce, the aging population, and a lack of basic and technical skills in younger workers—is a critical problem for manufacturing.

Companies are currently spending billions of dollars on training and retraining the workforce. More will need to be spent. However, strategies will have to be formulated to optimize the training effort and utilize existing resources as efficiently as possible.

Management: Management strategies that address the issues of changing demographics, changing cultural attitudes, and the need for more education and training will have to be developed. For example, new reward systems have to be devised, fringe benefits (such as day-care centers) have to be considered, career paths have to be identified, and a new corporate culture that builds company loyalty has to be developed.

Human assets will emerge in the twenty-first century as the most important resource of the company. World-class manufacturing companies will be those that recognize that fact and manage them accordingly.

Technology

Numerous benefits can be cited for those companies, both domestic and foreign, that have successfully implemented emerging technology in their manufacturing operations. While particular industry segments find themselves being set back at times by their competitors' latest application of technology, these setbacks represent the natural ebb and flow of perceived needs in response to competitive markets.

However, attempting to gain a long-term competitive advantage using technology alone will not succeed without an adequate infrastructure. Continuous improvement, addressing both structural and infrastructural elements of manufacturing, is the only proven approach to overcoming competitive barriers. There are no shortcuts around addressing basic issues.

This is not to say that keeping up with technology is not important. On the contrary, keeping pace with technology is a real concern. Ignoring the rapid changes occurring in such technologies as composites, precision machining, laser processing, and knowledge-based systems, is a sure way to erode competitiveness. Continuing to invest in modernization of both plant and equipment and looking for advances in technology that will improve competitiveness are important elements of long-term success.

World-class companies understand that the major impediments to the adoption of technology in manufacturing are not technological, but human. They understand that the implementation of technology is not complete simply because crates have been unpacked, equipment assembled and installed, and electrical wires connected. As a result, world-class manufacturers start with the human resources to produce awareness, buy-in, and the attainment of necessary skills to ensure implementation success.

World-class companies in the future will continue to apply technology to manufacturing environments strategically. They will work in meticulous detail to develop the infrastructures needed to support the technology. As companies move toward increased automation, they must develop a futuristic perspective concerning the strategic advantages of technology in the global arena. They must also recognize that their own management teams may be major hurdles.

Despite strong trends in applying technology to upgrade process communications on the production floor, the greatest payoffs reside in upgrading communication between people. The requirement for timely management information will increase dramatically in the 1990s, as time becomes a critical factor in competitiveness. Unfortunately, most companies are not prepared for the challenge. To be a world-class manufacturer in the twenty-first century will require superior communication and information management capabilities designed to carry information both vertically and horizontally throughout the organization. Goals will include real-time data transfer and information enhancement through artificial intelligence-based communications systems.

The External Environment

The framework does not function in a vacuum. Many elements in the external environment affect the operations of world-class manufacturers.

Federal laws governing how companies function are responsible for some of the most significant obstacles facing U.S. manufacturers. Current antitrust laws are woefully outdated. Their nationalistic perspective does not consider U.S. manufacturing in light of international competition. Tax incentives for manufacturers that invest in technology and education are limited and clearly inadequate for the challenges that lie ahead. International competition requires both state and federal governments to view national interests within the context of a global community and to become partners in the effort to help manufacturing companies compete.

The lack of national policies to support U.S. companies in their drive to become world-class is a critical issue today. Even the U.S. Department of Defense states that, "Policies restricting manufacturing are a national outrage."[11] That some manufacturers have achieved world-class status despite current government policies is a positive comment on the organizations' resolve to succeed. But needed resources are scarce. It is unrealistic to expect the bulk of U.S. manufacturers to attain world-class status, or even survive, when they are required to match individual resources against those of international consortia with billions of dollars at their disposal.

Over the years, the federal government has created a relationship with industry characterized by extensive restraints and requirements for documenting compliance with rules and regulations. Manufacturers invest significant effort in satisfying federal paperwork requirements and proving they are in compliance with the law. "Other countries, Japan in particular, are much more effective than the United States in achieving industry/government/labor cooperation on process and product development and . . . in implementing new ideas to make manufacturing more efficient, responsive, and technologically advanced."[12]

Current tax laws and trade policies do not provide an effective incentive for manufacturers either to train their workforces or to invest in new equipment and process technology. Relative to other industrialized nations, America's tax system provides little or no incentive for long-term investments. The U.S. taxes capital gains and earnings on savings and investments as ordinary income, at relatively high rates, while it either does not tax consumption or does so at very low rates. This results in reduced capital for industrial modernization and research and development incentives. The American tax system, through income, property, and labor taxes, places a tremendous burden on American products but not on equivalent products manufactured elsewhere and sold in the United States.

Establishing nationally-based programs to improve North American competitiveness rapidly is a critical component of an overall national policy agenda. Congressional passage of the National Cooperative Research Act of 1984 was a step in the right direction, and the establishment of the Malcolm Baldrige National Quality Award in 1987 is further evidence of growing concern for U.S. manufacturing capabilities. Additional steps which address the significant challenges raised by Japan's Ministry of International Trade and Industry (MITI) and the passage of the Single European Act are necessary, as is a response to the increasing challenges of other industrializing nations which are also organizing groups of companies to compete globally.

Contrary to popular belief, the global rules of competition have not changed. But, the basis for competition and the game plan have. In *The Competitive Advantage of Nations*, Michael Porter argues that,

> What is needed today in American industry is not less competition but more... Innovation and upgrading results, not from a comfortable home environment in which risks have been minimized, but from pressure and challenge from demanding home customers, from capable home base suppliers and most of all from local rivalry.

The challenge of foreign competition is very clear. All branches of the federal government, as well as state governments, the education community, and the manufacturing industry, must work together to act as a single entity. How this response is orchestrated will have a tremendous impact on the future of world manufacturing and America's aspirations to significant participation in world-class manufacturing.

With or without government support, the path for American manufacturing in the twenty-first century is clear. American manufacturers must pursue continuous improvement in products and processes and a hardy recommitment to competition. They must not become preoccupied with complaints of the unfairness of the practices of other nations; they must forge ahead.

Organization of the Remaining Chapters

This book follows the framework established for world-class manufacturers:

- Chapter 2 focuses on the critical developments in the area of quality and customer service. As the heart of the operating framework for world-class companies, this important chapter establishes the foundation for the remaining chapters of the book.

- Chapters 3 and 4 describe manufacturing strategy and the impact of manufacturing strategy on performance.
- Chapters 5 and 6 detail the approach being taken by world-class companies to measure manufacturing performance.
- Chapter 7 defines the best manufacturing practices being applied by world-class companies.
- Chapters 8 and 9 discuss the organization and management of the human resources within the organization.
- Chapter 10 presents a national challenge for reestablishing America's technological preeminence and reviews America's progress in adopting state-of-the-art manufacturing technology, and the problems that have been encountered along the way.
- Chapter 11 identifies the critical attributes that will differentiate world-class manufacturing in the twenty-first century and outlines the NCMS operating principles for companies aspiring to become world-class competitors.

A review of Chapter 11 will provide executives with a detailed list of the operating principles put into effect by world-class manufacturers. The reader should keep in mind that the principles for achieving manufacturing excellence are most effective if implemented in concert within the framework.

CHAPTER 2

THE QUALITY REVOLUTION

Recently, the corporate appetite for quality seems to have become insatiable. The pursuit of quality has led to substantially more than adjusting current systems. Radical overhauls have revolutionized manufacturing—from changing the way products are designed to instituting new ways of evaluating managers. The quality revolution has affected every facet of manufacturing, including the application of technology and management of the workforce. Quality has fundamentally reshaped management thinking about the importance of manufacturing functions for competitive advantage.

Leading companies are taking no chances that they will miss out on the strategic benefits of the quality revolution. Respondents to a recent study of automotive supplier competitive issues commented, "quality is a one-way street, and if you give a taste of it, you can't back off . . . you have to have it or someone who does will get the business."[1] With ever-increasing global competition, the demand for quality products and services will remain robust into the twenty-first century.

The path to quality improvement is long and difficult and has a variety of destinations. Many companies, such as the McDonnell Douglas Corporation, have taken dramatic steps to transform their culture into one based on quality.

McDonnell Douglas has made a specific commitment to transforming the company into a competitive organization. It started with a radical reshuffling of the transport-aircraft giant's organization chart. Four of nine management levels vanished, eliminating more than 2000 out of 5000 management slots. At the time, many insiders were convinced that unless McDonnell Douglas did something extreme, the company could be driven from the commercial aircraft market by Boeing Corporation and Europe's Airbus Industries.

The company now relies on a Total Quality Management System (TQMS) as a model for quality improvement and cost slashing throughout the corporation. As a result of TQMS, attention has turned squarely to manufacturing. The company has adopted a special brand of total quality on the shop floor, as empowered workers tell their supervisors what they will be doing on a given day and not vice versa. Vice President Smith reports, "The guys in the plants are the ones who know best how to make planes. Not us."

Progress toward quality goals can be slow. There is some question as to whether McDonnell Douglas will be able to achieve the quality transformation goals it has established. Within a year of its startup, only about 600 of the 45,900 blue- and white-collar employees had completed the mandatory 80 hours of Japanese-style training on quality techniques for building labor-management partnerships. Furthermore, the company is suffering from severe cutbacks in U.S. defense contracts. The verdict on McDonnell Douglas' efforts has yet to be returned, and the company continues to stress the need for patience among investors and employees.

One important lesson that corporate America has learned is that quality improvement takes time. Five to 10 years of constant effort, drive, and leadership appear to be the norm. And late-comers who try to leapfrog the cultural transformation and training processes may not gain substantive benefits.

Quality as a Way of Managing Business

Commitment to quality has become a fundamental way of managing business. To many companies, it has led to a sweeping overhaul in corporate culture, a significant shift in management philosophy, and a permanent commitment at all levels of the organization to continuous improvement. Although McDonnell Douglas' implementation of total quality managment through radical restructuring is atypical of the manner in the process is introduced, the company's team approach to TQM, relying upon an empowered and trained workforce for continuous improvement, is commonplace. More specifically, McDonnell Douglas' quality improvement program, which provides flexibility in work assignments, employee participation in decision making, and the building of trust between labor and management, is typical of the evolutionary approaches to quality progress.

Three American companies—IBM, the Nashua Corporation in New Hampshire, and the Tennant Company in Minneapolis—blazed the quality trail in 1980. Following their lead were Ford, Xerox, Baxter, Milliken, Eastman Kodak, Motorola, and a number of other American corporations that consciously chose to aggressively pursue formal programs for total quality management through *incremental* overhauls.

What is quality? Many definitions of quality exist, but these three statements identify the three levels of quality talked about today. They are cumulative, and the difficulty of achieving quality increases with each level.

- **Conformance Quality**: Quality conforms to specifications. Products meet specified tolerances limits or services meet specified standards, and products are characterized as "free from defects."
- **Requirements Quality:** Quality meets customer requirements. The perceived product and service attributes match the customer's expectations and fill the customer's needs.
- **Quality of Kind:** Quality is so extraordinary that it delights the customer. The perceived product and service attributes significantly exceed customer expectations, and customers are delighted with the value.[2]

This chapter covers the concept of quality, as it is interpreted today by manufacturing companies, and the approaches, techniques, and shifting strategies currently used in achieving quality. It examines the thinking of the quality gurus: Joseph Juran, W. Edwards Deming, Philip Crosby, Armand Feigenbaum, Genichi Taguchi, and Dorian Shainin. It reviews the traditional toolkit of quality improvement techniques, such as quality circles, Statistical Process Control (SPC), and Quality Function Development (QFD), and introduces seven new tools for enhancing quality. Finally, it examines practices in Japan to see where quality is moving and presents a plan for quality renewal in the 1990s and beyond.

Shifting Strategies for Achieving Quality

Every year from 1981 through 1988, quality ranked at the top of the list of key strategic competitive priorities identified by top executives in the North American Manufacturing Futures Project. Ironically, while strategic mandates for high quality held a tenacious grip on American manufacturing, the 1980s witnessed a fundamental shift in approach to quality, from *command and control* to *continuous improvement.* Figure 2-1 shows how key components of manufacturing strategies coincided with the American quality revolution.

FIGURE 2-1 Shifting North American Strategies Target Quality (The Top Five Manufacturing Action Programs in 1984, 1988, and 1990)

1984	1988	1990
Product/Inventory Control Systems	Vendor Quality	Continuous Improvement
Workforce Reduction Programs	Statistical Process Control	Work Safety
Supervisor Training	Worker Safety	Vendor Quality
Direct Labor Motivation	Visible Manufacturing Strategy	Labor/Management Relations
New Product/ Process Development	Worker Training	Enlarge Worker Responsibilities

Adapted from J.G. Miller and A.V. Roth, "Manufacturing Strategies: Executive Summary of the 1988 North American Manufacturing Futures Survey," Boston University Manufacturing Roundtable, 1988, and C. Giffi and A.V. Roth, "Making the Grade in the 1990s," Deloitte & Touche Third Annual Survey of North American Manufacturing Technology, 1989.

Along with North American strategies and the touted Japanese quality practices, to be detailed later in this chapter and in subsequent chapters, quality has affected European manufacturers as well. European businesses and consumers are buying Japanese goods because the products are well-made, reliable, and well-suited to the consumer's needs. Nonetheless, there is considerable market advantage to be gained by becoming the strongest, most quality-conscious manufacturing nation in Europe.

Like their American counterparts, European manufacturing executives consistently placed quality at the top of their competitive priority lists during the period 1983 to 1988. Quality-related programs have proliferated in Europe with positive outcomes. For the Europeans, too, quality results have taken years to obtain as they require a "coherent and encompassing program of activities inside the plant, improvements in the design engineering and upgrading of the supplier links."[3] The 1988 European high performers on quality have specifically emphasized the following action programs to achieve results:

- Statistical process control
- Vendor quality
- Quality circles
- Zero defects
- Standardization/narrowing product lines
- Value analysis/product redesign
- Group technology
- Flexible manufacturing systems
- Reconditioning of physical plants
- Giving workers more planning responsibility

Furthermore, quality improvement is seen as the basis for building the manufacturing capabilities described in Chapters 3 and 4. The Europeans argue that:

> A precondition to all lasting improvements are improvements in quality performance of the company . . . Once the company has reached a critical level of improvement in quality, it can tackle issues of dependability. However, this should not constrain further improvements in quality . . . Once a critical level of dependability is reached, the company can attempt to improve its manufacturing reaction speed . . . Once, and only once, the company has obtained a critical mass of improvements in these three areas, can it improve in a lasting way its cost position.[4]

Stiff international competition has made the quality of products and services a major determinant of market share. If one company cannot offer top quality at a competitive price, there is another company waiting to meet the customer's requirements. This new emphasis on quality is bringing changes to the traditional practices of quality control. In the past, quality meant inspection. Today, quality is integrated into the organization—from product conception, through design, manufacturing, marketing, and into the customer's hands. All operations and all organizational components contribute. This section addresses quality from several strategic perspectives:

- Changing perceptions of top management.
- The quality implementation continuum.
- The quest for a zero defects culture.

Top Management Imperative

The level of top management commitment to quality management is evolving. Top executives today pay attention to quality. Quality has become the most important strategic concern of American industry, as all companies try to account for the tremendously high cost of ignoring quality. In the 1988 ASQC/Gallup poll of top executives, 45 percent of the respondents attached the highest ratings to changing corporate culture and securing CEO leadership and involvement to improve the quality of products and services. In 1986, only 15 percent of the respondents gave CEO involvement the highest possible score.

A recent Organizational Dynamics, Inc. (ODI) Executive Opinion Survey found that 85 percent of executives in the largest U.S. companies agreed strongly that as leaders of their companies they were "personally committed" to improving quality. Yet, the typical factory in the U.S. invests 20 to 25 percent of its operating budget in finding and fixing mistakes. Major obstacles to quality improvement identified in that survey were:

- "Making quality happen, not just recognizing the need for it."
- "Management—not workers—is primarily to blame for quality problems."
- "Managers are too focused on short-term goals."[5]

The Quality Continuum

What is the attitude of companies toward quality today? The attitude of the 1970s, which proclaimed that the way for a company to succeed was to make products more quickly and cheaply and to push hard on sales, has been replaced. Now there is a recognition that, while products must be made quickly and cheaply, they must also be of high quality to remain competitive. For some companies, quality is still an option. For most, however, quality is now a mandatory part of continued business operations. Based upon trends over the past decade, most American firms lie somewhere on a continuum between a "white glove approach" and "quality nirvana" in their application of quality improvement (see Figure 2-2).

In Stage I, companies manage the manufacturing "indirects" to achieve quality. There is heavy emphasis on the application of control mechanisms, including production planning and control systems, development of new processes, and training of supervisors to oversee quality activities. Quality is managed by inspection and rules and procedures. If statistical process

Figure 2-2 The Quality Continuum

- **Stage I: White Glove Treatment** — A fundamentally authoritarian "command and control" approach to quality improvement which relies upon systems and supervisors to ensure quality.
- **Stage II: Quality Toolkits** — Focuses on tactical management of manufacturing "directs." These include people, processes, and materials which can be strategically dictated and delegated to middle managers and workers for implementation.
- **Stage III: Quality Nirvana** — Cascading from leadership, a quality culture fosters a total organizational commitment to continuous improvement of products, processes, and services through empowered employees, suppliers, and customers; the application of formal quality techniques; and a conducive operating environment.

Source: A.V. Roth, "A Vision for the 1990s: Gearing Up for the 21st Century through Intelligent Manufacturing," Duke University, 1990.

control exists, it is controlled by quality assurance professionals or supervisors. Workers have very little discretion.

Stage II companies are more aware of the need to focus upon the direct or controllable factors involved in quality. They employ vendor quality management programs to tackle problems with incoming materials, statistical process control to reduce variation in the production processes, worker training to develop skills, and safety programs to facilitate progress. While Stage II firms adopt basic quality management tools, the quality management function has been delegated; top level managers are not active participants.

In Quality Nirvana, the quest for quality permeates the organization. A key indicator of a Stage III company is fanatic attention to customers, products, and processes at all levels of the organization. Not only can a Stage III company quickly identify quality problems, it can fix the process to resolve them quickly as well. Stage III companies are "process solvers" rather than "problem solvers." They are capable of producing products and services that delight customers!

Robert Hall tells the Stanadyne story which describes a company well on the way to Stage III status. Stanadyne Operations makes two products:

diesel fuel injection equipment (FIE) and auto engine tappets. It holds 65 percent of the North American market for light diesel engine FIE. Stanadyne's six-year path to process improvement includes:

- A top-down dedication to total quality management, including active management participation with workers and building good labor relationships.
- Application of statistical methods for process controls and capability—stopping the process rather than inspecting defects.
- Ongoing operator training, multiskilled workforce, and employee ownership of processes.
- Daily supervisor checks of control charts to provide feedback information on tooling, maintenance, and other factors necessary to process improvement.
- Reorganized and guided quality circles designed not only to build employee skills for problem solving and presentation but to correct problems.
- Conducive atmosphere to make it happen, including facilitators, facilities, time, recording of ideas, visible presentation of results, and constant encouragement from management.
- Revised systems for incentives and measurement, including changes in the internal cost system to de-emphasize operation-by-operation costs and operation-by-operation buildups of the efficiency report; new management emphasis on total throughput time and overall effects; and supervisor scorecards kept on departmental improvements in quality and delivery given the available resources.
- Implementation of "quality cells," *quasi* cells which press management to actively prevent problems and correct maintenance and quality problems as they occur.
- Working with customers to develop new applications and improve understanding of customer requirements.

What was the payoff for Stanadyne? Robert Hall reports that after six years of continuous improvement in the manufacturing process, Stanadyne Diesel Systems showed:

- 30 percent total inventory reduction, from $27 million to $19 million.
- 75 percent reduction in plant lead times.
- 75 percent reduction in stockroom space.

- Reduction in aftermarket backorders from $1.2 million to $6000.
- 97 percent customer service level, up from 79 percent.
- Reduced production control staff from 32 to 3 people.
- Two-day lead time for finished goods for major customers and three-day lead time to change mix for customers.[6]

Along with improved efficiency and increased productivity, quality is a critical factor in increasing America's competitiveness in today's market. More corporations are recognizing the importance of ensuring not only the quality of the products they manufacture, but also the quality of service they provide along with their products. Companies like Stanadyne believe that no one can guarantee success; the best chance for achieving it goes to manufacturing firms that have high-quality capabilities and the ability to respond quickly to changes in requirements.

Stalking Zero Defects Cultures

At least some companies view world-class quality as a culture that must be installed in many areas of a company. This *quality culture* calls for high-quality, reliable products that meet customer needs; a responsible, consistent customer interface organization; a service organization that provides timely and accurate service; on-time delivery of products; and a participatory management structure that encourages employee involvement.[7] The old notion that a company has to expect its products to have a certain level of defects and errors is fading.

Motorola has become preeminent by making zero defects a fundamental part of the entire corporation. Xerox, through its *Leadership through Quality* program, Ford with *Quality is Job 1*, and Hewlett-Packard with its *Total Quality Control* efforts are close followers. Motorola is so fanatic about eliminating defects that it uses a statistical measure, six sigma, to express the quest for perfection. Simply stated, six sigma is a way of measuring quality control of only 3.4 defects per million parts. How important is Motorola's *Six Sigma* campaign? In 1987 alone, the company spent $25 million on a companywide education blitz, and, in 1989, it began linking most performance reviews and bonus incentives to Six Sigma requirements.

CASE STUDY

Motorola: Quality in the Fast Lane

Business Week calls this company the "rival Japan respects." In 1988, this firm was one of the first to win the famous Malcolm Baldrige Award for Quality. Company representatives often say that "total customer satisfaction is the challenge of the next decade." And they back this claim with the Six Sigma quality program.

The corporation is Motorola. With six divisions led by communications, semiconductors, and cellular phones, total profits exceeded $500 million in 1988. Motorola's Six Sigma quality program mandates that products be perfect 99.9997 percent of the time. But reducing every activity to the minimum time it takes to do something (Total Cycle Time Reduction) is also a passion at Motorola. Further, and less well-known, Motorola backs all of this up with a solid and consistent commitment to research and development (R&D).

Admired by Sony Chairman Akio Morita and focused on market share like other successful R&D-based companies, Motorola is a company that does many things right—not just one or two things correctly. Part of the philosophy that distinguishes Motorola from other companies is the knowledge that when you are the market leader in an industry you must work harder to stay there. Motorola's primary method of keeping its strong corporate culture focused on these issues is a massive program of education and support for its 105,000 employees. Courses include global competitiveness, risk-taking, statistical process control, and reduction of product life-cycles. The total price tag for this education effort: $60 million in 1989 alone. Corporate engineers are required to earn an additional advanced degree every six years.

Motorola spent 19 percent of its revenues—$1.8 billion—on training, capital improvements, and R&D in 1989. Some companies talk about commitment; Motorola truly invests in its future.

Although the Six Sigma Quality program and Total Cycle Time Reduction are the two key initiatives at Motorola, other things are important, too. The company intends to achieve and maintain product and manufacturing leadership. The firm has always been profit-oriented. Participative management was initiated early in the 1980s and continues. Finally, like other world-class manufacturing firms, Motorola knows how to make alliances with other organizations—not just other businesses, but institutions like universities.

Although Motorola's reputation for quality has become well-known since it won the Baldrige Award, its initiative in reduction of total cycle time is not as well-understood. Yet, it is equally important. The company culture says that Motorola can accomplish jobs in 2 to 4 weeks that competitors take 10 to 15 weeks to master. The firm places a high value on answering

Case Study, concluded

customer questions and order inquiries during the first telephone conversation, rather than the more typical seven to nine working days later.

The Motorola goal is to ship 95 percent of all orders complete instead of the more typical 50 to 70 percent. *Orders are delivered daily rather than weekly*. Motorola intends to help customers increase their inventory turns from 4 to 5 per year to between 8 to 14 per year. The Motorola operating goal is to grow three times as fast as the market and to be twice as profitable as the industry average.

The culture at Motorola supports the notion that the evolution of any quality strategy follows four stages:

1. Minimize negative impacts by fixing problems.
2. Maintain standards with competitors and customers.
3. Provide credible support to the business strategy.
4. Systematically align customer requirements with internal processes.

Total cycle time management and quality are not just two parallel programs at Motorola; they are assumed to be intimately related. One helps the other. To reduce cycle time, quality must be better. To get better quality, cycle time has to be reduced. Methods that solve one problem can help eliminate another. Both attack waste; the Japanese philosophy says eliminate all waste in manufacturing.

The final chapter of the Motorola story reveals four essential roles for general managers:

1. Commitment, including resource allocation to issues of importance for the future, such as education and R&D.
2. Establishment of clear goals with value statements to support them.
3. Communication of goals and value statements.
4. Linking of quality requirements to expectations of customers. This is consistent with the company philosophy that total customer satisfaction is the ultimate measure of quality. Consequently, there is no *finish line* in this business.

This case was prepared by J. Ettlie using material drawn from several sources including: Lois Therrien, "The Rival Japan Respects," *Business Week*, November 13, 1989, 108-118; and Carlton Braun, "Total Customer Satisfaction: The Quality Challenge of the '90s," presented at the 2nd annual Quality Conference, School of Business Administration, The University of Michigan, Ann Arbor, MI, April 27, 1989.

Hewlett-Packard's dozen-year efforts aimed at tenfold improvements in product reliability produced an estimated saving of $600 million in warranty repairs and $3 billion in manufacturing costs. Championing total quality at Xerox means routinely practicing prevention and benchmarking. The

Xerox staff routinely searches for best-in-class performers outside of their company. By doing that, Xerox increased copier market share to 15 percent and reduced defect rates 93 percent. In 1988, Xerox recorded a saving of $166 million from scrap reductions.

Quality Gurus

The key to quality improvement rests with management. Management must drive the quality system and match its behavior to the objectives of the organization. Some companies think that quality can be improved simply by sending people to the right quality training seminars—but that alone does not work. Quality must permeate everything a company does at all levels of the organization. You cannot buy a solution to all quality problems, nor is there a single, right approach to excellence. This section describes the more popular of the current approaches.

Juran

Joseph Juran focuses on product designs that are both high in quality and able to be manufactured to consistently high quality standards. Juran contends it is critical to obtain participation not only in engineering and manufacturing, but in all parts of the organization. He believes in a project team approach that raises standards continuously and solves problems. Managerial review systems should be set up to measure performance against quality goals, and on-site top management involvement in implementing a quality program is crucial.

Juran insists that quality goals be specific. The statement, "Quality is priority number one," is unacceptable. Instead, a properly specified goal would be "to raise quality levels to at least the levels of competitors" by a certain date or to "cut costs of poor quality by 50 percent in five years."

Juran's concepts are very similar to Deming's 14 points, except that he advocates other technical approaches instead of statistics. He focuses on the customer, an approach which has contributed to major achievements in marketing, and stresses a balance between customer wants (surface features) and customer needs (product use characteristics). Implementation of Juran's approach requires skills in team leadership.

Deming

W. Edwards Deming stresses simple, straightforward, quantitative methods that prevent quality slippage rather than after-the-fact repair. Although simple in concept, his process control statistics have the potential to revolutionize the management of any process, not just manufacturing and assembly operations. Deming's approach to improving quality includes 14 steps (see Figure 2-3).

Central to Deming's message is that competitive quality cannot be obtained with traditional inspection methods. He regards statistics as abstracted models of how all systems should function and feels that this theoretical base is essential for quality improvement.

Deming also believes strongly that "quality on the shop floor can be no better than the intent of management which is made in the board room," and

FIGURE 2-3 Deming's 14 Points

1. Create consistency of purpose toward improvement of product and service.
2. Adopt the new philosophy. Refuse to accept defects.
3. Cease dependence on mass inspection.
4. End the practice of awarding business on the basis of price tag. Require suppliers to provide statistical evidence of quality.
5. Find problems. Continually and forever make improvements.
6. Institute modern methods of training on the job.
7. Give all employees the proper tools to do the job right.
8. Drive out fear, so that everyone can work effectively.
9. Break down barriers between departments; encourage different departments to work together on problem solving.
10. Eliminate numerical goals, posters, and slogans that ask for new levels of productivity without providing specific improvement methods.
11. Eliminate work standards that prescribe numerical quotas; use statistical methods to continuously improve quality and productivity.
12. Remove all barriers to pride in workmanship.
13. Provide vigorous and ongoing education and retraining.
14. Clearly demonstrate management commitment to the above 13 points every day.

Source: W.G. Landon, "Kanban and Deming's 14 Points," *Quality*, September 1988, pp. 50-51. Reprinted with permission from *Quality* (Sept. 1988), a publication of Hitchcock Publishing, a Capital Cities/ABC, Inc., company.

maintains that until executives and managers actively adopt quality methods into their own decision making—in effect, model the correct attitudes and actions—a quality effort will not work.[8]

Changing the tangled web of interrelated behaviors and expectations is a much larger issue than changing statistics in most companies. Deming is eloquent in explaining what has to be changed but does not explain how to change behavior so that his methods will be accepted and adopted.

Crosby

Philip Crosby believes that by tracking the cost of quality, companies will be guided to the areas in which the greatest improvements in both quality and efficiency can be made. If it costs more to do something wrong than to do it right, the problem areas will show up when you audit for "quality fixing" expenditures.

Users of cost-of-quality programs often refer to his 14 steps (see Figure 2-4) as the framework for the implementation of an organizationwide effort.

FIGURE 2-4 Crosby's 14-Step Approach

1. Management Commitment
2. Quality Improvement Team
3. Quality Measurement
4. Cost of Quality
5. Quality Awareness
6. Corrective Action
7. Zero Defects Planning
8. Supervisor Training
9. Zero Defects Day
10. Goal Setting
11. Error-Cause Removal
12. Recognition
13. Quality Controls
14. Do It Over Again

Source: Philip B. Crosby, *Quality is Free* (New York: McGraw Hill, 1979), pp. 228-240.

Crosby's 14 steps are not as technically comprehensive nor do they address the cultural issues and management practices that Juran's and Deming's systems do. The process of implementing a Crosby-type program is much more a real-time method of identifying areas in need of basic management, leadership, and employee participation skills.[9]

Feigenbaum

Armand Feigenbaum is not as well known as is his Total Quality Control (TQC) system for quality improvement. This approach promotes the concept that all activities are synergistic. You cannot get quality improvement from a system that does not have all of its parts dedicated to quality.

TQC encompasses many different elements and includes essential elements from most other quality programs. To implement a TQC system, massive education is required. While all known systems and techniques are included, management practices and employee skills are not. His approach focuses on helping a plant or company design its own system.[10]

Taguchi

Genichi Taguchi's approach centers on a statistical method of rapidly zeroing in on the variations in a product that distinguish the bad parts from the good, to avoid endlessly testing for all possible defects. He uses an assortment of graphs, statistics, and tables to find the important variables and to then concentrate on reducing costs by reducing variation. Taguchi also promotes a concept of *robust design* which uses experiment control to optimize designs and manufacturing processes so that they become insensitive to factors beyond the manufacturer's direct control, resulting in forgiving designs and processes.[11] The Taguchi method extends beyond pass/fail or in/out of specification limits ways of thinking.

For example, a few years ago Ford asked Mazda to build some of the transmissions for cars it was selling in the U.S. Both Ford and Mazda used the same specifications but, it was found, customers with the Ford transmission had higher warranty costs and complaints about noise. Upon investigation, Ford found that while all their individual transmission parts were within "acceptable" tolerance limits, many of the component specifications were only in close proximity to limits. As a result, the summation of the randomly assembled parts interacted with greater friction than the transmission could withstand. Mazda, on the other hand, consistently worked to meet targets on the total project and minimized all variations from target.[12]

Taguchi's quality imperatives are paraphrased below:

- Quality losses stem mainly from after-sale failures due to poor design.
- Robustly designed products take into account both variations in the production process and variations due to customer use.
- Develop a system of trials that evaluates the impact of the overall system performance according to the average effect of change in component parts when subjected to varying parameter values, stress levels, and experimental conditions.
- In building robust products, first establish ideal levels of performance for individual components. Minimize the average (squared) deviations from that ideal when the components are functioning together and over various customer use conditions.
- Set acceptable product targets before manufacturing processes begin.
- Manage production to targets rather than to acceptable deviations from specifications.
- Persistently pursue robust designs that can be consistently produced, and diligently demand consistent performance from manufacturing.
- Attacking product failure problems due to customer use will reduce variation in the production system as a whole.
- Evaluation of alternate proposals for capital equipment should reflect the costs of the "average quality loss and of expected deviations from target."[13]

The significance of the Taguchi approach for American manufacturing is threefold:

- The design of experiments is Japan's real secret weapon in the product reliability war. The focus on robust design is critical.
- Quality is evaluated simultaneously from both an end-user and a process perspective.
- There is less reliance on Statistical Process Control for resolving quality problems. SPC is best used after the process or product is brought under control by design of experimental methods.

The bad news about Taguchi methods is that they are complex, expensive, and time consuming.

But relief may be on the way for true Taguchi followers. Advanced, menu-guided software systems, which are designed for both the novice user and the experienced statistician, are now available. The software can be used to construct Taguchi designs with Taguchi methodology and can significantly reduce the time and statistical effort required to perform Taguchi testing.[14]

Shainin

As a quality consultant, Dorian Shainin espouses three principles: establish realistic tolerances, reduce spread, and control the process by pre-control.[15] Shainin's approach centers on statistical engineering methods that focus on what he calls *variation research.* His methods are applicable to continuous, batch, and discrete parts manufacturing processes and are generally believed to be more cost-effective and simpler to use than Taguchi's or traditional SPC.

Shainin's trick is to use statistical methods to zero in very quickly on the root causes. Shainin attempts to isolate the largest cause of variation and make changes even prior to process capability studies. His pre-control procedures achieve almost immediate results; e.g., within hours and days versus months. With his pre-control procedures for ongoing processes, an operator without any calculations or charting can react much more rapidly than with the X-bar and R charts of the SPC method.

Shainin does not start with the executives to advance his techniques. He is usually called into a company to get rid of a particular glitch. Shainin works with the company's engineers to apply variation research. Charles Thierfelder, former vice president for Product Assurance at the RCA Electronic Division, reported that "Shainin helped RCA fix problems with its picture tubes that had seemed unsolvable . . . His techniques got (RCA) from a poor second to the Japanese to the best reliability in the world."[16]

Design of Experiments

While advocates of the Design of Experiments consider them to be superior to SPC techniques, a comparison among three general approaches, including classical methods, Taguchi, and the Shainin tools suggests that the Shainin tools run circles around the other two. Figure 2-5 illustrates the relative effectiveness of these methods.

Common Weaknesses

Although the approaches of Deming, Juran, Feigenbaum, Crosby, Taguchi, and Shainin are the most popular approaches today, failure to meet

FIGURE 2-5 Three Approaches to the Design of Experiments

Characteristic	Classical	Taguchi	Shainin
Technique	• Fractional factorials, EVCP, etc.	• Orthogonal arrays	• Multi-vari, variable search, full factorials
Effectiveness	• Moderate (20% to 200% improvement) • Retrogression possible	• Low to moderate (20% to 100% improvement) • Retrogression likely	• Extremely powerful (100% to 500% improvement) • No retrogression
Cost	• Moderate • Average of 50 experiments	• High • Average of 50 to 100 experiments	• Low • Average of 20 experiments
Complexity	• Moderate • Full ANOVA required	• High • Inner and outer array multiplication, S/N, ANOVA	• Low • Experiments can be understood by line operators
Statistical Validity	• Low • Higher order interaction effects confounded with main effects • To a lesser extent, even second order interaction effects confounded	• Poor • No randomization • Even second order interaction effects confounded with main effects • S/N concept good	• High • Every variable tested with all levels of every other variable • Excellent separation and quantification of main and interaction effects
Applicability	• Requires hardware • Main use in production	• Primary use as a substitute for Monte Carlo analysis	• Requires hardware • Can be used as early as prototype and engineering run stage
Ease of implementation	• Moderate • Engineering and statistical knowledge required	• Difficult • Engineers not likely to use technique	• Easy • Even line workers can conduct experiments

K.R. Bhote, "DOE-The High Road to Quality," Reprinted, by permission of publisher, from *Management Review*, January/1988, the requirements listed below in actual use can be the undoing of any program.

- Sound instructional method.
- Interpersonal communication.
- Team leadership skills.
- Interdepartmental communication.
- A vehicle to make sure the need for immediate quality actions is clear to everyone.
- A way to discriminate between activity that contributes to quality and action that does not.
- A simple message that will reliably trigger consistent and coordinated action.
- Individual initiative to do the quality activities without additional quality systems.
- A management willing to be active participants rather than cheerleaders.
- Integration of unions, vendors, shareholders, and customers in the effort.[17]

The major problem with most of the programs espoused by quality gurus is that they focus on the technical issues, or "hard," side of the problem and often ignore the human, or "soft," side of the problem. Unfortunately, the soft problems typically determine the short- and long-term success of a quality program.

Despite the differences between the various approaches, all programs do have a single central theme—focus on the quality of the process, not just the quality of the product.

Quality Toolkits

Whether a company adopts one of the highly publicized approaches described above or designs its own approach, certain other techniques used by U.S. companies bear mention:

- Employee involvement teams
- Cost of quality

- Statistical Process Control
- Quality Function Deployment
- Seven new tools

Organizing for Quality

By the twenty-first century, quality breakthroughs will emanate from the ways in which the company organizes its human resources for quality improvement and prepares its workforce for technological advancement. Quality improvement and technological advancement are two sides of the same coin. While it has been generally assumed that quality problems are "traceable mainly to a poor workforce, poor workmanship, and poor maintenance of production facilities and equipment . . . (it is becoming increasingly evident), that you can't get quality through fixing the wrong part of the whole system."[18] Quality improvement must be designed into every aspect of the organization.

Much has been written regarding the necessity of involving all employees from top management through production floor employees in a participatory management environment. World-class companies are finding that customer excitement about products and services exists in direct proportion to the excitement and commitment of the workforce. Employee Involvement (EI) is a method for providing employee enrichment and a challenging environment in which people want to work and strive for continuous improvement. One potential benefit of EI is improved organizational flexibility. Workers are in a better position to anticipate problems and to experiment and innovate.

Even so, there are no quick fixes in managing worker-related problems. EI is not intrinsically good or bad. The organizational readiness for EI and the manner in which it is executed and responded to determine the relative usefulness of the approach in implementing quality programs. Several factors appear to be critical to successful implementation of EI teams.

The degree and form of EI participation varies widely from company to company. There are two general forms of EI: quality circles and cross-functional teams. Either may become a superteam in which the management and control function is largely delegated to team members.

Quality Circles

The basic definition of a quality circle is a "group of workers from the same work area who voluntarily meet on a regular basis to identify, analyze, and solve various work-related problems."[19] Participation in quality circles usually does not coincide with assumption of responsibility by workers. Even worse, employees find problems that others are expected to fix.

The quality circle group members meet for a few hours every week or two to discuss, analyze, and propose solutions to quality problems and other common concerns. Quality circles are a parallel structure separate and distinct from the on-going activities of the company. Most groups are led by a special type of facilitator, and to produce change, the groups must sell their ideas to the management of the regular work organization.

The Japanese have utilized quality circles successfully for nearly 20 years. Companies in the United States that are concerned about international competitiveness and the level of U.S. manufacturing quality and that want to move to a more participatory management style have recently adopted quality circles nearly *en masse*. In fact, by 1985, quality circle activity in North America had peaked at a point where more than 90 percent of Fortune 500 firms—an increase of more than 100 percent since 1982—reported having them.[20]

However, research on the success of traditional quality circles in the United States indicates mixed results:

- The active participation rate of employees is under 15 percent.
- The attrition rate exceeds 70 percent.
- Blue-collar circles have an average life-span of six months. The average life-span of a white-collar circle is only 10 months.

Three basic factors have led to the poor showing of quality circles in the United States:

- Organizations were ill-prepared for quality circles.
- The quality circle concept was misunderstood and its successes exaggerated. As a result, it was superficially applied.
- Execution was ineffective because of management's lack of conviction regarding the process.[21]

In addition, shop floor blue-collar quality circles have different problems from white-collar workers' circles (see Figure 2-6).

The long-term future of quality circles is unclear. Some management observers contend that quality circles are unstable structures that face one of two possible fates—extinction or evolution into another form. Evolution into other forms is occurring (see Figure 2-7), given the current macropolitical trend toward a more participatory management style by many companies and the emergence of cross-functional teams and superteams.

FIGURE 2-6 Factors Associated with Quality Circle Failures

Blue-Collar Failure Factors	White-Collar Failure Factors
• Skepticism by employees	• Immediate supervisors' poor attitude
• Fear loss of job	• Lack of projects
• Threat to union	• Staff movement
• Management claims the savings	• Leaders' lack of time
• Undermining by shop stewards	• Disillusionment with the program

Adapted from Michael W. Piczak, "Quality Circles Come Home," *Quality Progress*, 21 (December 1988), pp. 37-39.

FIGURE 2-7 After Quality Circles

SCOPE / SPAN	SUGGESTIONS	INVOLVEMENT
ORGANIZATION WIDE	- ADVISORY GROUPS - TASK FORCES	- MULTILEVEL COUNCILS - BUSINESS TEAMS
WORK GROUP	- QUALITY CIRCLES	- SEMI-AUTOMATIC WORK GROUPS

Source: Michael W. Piczak, "Quality Circles Come Home," *Quality Progress*, 21 (December 1988), pp. 37-39.

Cross-functional Teams

Members of cross-functional teams represent employees from different disciplines and functional areas. Different levels of the organization may be represented as well. Members work on specific problems or on cross-functional business unit issues. Often, the teams disband when their projects are completed.

Superteams

The management innovation breakthrough of the 1990s is the deployment of self-managed superteams. Self-managed teams may be comprised of workers in the same area, like the quality circles of the 1980s, or they can be formed around cross-functional members. The important distinction is that they manage themselves; no boss is required. Corning CEO Jamie Houghton, whose company has roughly 3000 teams, says, "If you really believe in quality, when you cut through everything . . . it's empowering your people; that leads to teams."[22]

Superteams perform their own management and control functions, including determining profit targets and strategies, arranging work schedules, ordering equipment and materials, and hiring and firing team members. They may be permanent or temporary. Superteams are best used when there is a high level of dependency among three or more people and complex manufacturing processes to deal with.

While almost half of the Fortune 1000 companies sampled reported that they will place significantly more emphasis on self-managed teams in the future, only seven percent currently have them in place. The results, however, have been impressive:

- A General Mills cereal plant in Lodi, California, runs its factory with no managers present on the night shift.
- A team of Federal Express clerks spotted a $2.1 million-a-year problem.
- A team of millworkers at Chaparral Steel selected and installed machinery after traveling the world to find the best. Their mill is one of the most efficient in the world.
- Cross-functional teams at 3M tripled the number of new products.
- At Johnsonville Foods of Sheboygan, Wisconsin, productivity has risen by at least 50 percent since 1986.

- Rubbermaid sales on a new product developed by superteams showed sales running 50 percent greater than projected.[23]

Problems of Teams

Organizing superteams also presents some problems.

- **Require consensus.** People must be trained to feel comfortable with airing their opinions. To add to the communication difficulty, the skill mix and demographic characteristics of team members are expected to vary greatly in the next decade. Forecasts of the educational and racial composition of the twenty-first century workforce are given in Chapter 9.
- **Provide less opportunity for advancement.** With leaner organizations, there are fewer middle management positions available for workers to aspire to. Chapter 8 discusses future trends in organizational size and the implications of restructuring.
- **May become ineffective due to facility layout.** Traditional functional layouts may pose barriers to communication and collaboration. Team furniture creates "homey neighborhoods" of workers. Teams work in small areas with a central work table for team meetings; nearby desks are arranged for privacy.
- **May cause resistance by middle management.** Traditional powerbases are eroded. Internal blockers and bureaucratic resistance can cause an uneven commitment to team development and implementation.
- **Are not a panacea.** This is a long, hard process that requires changes in management and employee roles. It is easier to implement superteams in a "greenfield" facility than to adjust an old one.

Meeting Information Requirements Boosts Team Productivity

Regardless of whether the type of Employee Involvement (EI) being deployed is a quality circle, a cross-functional team, or a superteam, information plays a critical role in shaping the quality improvement strategy. The nature of the information and control systems can substantially affect the organization's strategic and economic performance. As will be described in Chapter 5, accounting systems play a custodial role and are the main determinants of information flow in business. As closed systems, accounting systems separate management from ownership and keep tabs on assets rather than contribute to organizational effectiveness. They can stifle initiative and creativity because they place inordinate pressure on the organization to standardize to meet mechanistic measurement demands.

Company-level information systems are seldom neutral. The intrinsic nature of the company's information processing characteristics can either bolster or impede EI team productivity. To boost team performance, follow the guidelines below.

Put decision-making where the information is. Ordinary workers have a lot more information in their heads than management usually gives them credit for.

Simplify information systems. Organizations typically suffer from information overload. Elaborate and expensive control systems either collect more information than the organization can absorb or collect superfluous, biased information that is nonthreatening to the basic values and power bases.

Focus information systems on providing teams with opportunities for problem solving, preventive maintenance, and identifying areas in need of improvement. Push quality data down to the lowest possible level of the organization. Monitoring information which shows bottom-line results, such as cost of quality, percent productivity increase, headcount reduction, and ROI, does little to help teams resolve problems or find opportunities.

Accept data gathered by EI teams as legitimate and not as a "tool of suspicion." Often, overcontrol occurs because management distrusts employees. Managers feel constrained by the measures that are acceptable and valid in their firms.

Get information into the hands of the people who need it. ". . . Without making sincere efforts to increase information flow to workers (companies will) face tough sledding." Data should be in a useable and easy to understand form.

Facilitate information sharing and feedback, both horizontally and vertically. "The technical and nontechnical system planning must take place together, and . . . management must arrange the process so that everyone involved sees it as a continuing, interactive process."[24]

Macropolitical Influence on Teams

The institutionalization of small-group teams has become widespread internationally. The relative success of team activities is influenced by micropolitical factors within the organization that inhibit or enhance the probability of success, such as those cited above concerning quality circles, cross-functional teams, and superteams. However, recent research analyzing team activities in three national settings, Japan, Sweden, and the United States, highlights a macropolitical process that helps explain the relative spread and outcomes of small-group activities. Macropolitical processes include the political dimensions of an entire industry or nation. These processes are outside the control or direct influence of the organization and include elements such as the direction of national business leadership, the orientation of organized labor, and the national infrastructure in place.

This comparative research highlights three important factors that contribute to the spread of small-group activities in a nation: "The incentives embedded in the national labor markets for management to innovate; the establishment of well-funded industry or national level organizations supported by management to communicate methods and support change; and the disposition of organized labor toward these changes and its ability to enforce its preferences."[25]

Figure 2-8 summarizes the patterns of national experiences and the contrasting macropolitical climates. Striking findings suggest that much of the adoption and diffusion of managerial innovation concerning small-group activities are a function of macropolitical influences from outside the organization. Furthermore, Figure 2-9 suggests that survival rates of small-group activities may also be explained in part by macro forces.

The basic requirements for achieving a national infrastructure to legitimatize and promote team activities are:

- National business leadership must be committed to small-group activities as a solution to perceived problems.
- This (small-group activity) must be uncontested by labor.

The seeds of a national movement toward small-group activities have been planted, as evidenced by the awkward evolution from quality circles to superteams. A national movement toward teams could become a normal part of organizational change in the same sense that scientific management, Taylorism, large-lot production, and scale economies took hold and transformed industrial practice in this country.

The Shocking Cost of Quality

Many companies joined the quality revolution because executives were astonished to learn about the impact of quality-related costs. It is unlikely that world-class manufacturers will develop elaborate systems to keep track of the total cost of quality for an organization. However, it is important for managers to have an intuitive grasp of the fact that poor quality drives costs

FIGURE 2-8 Summary of National Experiences

	Timing and Scope of Innovation	Management Incentives to Innovate	Characteristics of Mobilization	Political Dimensions
UNITED STATES	Piecemeal experimentation in 1960s and 1970s. Limited primarily to blue-collar workers at the shopfloor level.	Few incentives perceived.	Many piecemeal efforts, absence of well-funded and centralized infrastructure for diffusion.	Neither government, union, nor management leadership committed to innovation. Unions often hostile.
JAPAN	Began early 1960s. Now widely diffused. Limited initially to blue-collar workers at the shopfloor level.	Significant labor shortage and rapidly accelerated education levels with raised expectations.	Management organizations provide well-funded and organized infrastructure for diffusion.	Top management develops commitment to innovation. Unions in private sector passively accept.
SWEDEN	Began late 1960s. Now widely diffused. Spread rapidly to white-collar and public sector. Operative from shopfloor level to board of directors management.	Severe labor shortage and rapidly accelerated education levels with raised expectations.	In early years, joint union-management efforts develop effective infrastructure for diffusion. Later management organizations provide focus for shop- and office-floor efforts.	Top management, union leaders, and government all develop commitment to innovation. Union committed but sets its own agenda.

Reprinted from "The Macropolitics of Organizational Change: A Comparative Analysis of the Spread of Small-Group Activities" by Robert E. Cole, published in *Administrative Science Quarterly*, Vol. 30, No. 3 by permission of *Administrative Science Quarterly*.

FIGURE 2-9 Macropolitical Dimensions of the Process of Change

	Traditional Union Orientation	Union Perception of Small-Group Activities	Union Action	Union Means to Enforce Preference	Degree of Conflict Consensus with Management	Outcome
UNITED STATES	Control of job opportunities; vested interest in existing job definitions and related pay scales.	Suspicion; threatens structure of bargaining at industry level; disinterest.	Little union action.	Industry-level agreements; factory-level negotiation.	Did not become an issue.	Very limited spread of new structures.
SWEDEN	Centralized national collective bargaining and political action through Social Democratic Party.	Initially suspicious; later saw as a means to satisfy rebellious rank and file, still later as one prong of strategy for work control.	Began own efforts, initially in cooperation with employees, later on own.	National-led labor agreements and passing of legislation through SDP.	Initial active cooperation, later contention of direction and extent of changes.	Initial rapid change, later halted due to conflict over course.
JAPAN	Stress job security; allow management prerogative of organizing workplace.	Neutral; did not threaten vested interests.	Uninvolved, although consulted at plant level.	Industry-led agreements; factory-led negotiation.	Consensual process of change.	Thorough, rapid process of change.

Reprinted from "The Macropolitics of Organizational Change: A Comparative Analysis of the Spread of Small-Group Activities" by Robert E. Cole, published in *Administrative Science Quarterly*, Vol. 30, No. 3 by permission of *Administrative Science Quarterly*.

up. Conceptually, the way to communicate this fact is in the language of money. While the concept was popularized by Phil Crosby's *Quality is Free*, the basic components of quality costs were first identified by Juran in his 1951 book, *Quality Control Handbook*. The notion of the cost of quality can be used to build awareness of the impact of quality on business performance and to help operating units prioritize opportunities for quality improvement and assess their own programs.

The traditional costs of quality are divided into the following four categories described below.

Cost of Prevention: What does it take to prevent defects, failures, and breakdowns? Costs include job training, quality improvement teams, expenses associated with in-process quality control methods (SPC), simulation tools, and adaptive control systems.

Cost of Appraisal: What does it cost to ensure that products conform to specifications? Typical costs are associated with inspection and testing. Costs include Computer-Aided Testing (CAT) systems, Coordinating Measurement (CMM) systems, machine vision systems, and incoming material inspection costs and the costs associated with inspectors and QC engineers and their supervision.

Cost of Internal Failure: How much does it cost to correct processes and procedures before the products reach customers? Internal costs pertain to rework, repair, downtime, scrap, retest of repaired parts, and yield loss.

Cost of External Failure: How much does it cost to correct failures after products have reached customers? External costs are discovered by the customer which results in warranty charges, complaint adjustments, and service groups.

For each of these categories, the costs of quality are typically expressed as a percentage of sales and are regrouped into *the cost of conformance* and *the cost of nonconformance*. Why is quality so important? What is the shock value of the cost of quality? As shown in Figure 2-10, it is not uncommon to see the total cost of quality be 15 to 25 percent of sales, with the following breakdown:

In addition to the quality costs cited above, Xerox uses two others in strategic planning for quality:

- **The cost of exceeding requirements:** What is the cost of doing too much?
- **The cost of lost opportunities:** What is the cost of producing poor quality in the marketplace? This would include the cost of lost sales and reputation due to producing low-quality products.

FIGURE 2-10 Typical Cost of Quality

Cost Components			%Sales
Prevention/Appraisal Costs	==> COST OF CONFORMANCE	==>	5 %
Internal/External Failures	==> COST OF NONCONFORMANCE	==>	10-20%
TOTAL QUALITY COSTS ===>			15-25%

Source: P. Crosby, *Quality Is Free* (New York: New American Library, 1979).

Of these costs of quality, many believe that the external cost of quality is the highest. Others add to the list the cost of keeping buffer stocks and excess capacity as protection against quality uncertainties (see Figure 2-11).

Less traditional advocates of quality improvement efforts suggest that the cost of prevention should not be counted, because prevention is everyone's job. Quality cost information can serve as a prototype cost management system because:

- Quality cost systems incorporate measures of significant organizational success factors such as defect rates, rework costs, and warranty claims.
- The development of quality cost systems necessitates an understanding of the drivers of quality costs.
- Quality cost systems cut across organizational boundaries to accumulate costs for a specific purpose.
- Non-accounting managers often take the initiative to develop quality cost systems; as a result, the systems are better received in operation areas.[26]

Research indicates that total quality costs decrease as companies shift their quality expenditures from failure costs to prevention and appraisal costs. There is growing evidence, however, that firms embarking on quality improvement programs are not developing large-scale tracking systems to measure the cost of quality. In fact, the current wisdom is that the cost of prevention so far outweighs the cost of nonconformance, that there is little

FIGURE 2-11 Quality Cost Distribution

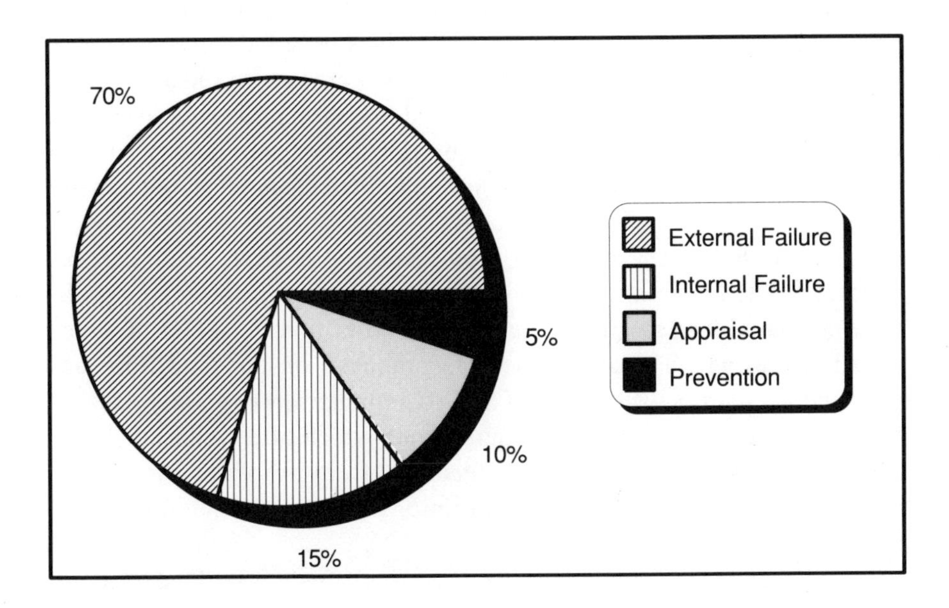

Source: W.J. Morse, "Accounting for Quality Costs—A Critical Component of CIM," *CIM Review* (New York: Auerbach Publishers). © 1986, Warren, Gorham & Lamont Inc. Used with permission.

need to add to nonvalue-added transaction processing of these results. Care must be taken that measuring the cost of quality does not become an end in itself. Rather, quality measurement efforts, like the Taguchi methods to find the optimal cost performance combinations, may yield more practical and cost-effective use of resources.

Statistical Process Control

SPC is not new, but until recently it was not well-accepted in plants making discrete products. Conversely, many process plants have practiced a form of SPC for years, calling it *feedback control*, in which a parameter variation from a set-point value known to be best for the process sends a signal related to the deviation to a control component such as a valve, pump, or heater to effect automatic adjustments to bring the parameter back to the desired reading.[27]

Unlike continuous-product feedback control plants, discrete-product plants have historically been labor intensive. Operators and line supervisors were relied upon to monitor production. If quality problems occurred, personnel either adjusted the process themselves or called a quality person for help. Statistical process control techniques were known and understood in some plants. However, the time required to collect and manipulate data sometimes allowed large numbers of questionable or poor quality parts to pass through manufacturing operations and limited the application in many plants.

Recently, however, increasing competitive pressures, customer demands, and the lowered tolerance of production systems for quality problems have combined to greatly increase the utilization of SPC in all types of manufacturing. The appearance of computers on the shop floor has enhanced the increased adoption of SPC. Computers have greatly reduced the data collection and analysis requirements which often fell to the shop production personnel. Today, SPC feedback from processes can be collected in something approaching real time, with minimal interruption of production efforts.

A final step in the adoption of SPC is currently underway. The injection of advanced technology into the production environment is increasingly automating both quality assurance and production processes. What this means for SPC adoption is that the loop from data acquisition through data manipulation and storage, to decision making, to process control inputs, can be completed without human intervention. It is important to note, however, that SPC is not simply a process control tool. It can also identify the causes of variability and separate those causes in such a way that workers, engineers, and management can change the process and reduce variability. A major reason the United States has not been as successful with SPC as Japan has is that it views SPC as a purely process control/monitoring aid.

There is mounting evidence that Design of Experiments, such as those of Taguchi and Shainin, may push the U.S. further along in quality improvement than SPC. However, SPC is significantly easier to conceptualize and apply and, subsequently, more popular. SPC captures the heart of statistical thinking about variance and randomness. Many companies are using both Design of Experiments and SPC.[28]

In Chapter 7, Just-in-Time (JIT) production techniques and delivery schedules are described. Quality tools like SPC are often used with JIT practices, because successful application requires high incoming quality, known process variation, and overall system reliability and synchronization. The JIT philosophy of waste elimination as an approach to quality will become more widespread.

QFD – Discovering the Missing Link in Quality Management

The quality revolution in America involves many transitions in the definition and application of quality management. Total quality management extends beyond total quality control – beyond producing – to specification. It extends to clearly determining customer and design requirements which can be fed into the manufacturing process. Two technologies adopted for determining requirements are the Taguchi methods, already summarized, and Quality Function Deployment (QFD).

QFD is a broad product development tool that combines aspects of value analysis/value engineering with market research techniques. In SPC, the primary objective is to build the product right the first time. The primary objective of QFD is to specify both the product and the process correctly from the outset. QFD translates the voice of the customer into product specifications by associating customer requirements with the appropriate technical requirement for each stage of product development and production (see Figure 2-12).[29] Furthermore, QFD may bridge a gap between customer-driven specifications and concurrent engineering, as described in Chapter 7.

QFD has been used extensively in Japan since the 1960s. It is a natural for product development planning. Multifunctional teams are often formed to integrate critical product and process decisions. A number of tools, such as the House of Quality and relationship matrix, are used to facilitate QFD applications. Successes with QFD include:

- Toyota's design cycle has been reduced by one third; over a seven-year period with four startups, total ramp-up costs on one product line were cut by 61 percent.
- The number of Engineering Change Orders (ECOs) and design cycle times at Aisin Warner, a subsidiary of Aisin Seiki Co., Ltd., were cut in half.
- Komatsu MEC introduced 11 new products, 5 of which were concurrent, within a two-and-a-half-year development cycle.
- Tokyo Juki Kogyo Co. rejuvenated a mature and declining line of sewing machines, gaining a significant competitive advantage.[30]

FIGURE 2-12 Addressing the "Whats" through QFD

- **Customer Requirements:** What do customers want and need from the product/service? Many customers view the same product/service attributes as important (convenient to use, lightweight, portable, long-lasting, clean design, durable, easily assembled), but it can be difficult to quantify these attributes.
- **Design Requirements:** Loose customer specifications are translated into broad-based "hows" which are often measurable product/service specifications; e.g., "portable" is translated into design requirements for thickness, width, and weight.
- **Parts/Service Characteristics:** Key components of the product/service are designed to meet the critical design specifications.
- **Manufacturing Processes:** The manufacturing processes that are deemed most critical to creating the desired part/service characteristics are specified.
- **Production Requirements:** Shop floor requirements, including statistical process control, worker training and making failsafe, are detailed.

The application of QFD is a fairly recent innovation in the U.S. The first recorded case studies were in 1986. Kelsey Hayes used QFD for a new electromechanical product—a coolant sensor—which fulfilled important customer requirements: "easy-to-add coolant, easy-to-identify unit," and "provide cap removal instructions." Now, Ford, General Motors, and Chrysler have begun using QFD. Others applying the technique are Baxter Healthcare, Digital Equipment, and Procter and Gamble. Due to the newness of QFD in the U.S., detailed evaluation of its impact is not possible. But success stories are beginning to trickle in.

Tools for Managers — The Seven New Tools

Seven problem-solving tools are commonly used by quality circles and cross-functional teams. Many groups are trained in a variety of the KAIZEN problem-solving techniques, including the following seven tools:

- Pareto charting
- Histograms
- Fishbone techniques/cause-and-effect diagrams

- Control charting
- Scatter diagrams
- Graphs
- Check sheets

In Japan, seven new KAIZEN tools for comprehensive problem solving are being taught by quality management consultant groups. A detailed description of the tools is given in Mizuno Shigeru's *Management for Quality Improvement: The 7 New QC Tools.* The tools are outlined below:

1. **Affinity Diagram:** Using essentially a brainstorming approach, many ideas, opinions, and issues are grouped by subject matter.
2. **Relationship Diagram:** This tool is used to clarify the interrelationships often found in complex, multivariable problems and to display graphically the logical, if not causal, relationships among factors.
3. **Tree Diagram:** As an extension of the value engineering concept of functional analysis, this tool systematically maps out, in increasing detail, the full range of paths and tasks that need to be accomplished to reach every goal.
4. **Matrix Diagram:** A versatile tool that is useful in clarifying the relations between two different factors. For example, it is one of the tools used in linking QFD requirements of whats and hows.
5. **Matrix Data Analysis Diagram:** A tool used when further expansion of tasks, issues, and actions from the matrix diagram is required.
6. **Process Decision Program Chart (PDPC):** This tool is used for contingency planning so that the optimum conclusion is reached and surprises are prevented. It identifies all conceivable events that can occur from problem statement to solution.
7. **Arrow Diagram:** This is a (CPM/PERT) network approach used to plan the most appropriate schedule for any complex task and all of its related subtasks. It shows the most likely completion time for each activity and the project completion time.

Most of these tools have their roots in post-World War II Operations Research. They were refined and tested by the Japanese committee for the Society for QC Technique Development and introduced into the U.S. in 1984. Typical applications for the new tools include:

- New product development/R&D

- Safety improvements
- Competitive/market analysis
- All areas of quality improvement
- Quality deployment
- QA systems improvements

Practices in Japan

U.S. industry is attempting to emulate the success of the Japanese in providing a consistent and extremely high level of product quality. Many times in active practice, as well as in much of the current literature, the tendency is to seek the visible tools and techniques the Japanese use and inject them into U.S. manufacturing operations. For example, U.S. manufacturers are now widely utilizing SPC, quality circles, and JIT. However, the implementation in the United States of techniques proven effective in Japan often overlooks the basic foundation from which the Japanese success has been achieved. There are differences between Japanese manufacturers and U.S. manufacturers which are important to note. This section highlights some of those differences discovered in a 1986 research mission to Japan.[31]

Training

Japanese industry has made a major commitment to training its employees in quality improvement methods. More than 100,000 engineers are trained each year in a quality curriculum that covers a broad range of topics. This in itself indicates that all parts of Japanese companies are participating in quality improvement activities.

A major difference between the Japanese quality training programs and those in the United States is the length and format of the courses. In Japanese companies, courses meet several days per month over a period of many months, students work on homework between class sessions, and the total class time may be as much as 150 to 200 hours (see Figure 2-13). In contrast, U.S. training programs are usually built around a short-course format in which the total class time is limited to a single one- to five-day period.

FIGURE 2-13 Typical Quality Courses Offered in Japan

COURSE	LENGTH (HOURS)	CAPACITY
QC Seminar for Management	48	70
Total Quality Control Seminar for Management	24	40
QC Basic Course	188	70
Elementary QC Seminar	52	70
QC Seminar for Foremen	40	50
On-line QC Seminar for Manufacturing	36	30
QC Circle Course for Administrators	22.5	60
Elementary QC Circle Seminar	19	50
QC Seminar for Sales and Clerical Area	24	40
Quality Engineering for Management	12	30
Elementary Design by Experiments	37.5	40
Design of Experiments	130.5	50
Reliability Seminar for Management	24	35
Elementary Reliability	24	30
Reliability	60	40
FMEA/FTA Methods and Case Studies	12.5	60

Source: G.E. Box, N. Raghu, N. Kacker, N. Vijay, M. Phadke, A.C. Shoemaker, and C.F.J. Wu, "Quality Practices in Japan," *Quality Practices*, March 1988, pp. 37-41.

Statistical Methods Emphasize Design of Experiments

The Japanese make wide use of statistical methods. In one company, more than 6000 experiments were conducted in one year. In another, design engineers conducted 48 multifactor experiments over a three-year period during the design and development phases of a product. The Japanese are making significant strides in reducing process variation and problem solving using methods like Taguchi, Shainin, and other classical approaches to the design of experiments.

One area in which the Japanese are lagging behind the U.S., however, is in statistical computing and the use of software for data analysis. The few statistical software packages available in Japan are not comparable to the leading ones on the market in the U.S. Very little attention is given in

Japan to exploratory data analysis, diagnostics, graphics, and other methods that have become routine in the United States.

Smooth Transition for New Products

After a successful market feasibility study in Japan, a chief engineer is appointed to oversee the entire product realization process, from production planning, design, and manufacturing, to sales and customer service. The chief engineer's responsibility and accountability extend to all phases of the process to ensure continuity and a smooth transition from one phase to the next. In addition, some team members from each phase go on to form part of the team for the next phase to smooth the transition and provide for better team work. This methodology contrasts with the U.S. practice of passing a product from engineering group to engineering group sequentially.

Quality Improvement Activities

Many of the quality promotion activities in Japan are aimed at motivating people, improving their skills, and providing greater job satisfaction. Quality in Japan is promoted at the national level and at the regional level through use of recognition awards, commendations, and ceremonies. The Deming prizes, initiated in 1951, have had the greatest effect. These awards can be given to individuals (Deming Prize) and to companies (Deming Application Prize). So many companies have won Deming Application prizes that the Japan Quality Control Medal was set up in 1970 to spur continued improvement in Japanese companies. A company can only be considered for the Japan Quality Control Medal when at least five years have passed since it won the Deming Application Prize.

The Japanese began their focus on quality in the 1950s with statistical quality control, but soon realized that effective quality control (QC) is a companywide activity. They started promoting quality control in all functional areas of the business. Due to the strategic importance of quality, QC activities are run by top management. The general manager of the QC department is usually equivalent to a vice president.

Quality circles are popular in Japan. Toyota Motor Company, with 65,000 employees, has about 6000 quality circles. Top management encourages and actively promotes the activities of the quality circles by arranging periodic meetings of the different circles and giving commendations for outstanding achievement. Quality circles have had a significant impact on performance in Japan.

- Toyota Motor Company and Nippon Electric Corporation each report $10 million in savings attributable to quality circles.
- Nippon Kokan Steelworks claims as much as $66 million in cost savings.
- Overall, the Japanese quality circle movement has been credited with contributing a total of $25 to 30 billion to Japan's economy in less than 20 years.[32]

Design for Customer Satisfaction

The survival of any business requires that the business establish quality for its present and prospective customers so that the value of the product and service delivered is as good as, if not better than, that offered by the competition. Spencer Hutchens, Jr., senior vice president of Intertek Services Corporation and president of the American Society for Quality Control, stated in *Fortune*:

> To survive and prosper these days, a company must find ways to add value to its transactions with customers. The winning strategy is not simply to satisfy your customers' basic requirement but to go beyond their expectations so they'll become loyal buyers who keep coming back.[33]

Some of the quality strategies used by the best companies to go beyond customer satisfaction and to build loyalty were:

- Cultivate a partnership with your customers.
- Enlist employees as colleagues in business.
- Actively engage your suppliers.
- Manage quality by facts, not by guts.
- Prioritize your quality improvements.
- Encourage innovation and experimentation.
- Close the loop through after-sales service.

These strategies are discussed below.

Cultivate a Partnership with Your Customers

Help your customers become winners, too. Baxter Healthcare Corporation shares its corporate Quality Leadership Process with customers. An example is Baxter's partnership with Duke University. The hospital's transfusion service has applied Baxter's quality leadership process to its operations and, as a result, has significantly improved service to patients. In addition, under the aegis of David Auld, vice president of Quality Leadership, Baxter has supported the hospital's extended family. Baxter shared in the development of quality management course material for the Department of Health Administration at Duke Medical Center, so that future health administrators can be adequately prepared in total quality management. Baxter has forged a number of other joint efforts on quality leadership with other leading hospitals. Baxter's formula for success is to know and understand customer requirements and to go all the way to meet those requirements.

The relationship between DuPont and Milliken Company, the giant textile manufacturer which won the 1989 Malcolm Baldrige Award, exemplifies a successful partnership. Only several years earlier, Milliken purchased fibers from a number of suppliers based upon price. DuPont forged a new partnership with Milliken on the principle that greater interdependence between them would benefit both. In most new product areas, the two companies worked together on both product and markets.

Chairman and CEO, James R. Houghton, said that at Corning Glass Works quality is defined as understanding customer requirements and meeting them 100 percent of the time. Corning workers at the State College, Pennsylvania, plant are visiting their customers to learn more about their processes and to see what they can do to meet requirements. Corning boasts that its State College plant's glass is so good that it saved one domestic customer $1 million a year. Houghton further indicated that when Corning sends its glass for qualification by Japanese tube makers, the Japanese customers do not believe the numbers because they are too good.[34]

Digital Equipment has scored big using its service organization as a way of building lasting relationships. When Chase Manhattan Bank's computer system was downed by an electrical fire at 3:00 a.m. on January 4th, 1990, Digital put two field service crews on-site before dawn and, simultaneously, a Digital team from manufacturing, sales, and field service coordinated overnight delivery of additional parts and peripherals, enabling Chase to successfully complete 95 percent of its business transactions the same day.

Enlist Employees as Colleagues in the Business

Total employee involvement in quality improvement goes far beyond quality circles.

- At General Motors, this translates into the new UAW/GM Quality Network. The Quality Network is a uniform program to promote quality improvement in all U.S. operations. Every employee participates in the search for the best strategies for dealing with people, material, equipment, customer satisfaction, environment, and methods and systems.
- Perkin-Elmer, the world's largest producer of analytical instruments, uses a team-based strategy to give employees a greater share in decision-making processes. Teams representing manufacturing, quality assurance, and service participate in the product development process, ensuring that all departments' needs are met.
- At the Corning Glass Works State College plant, people design their own machines, and teams of people completely redesign processes from the office to the factory floor.[35]

Changing an organizational culture is not easy; the transformation is an ongoing process. Jack Welch's Work-Out program, reportedly GE's most far-reaching vehicle for employee involvement, has penetrated only three layers of management at best. GE's quality and productivity gains in the mid-1990s are expected to come from voluntary behavior of the rank and file. Quoting Welch, "Liberating and empowering middle managers and reaching the 300,000 souls still on the GE payroll . . . It has nothing to do with whips and chains and taking heads out . . . We're trying to unleash people to be self-confident, and so to take on more responsibility."[36] Clearly, GE is just beginning its transformation, and persistence will be required up to the next century.

Actively Engage Your Suppliers

Companies have been narrowing their supplier bases over the past decade in efforts to work more closely with a limited number of suppliers. Why? Quality goes up and the price goes down. With its significant reduction in the number of suppliers, Xerox was able to make hefty investments in training for statistical quality improvement and just-in-time manufacturing techniques. In addition, Xerox involved suppliers as partners in product development. Xerox's reported payoff, measured in terms of parts per million (ppm) defects, went from 10,000 ppm in 1983 to 350 ppm in 1988.

Motorola's Partnership Growth Advisory Board proactively breaks down adversarial relationships with suppliers that have existed for decades. The board includes 15 supplier executives who are encouraged to advise

Motorola about what it needs to be a world-class company, from their perspective. After winning the Malcolm Baldrige National Quality Award in 1988, Motorola urged all suppliers to start implementing *Six Sigma*-like programs and apply for the Baldrige Award themselves. Building a world-class supplier base demands the formation and cultivation of such relationships and open communications. Motorola, like Xerox, has devoted considerable energy and resources to educating its suppliers and developing their capabilities through training, hands-on technical assistance, and demonstrations.

Ford's *Quality Is Job One* program turned its zero defects crusade outward to its suppliers. Ford is working with suppliers to meet higher and higher standards. For example, any North American supplier who wants to obtain new business from Ford Motor Company after January 1, 1990 must have a Q-1 award status from Ford. This means that approximately 66 percent of Ford suppliers must become qualified. Supplier partnerships push excellence to the limits.

Manage Quality by Facts, Not by Guts

Being a world-class manufacturer demands significant cultural and behavioral changes at all levels of the company. Management must be able to manage by fact and have the information that taps into customer satisfaction. As will be seen in Chapter 3, perceived levels of customer satisfaction are correlated with business success. Companies must use measurement systems to identify customer needs, expectations, and satisfaction and to measure internal and external customers' attitudes. Methods such as QFD and Taguchi are powerful in determining measurable requirements. Other important measures are the level of service quality delivered and gaps between customer expectations and the service delivered.

The Forum Corporation, a leading training and research firm, conducted an extensive survey of customer service practices at 15 distinct business units within 14 companies. It found that customers were more likely to switch vendors because of perceived service problems than for price or product quality issues and that employee attitudes about how well the company was doing with respect to serving its customers were highly correlated with attrition. The number one reason for employee turnover was the perception that the company was not doing a good job or caring enough about serving its customers well.

Prioritize Your Quality Improvements

Not all improvement efforts are equal. Some are likely to produce higher payoffs than others for the resources committed. *Fortune* reported that, in the early 1980s, Johnson & Johnson, Xerox, and AT&T sponsored Technical Assistance Research Programs (TARP) to devise a methodology that addressed quality priorities. The TARP method is summarized below:

1. Identify the frequency of major sources of customer dissatisfaction, from pre-sale to post-sale product performance and service.
2. Identify the impact on the market resulting from a single occurrence of the problem such as the degree to which an unhappy customer will not repurchase or recommend the product or buy other products made by the same company.
3. Determine the overall market damage by multiplying the overall customer base by the problem frequency and by the percentage of customers lost if the problem is encountered.
4. Rank order the problems in terms of market damage.
5. Determine the cause and evaluate action alternatives to eliminate each problem.
6. Perform a cost-benefit calculation to prioritize the actions.[37]

Encourage Innovation and Experimentation

Encourage innovation and risk taking as an opportunity to better meet customer requirements. Quality improvement has rekindled the entrepreneurial spirit of Eastman Kodak which renewed its emphasis on quality in 1983-84. Chairman Colby B. Chandler states:

> There is a new sense of urgency at Kodak. The future is now. We can no longer afford to take years to bring new products to the marketplace. By emphasizing quality at every level of the company we've been able to speed up the process without losing sight of our commitment to quality products and services at competitive prices.[38]

At 3M, the strategic management imperative today is to go beyond customer satisfaction. The vehicles for doing so are innovation and quality based upon real demands made directly by customers.

Innovation by experimentation is one way in which America can leapfrog the Japanese in their own game of concentrating on design quality as well as production quality.

Figure 2-14 shows that the Japanese rode the crest of the SPC wave until the 1970s, just when the U.S. started to climb on the SPC bandwagon. In one U.S. company, a turret lathe, turning out cylindrical rotor shafts with a diameter requirement of .25 inches .001 could not hold tolerances. The foreman recommended that the lathe be junked in favor of a new one costing $70,000. Statistical experiments were performed with the old lathe to determine the causal factors. The bottom line was a $450,000 cost reduction due to quality improvement and purchase cost saving accrued in the first year alone, because the reject level was reduced to zero.

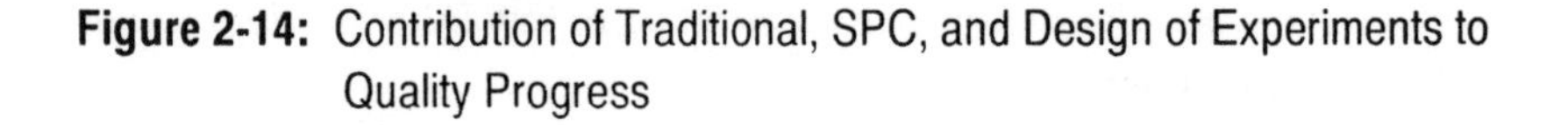

Figure 2-14: Contribution of Traditional, SPC, and Design of Experiments to Quality Progress

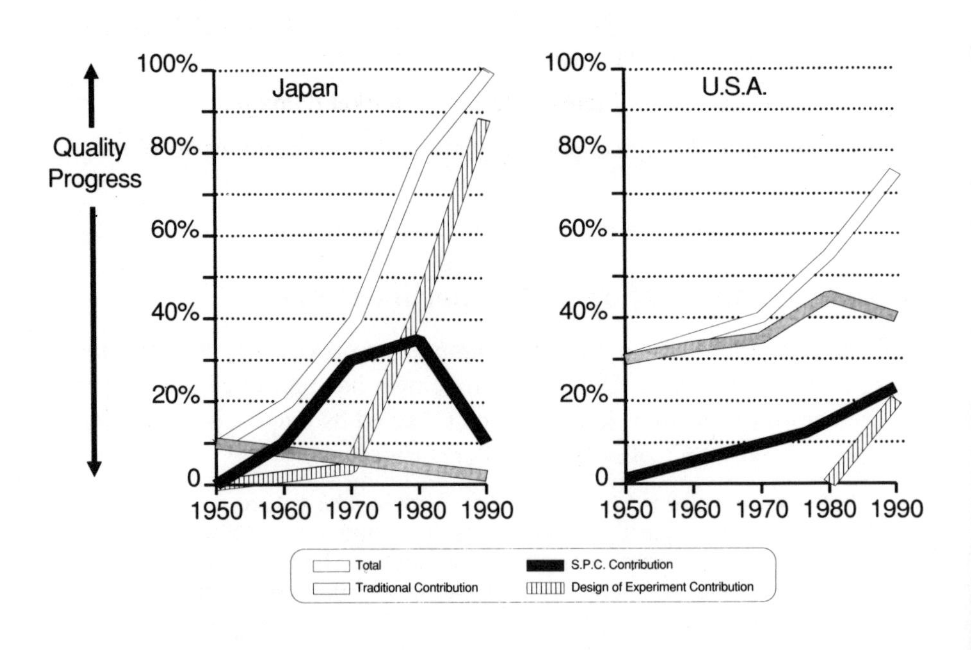

Source: K.R. Bhote, "DOE-The High Road to Quality," Reprinted by permission of publisher, from *Management Review*, January/1988, © 1988. American Management Association, New York.

Ironically, the Japanese secret weapon of today is the application of Design of Experiments to reduce variation:

- The perception of quality is enhanced. Customers want products that consistently and uniformly cluster around target values, not products that merely meet specifications.
- Focusing upon target values reduces field failures which represents huge profit potentials.
- Improve quality by factors of 10:1 to 100:1 in a few years.

Close the Loop through After-sales Service

The after-sales service organization has traditionally been viewed as the means by which the manufacturer installs the product and ensures reliability of the product in the field, possibly as a strategic substitute for better quality. It has been treated by many practitioners as a necessary evil, rather than as an opportunity to cement a relationship with the customer.

Customer bonding, the process of tying a customer to a firm's particular product or service, improves sales potential, obtains feedback, and builds synergies. After-sales service provides a number of unique opportunities to better serve customers and to pursue "close the loop" discussions with customers that can feed into product design, distribution, and pre-sales.

IBM sales and services forces are taking almost unbelievable measures to please their customers. They are comparing sales representatives with total revenue, not commission, so that they can concentrate on the customers' needs. They are even tying in other companies' products if that is what the customer must have. IBM's redeployed field operation is learning to embrace customers who mix and match brands. Often, they are an integral part of the customer's team. Eager to please customers, IBM is getting technologies out of the lab faster and closing the loop.

To close the loop, two critical factors must be considered (see Figure 2-15):

- The role of technology.
- The role of customer contacts.

Advances in technology, particularly complex telecommunications and information technology, are creating new opportunities for firms to exploit after-sales service as a competitive weapon. This enabling technology extends the scope of after-sales service significantly beyond that of traditional field service. This potential is especially high for capital goods.

FIGURE 2-15 Closing the Loop

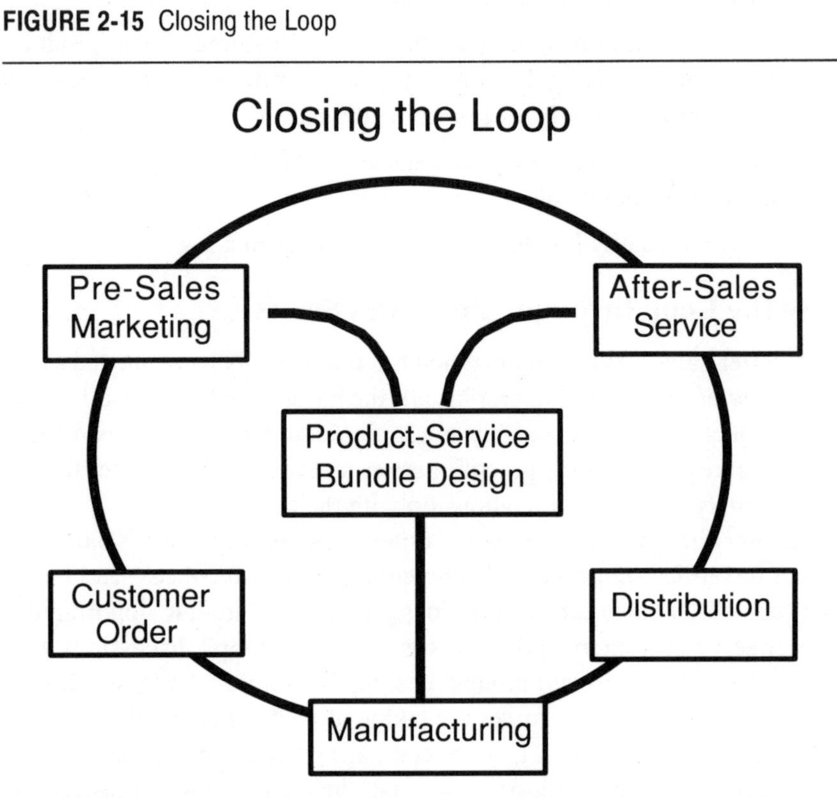

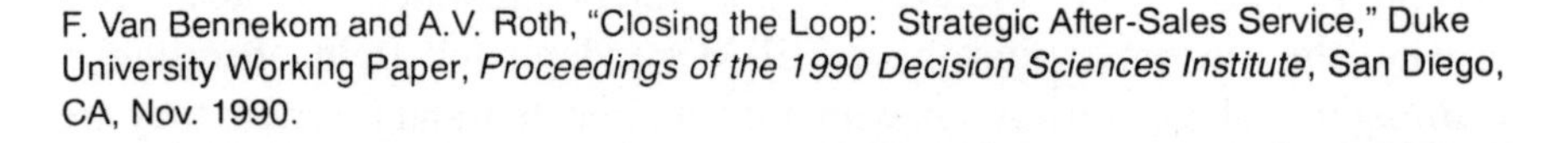
F. Van Bennekom and A.V. Roth, "Closing the Loop: Strategic After-Sales Service," Duke University Working Paper, *Proceedings of the 1990 Decision Sciences Institute*, San Diego, CA, Nov. 1990.

The linking of whole organizations with one company through the use of information technologies has, in essence, created one long-linked technology where the information exchange is substituted for traditional buffers between functions, and even with customers. But technology will never totally replace people. What people really want is service—personal service, with a personal touch. Instead, the role of technology is to turbocharge workers by making them smarter and better service providers.

After-sales service support can build customer relationships by improving the capabilities of the company's customers as well. As in the Corning, Baxter, DuPont, and IBM examples previously presented, customers are better able to compete because of the relationship. Therefore, understanding the role of contact in this interplay between customer and service provider becomes even more imperative in developing an after-sales service as a component of an operations strategy for quality.

The potential role of the after-sales service organization as a contact mechanism for client relationship development and improving quality in manufacturing has not been developed. A basic conceptual model for the strategic roles of after-sales service is shown in Figure 2-16. Note that a high level of customer contact can be provided through either technology or face-to-face contact.

FIGURE 2-16 Strategic Positioning of After-sales Service

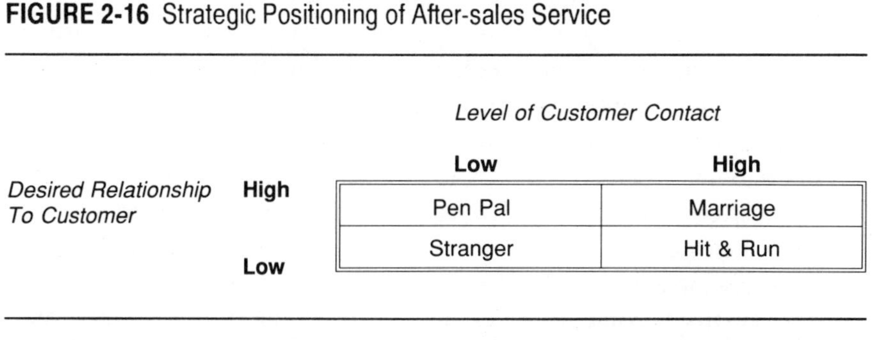

Adapted from F. Van Bennekom and A.V. Roth, "Closing the Loop: Strategic After-Sales Service," Duke University Working Paper, *Proceedings of the 1990 Decision Sciences Institute*, San Diego, CA, Nov. 1990.

These are the kinds of after-sales relationships customers and companies have:

Stranger: After-sales service that is designed for low levels of customer contact and shallow relationships is likely to be treated as providing *ad hoc* repairs to products. Customers generally remain strangers to the firm and contact is avoided as a cost. The strategic role of after-sales service is one of cost minimization. There are few opportunities to increase sales directly. Stranger services may be subcontracted out without much opportunity cost to the firm.

Hit & Run: Hit & Run services are those designed for a high level of customer contact with a shallow relationship. Appliance repair done at the customer's home, not involving a warranty or service contract, is an example of this type of service.

For both Stranger and Hit & Run services, being competitive means designing reliable products with low failure rates that sell at a low cost and can be repaired quickly. It also means that orders are processed quickly and products are manufactured with a high degree of conformance quality and delivered in a timely manner.

When companies desire to build relationships with customers, however, they have the opportunity to turn those customers into clients. The competitive advantage now becomes proactive in that both sales support and ongoing user support are enhanced. Two relationship-building strategies for after-sales service are possible:

Pen Pal: In Pen Pal services, customer contact is minimal but a membership or relationship exists with the company. For example,

- American Express cardholders have a significant membership relationship with the company but relatively low contact.
- Maytag's new 10-year/money back warranty program on its refrigerators is an attempt to forge a relationship with little or no contact.
- At the GE customer service center in Louisville, the customer service representatives are empowered employees equipped to be responsive to customers. Data is fed back to product design and manufacturing.
- Armstrong created a toll-free service center to tell customers how to keep their no-wax floors clean and shiny. The center has representatives available 12 hours a day. As it turns out, this after-sales service has actually become pre-sales support for prospective customers and, hence, this service closes the loop.

Not only are Pen Pal customers satisfied, but their knowledge is captured by the service and fed back to the company. This membership relationship allows the business to use direct mail and other forms of sales more effectively.

Marriage: Marriages offer economies of scope for the business. Customers have high contact with the service provider and build intimate relationships. Marriages are important, because both customers and manufacturers derive significant synergy from a relationship that is mutually supported. Marriage services develop the customer's knowledge base, help draw new sales, increase the opportunity for introducing new ideas for products, and provide dynamic feedback to the manufacturing firm.

Marriages occur when providers have resident employees in the customer's business and/or have an ongoing technological interface that bridges the direct contact gap.

Pioneer International builds marriage relationships with farmers. In *The Service Edge*, Ron Zemke reports that Pioneer's agronomists not only produce seed but become specialists in the conditions of a local area. They meet regularly with their customers, send out timely publications, and provide proprietary software to help farmers manage their own businesses.

In practice, many firms employ hybrid service strategies. The point here, however, is that the operations designed for each of these pure strategies may vary considerably.

- The after-sales service operations design must be consistent with its purpose in the strategic bundle of products and services.
- After-sales service offers a tremendous opportunity to close the loop between manufacturing and other functional areas of the business.

Conclusion

Commitment to quality is one of the most prominent initiatives that a company can undertake. Its objective is to arrive at an elusive goal — to go beyond total customer satisfaction to win trust and loyalty. Quality must be designed into the entire organization. Some key questions senior managers should ask:

- How is quality defined and measured?
- How is quality communicated?
- What is the cost of quality?
- How do quality and service objectives fit into the strategic plan? What are the functional strategies?
- How are quality policies being adapted and supported?
- Are the employees capable of delivering quality; e.g., are they equipped to do so and are they knowledgeable?
- Is the company's quality toolkit complete?
- How does the company's product and service mix build trust and loyalty?
- What percent of management's time is devoted to quality improvement?

The answers to these questions cannot be left to the experts. Nor can quality be delegated. It must be embodied in the culture of the organization, the design of the processes, and the preparation of the workforce. Quality is a moving target that allows for no veering off course or self-congratulation. The modern approach to continuous improvement requires that all employees be involved in the process.

Management behavior reinforced by current financial measurement systems is the root cause of American quality problems. A primary job for twenty-first century managers is to keep all employees motivated for quality improvement. Unfortunately, many American managers do not perceive this challenge as their primary job. "For macho manufacturing managers to push for deep change, they have to become convinced that the company has serious problems. Unless they are convinced they are in trouble, they won't search for a solution. And unless they continue to believe that, they will not continue to hunt out the best. And of course, that is often a bitter pill to swallow."[39]

Quality requires that the critical elements of manufacturing strategy be covered: management of human assets, product and manufacturing leadership capabilities, cycle-time reduction, and performance management systems. Each of these areas will be discussed in subsequent chapters. Clearly, quality improvement is at the heart of world-class manufacturing. And continuous improvements in product and processes are essential to survival in a global economy.

CHAPTER 3

MANUFACTURING STRATEGIES AND AGENDAS

Know the enemy and know yourself; in a hundred battles you will never be in a peril. When you are ignorant of the enemy but know yourself, your chances of winning or losing are equal. If ignorant of both your enemy and yourself, you are certain in every battle to be in peril.

Sun Tzu
Chinese General (440-30 B.C.)

While Sun Tzu's experience as a general in the service of the King of Wu in the fourth century B.C. may seem rather far removed from the commercial and industrial wars of the late twentieth century, his thinking is remarkably fresh. Much of our understanding of present-day business strategy is based on the writings of military and political strategists over the last three thousand years.

Consider Komatsu Limited's assault on Caterpillar Tractor Company. With the rallying cry of "Encircle Caterpillar," Komatsu carefully evaluated Caterpillar's weaknesses and then set out to build into its operations the capability to exploit those weaknesses. Caterpillar did not foresee the potential threat of Komatsu and was relatively unprepared for the breadth and depth of the attack.

Caterpillar now knows its enemy and the complexity of the battle. From 1983 to 1987, it revamped its manufacturing strategy in an effort to bring down costs. Its 1986 *Plant with a Future* program called for investments of $1 billion a year over five years to overhaul its manufacturing operations.[1] It is also seeking government support for a policy on currency squeezes. As of 1990, Caterpillar has made great strides in knowing and improving itself. However, executives at Caterpillar's world headquarters still fret that a force beyond their control is undermining their efficiency drive. The difference in the exchange rate alone has slashed Komatsu's costs by a greater percentage than the cost reductions Caterpillar expects from its overall modernization program.[2]

Caterpillar's experiences illustrate the problems experienced when many adverse situations combine with a company's sluggish reaction to encroaching global competition and exchange rate instability. Similarly, North American automakers missed the threat of Japanese competition. They failed to notice as Honda built into its motorcycle manufacturing many of the capabilities in engine design and production that later allowed it to enter the automobile business. Today, Honda's core capabilities in engines form the center of its competitive advantage in motorcycles, automobiles, and electric generators.

The changing market situation requires constant vigilance and accurate surveillance. The lessons for American manufacturing are clear. Product superiority is the basis for global leadership; manufacturing costs play a critical role in the competitive battle; floating exchange rates require flexibility and timely changes in sources of supply. To win the battle, manufacturers must design and execute strategies that are resilient to currency fluctuations and consistently meet customer needs.

John Deere & Company's Consumer Products Division foresaw that the competition in the mid-1970s was not domestic but from the Pacific Rim. It decided that the best way to meet the formidable challenge ahead was to anticipate it and prepare for it before it became insurmountable. First, Deere evaluated its purpose as a manufacturer to meet the needs of customers, broadly defined as workers, John Deere dealers, and end customers. Second, it laid out a strategic course of action that focused the factory on quality and reliability. For six consecutive years, sales have increased by more than 18 percent a year.[3] Strategies may never work perfectly, but to neglect them is perilous.

The Birth of Business Strategy

Although a distinction was made between corporate and business unit strategy early on, it was not until the 1960s that manufacturing strategy was recognized as an organized field of inquiry.

- Corporate strategy defines what businesses the company should be in, how resources are to be acquired and allocated among different businesses, and the corporate objectives.
- Business unit strategy defines how the business is to compete in terms of markets and products and the distinctive competencies required.

- Manufacturing strategy is the blueprint that defines the manufacturing choices that will give the business a distinct advantage in the market, and that provides for the coordination of manufacturing support so that products *win orders* in the marketplace at a better level than competitors.

Terry Hill, in *Manufacturing Strategy*, indicates two strategic goals of the manufacturing function. First, that the distinct, unique technology developments in manufacturing processes and operations offer a marketing advantage that competitors are unable to match. For example, Northern Telecom and NeXT Inc. have developed sophisticated, flexible, and difficult to replicate software systems to run their factory automation. The North American Manufacturing Futures Project data indicates that successful firms have more internal control over their core process technologies.

Second, manufacturers must make process and infrastructure design choices (for example, procedures, information systems, and human resource policies) which are consistent with the way that products win orders. The specific nature of the manufacturing choices will be discussed in the next section. How leading manufacturers orchestrate their strategies to win orders is the subject of Chapter 4.

Wickham Skinner in *Harvard Business Review* reports that "architectural skill is the key to change. We have generations of managers who are good at running operations. That's housekeeping. But what the United States needs now is managers who can design their own manufacturing systems — not just stabilize and coax production out of them. If the structure of a plant, like the character of a building, is not designed for a specific, focused objective, it can only fail."[4]

Manufacturing strategy is really about specifying the architecture of the manufacturing function so that it provides strategic strength to the company. Manufacturing strategy is a set of design choices that are made to reflect and support:

- How the business will compete.
- The specific competitive product lines and markets.
- The competitive priorities of the business.
- The core process and product technology.

The nature of the manufacturing task is new; the key issues are no longer confined to operational control, and a whole new perspective of customer is required. For these reasons, the world-class companies have evolved customer-driven manufacturing strategies.

Figure 3-1 shows the development, or metamorphosis, of manufacturing strategy from unfocused manufacturing activities to a fully integrated customer-driven strategy.

This framework is important in describing the conceptual aspects of manufacturing strategy and for evaluating and exposing the implications of manufacturing in the business. While the exact number of firms in each phase is unknown, one-third of the respondents to the 1988 North American Manufacturing Futures Survey indicated that they had no current programs or activities to define a manufacturing strategy, leading to the presumption that this group has either no formal strategy or, at best, an articulated strategy.

FIGURE 3-1 Metamorphosis of Manufacturing Strategy

Phase 1: No formal manufacturing strategy: Unfocused manufacturing activities *push* production.

Phase 2: Articulated manufacturing strategy: Charismatic leader heads the manufacturing function.

Phase 3: Written manufacturing strategy: Blueprints for action.

Phase 4: Communicated and visible manufacturing strategy: Everyone knows his or her part within the strategy.

Phase 5: Customer-driven strategy fully integrated within the business: Value-added is maximized throughout the organization. Customer requirements *pull* production.

Source: A.V. Roth and H. Schneider, "Customer Driven Manufacturing Strategy: A Paradigm Shift," Duke University, 1990.

Phase 1: No Formal Manufacturing Strategy

Throughout much of the post-war period, top managers able to turn out product and control costs under strong growth pressures were regarded as heroes. Limited competitive pressures permitted American manufacturers to divert their attention from the details of production to the more glamorous fields of marketing and finance. The typical role of the manufacturing function was performing short-term tasks to meet day-to-day targets and make products designed outside the manufacturing area. Manufacturing was centered largely on efficiency and productivity of operations. Many manufacturing managers did not know the business objectives of the factory. They were there to make product. There was little coordination of manufacturing with the business plan. Customers were forced to make tradeoffs between product and service attributes, such as cost and quality or delivery and quality. The presence of a clearly defined manufacturing strategy was virtually unknown and the utilization of the manufacturing function as a competitive weapon was unfocused at best.

On the other side of the globe, Japanese manufacturing firms were creating unprecedented value as part of their global strategy. The Japanese strategy featured price, quality, and service concurrently. The Japanese focused a tremendous effort on achieving product quality, lowering product cost, and developing a service network. The Japanese pulled demand. From the start, they created high-value products that filled needs and created a sense of confidence. The role of manufacturing in achieving these capabilities was significant.

American profitability and market share dwindled. American manufacturers did not understand that customer requirements drive profitability. Nor did they understand the critical role that manufacturing plays in creating the order-winning capabilities by which they could offer a bundle of product attributes simultaneously and build market share.[5] American firms displayed an inability to effectively deploy the manufacturing function for competitive advantage. Manufacturing was relegated to a low status. Conventional financial justification techniques often hindered renewal of technology. Organizational learning and readiness were stifled by traditional performance measurement systems.

Phase 2: An Articulated Manufacturing Strategy

The global competitive shakeout of the 1970s and 1980s transformed the American manufacturing function. Leading companies found that manufacturing mattered. Charismatic leaders surfaced who, many believed, would elevate American manufacturing performance to an unprecedented status. Lee Iaccoca ran Chrysler by his charismatic style and vision. Steven Jobs at Apple Computers created the revolutionary Macintosh personal computer. These leaders and their management teams shaped the manufacturing functions of their organizations through instinct and natural leadership.

Phase 3: A Written Manufacturing Strategy

As the 1980s progressed, leading manufacturers developed written manufacturing strategies as blueprints for action. These firms found that the discipline required to put onto paper the strategic actions being considered not only clarified the entire thought process, but resulted in a type of formal contract. Once on paper the plan is real. This process of committing thought to paper has been found to be critical in the successful execution of planned activities.

Phase 4: A Communicated and Visible Manufacturing Strategy

Other firms not only documented their plans, but communicated them to everyone in their organizations. Researchers found that as understanding of the basic manufacturing strategy permeated the manufacturing function, profitability increased. According to the 1988 North American Manufacturing Futures Project, 10 percent of the responding firms provide their employees with a broad understanding of the basic goals and plans of the manufacturing enterprise. The manufacturing strategy diffusion process appears slow, as 4 percent of the firms surveyed in 1981 shared their plans and goals with all employees.

Best manufacturing practices of the twenty-first century are expected to accelerate the rate at which manufacturing strategies will be shared with employees. This is expected to be a result of changes in manufacturing performance systems (see Chapter 6).

Phase 5: A Customer-driven Strategy

Developing and implementing a manufacturing strategy is obligatory today and will continue to be important. The focus, the content, and the process of implementation, however, are rapidly being redefined. A major evolutionary shift in the manufacturing function is occurring. It will change the character of manufacturing strategy and intensify over the next decade. Manufacturing firms are entering Phase 5, the development of customer-driven strategies.

Customer-driven strategies pull requirements through the firm and serve as the focal point of the total business effort. In coming to grips with the

CASE STUDY

The GE Motors Market-back Transformation Strategy

Having customers drive the total business is the essence of the GE Motors "market-back business transformation approach," according to Herb Schneider, manager, GE Corporate Business Development and Planning. Meeting or exceeding customer requirements is paramount at GE Motors, a division of GE. Cross-functional business teams include engineering, sales, employee relations, finance, marketing, and manufacturing personnel and are headed by a cross-functional steering committee.

Each team member visits three key customers to determine what issues exist, what their perceptions are of GE Motors, and what their needs are. Customers are selected to represent all product/market segments. Teams summarize what they hear from customers and contrast it with their performance. All managers are required to take a total business perspective and become totally involved.

The manufacturing process at GE Motors is completely integrated and defined by customer requirements; it is customer-driven. When customers identified 3 to 18 week cycle times as an obstacle, the problem was pushed down through the entire business, not just to production. Cross-functional project teams were created to deal with reducing cycle time. The GE Motors multifunctional approach reduced the time from order placement to delivery to less than 14 calendar days. Production's role in reducing the production cycle time was aided by changes in product design and pricing strategies, implementation of JIT, and redefinition of supplier relationships.

This case was prepared by A.V. Roth based on an interview with Herb Schneider of GE Corporate Business Development and Planning.

competition, companies such as Ford, Xerox, and GE have made a concerted effort to understand and be responsive to consumer needs.

Customer-driven manufacturing requires strategic management of the interrelationships between manufacturing and other functional areas and direct contacts with customers. The new role of manufacturing is relating to and working with customers in new and different ways. The traditionally buffered technical core is being unveiled for meeting total customer needs.

"The factory of the future is not a place where computers, robots, and flexible machines do the drudge work. That is the factory of the present which, with money and brains, any manufacturing business can build. Of course, any competitor can build one too—and that's why it is harder to compete on manufacturing excellence alone." The factory of the future is a service factory.[6]

The *service factory*, as the hub of new activities, will provide a full range of services that extend beyond the traditional focus on material transformation. Manufacturers, such as Harley-Davidson, are showcasing the factory floor and using it as a way to build customer relationships. The factory will also serve as a consultant to the field and customers and as a laboratory in which to experiment with better products and processes. In the quality revolution, close cooperation between manufacturing and suppliers is common, as is the role of after-sales service in closing the loop formed by manufacturing, marketing, design, sales, and research and development.

Manufacturing Strategy for the 1990s

Before launching into a full-scale discussion of manufacturing strategy, it is worthwhile to note that, except in very special circumstances, excellence in functional strategies does not necessarily translate into excellence in overall business strategy. *A manufacturing strategy cannot be successful without the integrated support of all other functional areas.* Manufacturing excellence in the 1990s requires that the manufacturing function have a clearly articulated customer-driven manufacturing strategy and the means to execute it.

Viewing a manufacturing strategy as a pattern of choices produces an operational yardstick where behavioral comparisons of strategic choices among manufacturing companies can be systematically gauged. For example, when low price is viewed as the competitive weapon in the marketplace, the manufacturing capability to produce at low cost is key. Competition on the basis of price requires that manufacturing choices concerning location, product design, job design, equipment, and process technology be made on the basis of efficiency-related criteria to lower the production costs.

Henry Ford's goal of putting an automobile within the reach of the average American family meant that the Ford Motor Company had to develop a manufacturing strategy consistent with this objective. Assembly line production processes and specialization of labor workforce policies were developed. It was said that you could purchase a Model T in any color – as long as it was black. And black Model Ts were produced for an eager populace at a low cost. The pattern of manufacturing choices made by the Ford Motor Company at that time can be called its *manufacturing strategy*.

The 1990s offer manufacturers opportunities for organizational renewal and creation of a competitive advantage, and the chance to restructure to develop new capabilities. Companies are rediscovering manufacturing and its importance in achieving their overall strategic goals. Often, this is expressed as a company goal to achieve world-class manufacturing status.

Manufacturing Strategy Is a Set of Choices

A manufacturing strategy begins by identifying the importance of the internal capabilities which the production unit possesses or must develop for the business unit to compete effectively. Determining the appropriate manufacturing strategy for a company requires an honest appraisal of the manufacturing capabilities of the company in terms of its competitive priorities.

Manufacturing capabilities enable a company to exploit technological and market opportunities to win orders and maintain share. Clearly, the primary forces driving manufacturing excellence are management decisions concerning how the company will invest its resources to develop and maintain the required competitive capabilities.

Whether or not a formal manufacturing strategy is written, every manufacturer chooses a set of key action programs that are to be undertaken and are to be the focal point of management attention and resource deployment. Each manufacturer's specific portfolio of key action programs defines its manufacturing strategy. Chapter 4 describes the manufacturing strategies of leading manufacturing firms. Remaining chapters highlight important aspects of manufacturing decisions.

A useful conceptual framework for organizing the diverse sets of manufacturing choices is shown in Figure 3-2. Within this framework, the manufacturing strategy is comprised of a series of structural, infrastructural, and integration choices. The structural components are the hard, or brick-and-mortar, choices concerning technology, capacity, facilities, and vertical integration. Infrastructural, or soft, aspects of the operations are comprised of the management policies and systems that are linked to structural components. Integration choices describe how manufacturing will strategically interface with internal and external functional boundaries.

FIGURE 3-2 Manufacturing Strategy as a Pattern of Choices

STRUCTURAL STRATEGY CHOICES*

Facilities

- Where should facilities be located?
- What are the number and sizes of the plants?
- How should they be focused?

Capacity

- How much capacity should be available?
- How should the capacity be utilized?
- What is the correct mix of permanent vs. temporary capacity?

Process Technology and Equipment

- What technology should we employ?
- How much of our process technology should we develop internally?
- How should we service and maintain the equipment?

The Degree of Vertical Integration and Control

- Should we make or buy our materials?
- How many suppliers should we have?
- How should we manage and control our supplier network?

Materials Systems

- How should we store and transport our materials?
- How should we control and secure our materials?

INFRASTRUCTURAL STRATEGY CHOICES*

Human Resource Policies

- What skills are required?
- How should we train our employees?
- What should we train our employees on?

Quality

- What is our quality policy?
- What approach will we use to achieve our quality objectives?
- What is the optimal quality position for our company?

Production Planning and Control

- How should we plan our material needs?
- What system of production control will we employ?

Figure 3-2, concluded

Performance Measurement and Rewards

- How will we measure performance?
- How will we define superior performance?
- How will we reward desired performance?

Organizational Structure and Design

- How will we organize?
- How many layers in management do we need?
- Should we organize around product or process?

INTEGRATION STRATEGY CHOICES**

External Integration

- What types of relationships will be developed with suppliers?
- What types of relationships will be developed with customers?
- What will be the degree and nature of customer contact?
- What types of strategic alliances and partnerships will be employed?
- How will the global networks be established and communication maintained?

Internal Integration

- How will the interfaces with other functional units—R&D, marketing, engineering, services, finance—be bridged?
- How will strategic fit with the business unit be linked with manufacturing?
- What mechanisms will be employed to integrate product/process choices?
- How will manufacturing technologies be shared between the business unit and corporation at large?

Adaptive Mechanisms

- How will organizational learning be hastened?
- How can the manufacturing function work smarter?
- Where can system synergies occur?
- How will technology be transferred?

(*)Adapted with the permission of The Free Press, a division of Macmillan, Inc. from *Dynamic Manufacturing: Creating the Learning Organization* by Robert H. Hayes, Steven C. Wheelwright, Kim B. Clark. © 1988 by The Free Press.

(**)Adapted from A.V. Roth and H. Schneider, "Customer Driven Manufacturing Strategy: A Paradigm Shift," Duke University, 1990.

Boundary Management

The competitive environment and the strategic manufacturing task have profound implications for the types of knowledge that a manufacturing organization requires. New knowledge requirements often come from outside the manufacturing function as companies competing to design successful products contend with different types of process technologies, materials, and skill requirements.

The more traditional approach to manufacturing structural and infrastructural choices follows a closed systems model. The manufacturing core and, concomitantly, the architecture of manufacturing strategy, has been designed to be buffered from other departments and the external environment. Boundary information had been filtered by management, specialists, and support functions. However, as organizations become flatter, line and staff positions are combined; workers are empowered, and the traditional boundary spanning roles fall to employees.

This evolution suggests that an open systems model of manufacturing, which requires an ongoing information exchange between the manufacturing functions and the external environment, may be more appropriate for operating in the twenty-first century. When it comes to integration, the set of manufacturing choices becomes an arsenal of new strategic weapons. *Boundary management* is the manner in which integration strategy choices are implemented.[7]

Current manufacturing innovations, discussed throughout this book, are pushing manufacturing toward an open systems perspective. Among the most prominent or these innovations are concurrent engineering of product and processes, after-sales service, vertical control of suppliers and distribution (JIT/TQM), interfacing marketing and manufacturing (QFD), globalization of manufacturing, and participatory human resources policies (EI). Each of these action areas involves the coordination of various functional plans in which the management of cross-functional and cross-firm linkages is vital. Performance measurement on the quality of interfaces is presented in Chapter 6.

For example, manufacturing, finance, marketing, and product development may all have different plans for a new product, which when taken alone or combined without an overall structure, may be incompatible. Boundary management defines the way in which the strategy instruments "sound" together. Boundary management will reshape the architecture of the manufacturing organization to an open systems design.

Boundary management places a premium on exploration of interface choices at the manufacturing perimeters. Its objective is to open communication channels, to search actively for new opportunities, and to assimilate appropriate new knowledge into manufacturing and vice versa to the external parties. Therefore, a whole new set of interactions that were not contained in the existing manufacturing strategy must be exploited for competitive advantage. Boundary management extends beyond the mere reinforcing of current manufacturing capabilities. It is about creating new capabilities that pose significant barriers to entry.

Boundary management forces manufacturing to ask a new set of questions, to employ new problem-solving approaches, to develop new manufacturing capabilities, and to continually redefine the way in which the choices are to be linked together. Usually, the core technology of the manufacturing function remains the same, but refinements in the integration and the linkages or reconfigurations of existing resources are made. At the same time, the manufacturing function must change its orientation from one of containment within a well-defined area to one of active searching for new solutions within a constantly changing business context.

Attaining world-class manufacturing status requires the development and implementation of a customer-driven manufacturing strategy. A major hurdle is that manufacturers must simultaneously consider all elements of boundary management, structure, and infrastructure as they formulate their strategies. A second hurdle is the careful execution of strategy which requires small improvements on a day-to-day basis.

Focused Factories

Perhaps the greatest advancement in manufacturing strategy during the past two decades emanated from Wickham Skinner's notion of *focused factory*. The focus of the factory is defined by the production task—that is, those performance criteria in which the factory must excel to compete. The set of manufacturing choices is then organized to carry out this task.

Quality, cost, flexibility, and delivery capabilities are affected by those choices. The concept of focus provides a rationale for plant within a plant design and, subsequently, the rapid growth of manufacturing cells. A focused factory can provide many advantages, because its entire operating environment is focused on accomplishing the particular manufacturing task demanded by the company's overall strategy and the realities of its technology and economics. Focused factories offer the opportunity to stop

CASE STUDY

Strategy as Capability Building: How Toshiba Did It

Much of manufacturing strategy involves the building and leveraging of capabilities of the production function. While the exact contents of a manufacturing strategy will vary from industry to industry and company to company, manufacturing strategy as a pattern of choices should be linked to the business strategy which must be accomplished to win orders.

Figure 3-3 illustrates how Toshiba successfully deployed a manufacturing strategy to produce the breakthrough T-3100 laptop personal computer. Toshiba first introduced the T-3100 laptop computer at a trade show in London in January 1986. Since that time, Toshiba has been the market leader of laptop computers in the U.S., Europe, and Japan. In developing this product, Toshiba successfully utilized the target costing method described in Chapter 5. This method evoked a customer-driven manufacturing strategy in which a major innovation in personal computers occurred. The stages in Toshiba's customer-driven manufacturing strategy development are given below, defined by business, marketing, and manufacturing strategies.

Business Strategy

Toshiba's strategic intent was to penetrate the desktop computer market at a time when the IBM PC dominated. As a late entry, it sought to develop a highly differentiated product from the IBM desktop, yet maintain IBM operating standards. The initial business strategy was to develop a computer that would compete on price and performance. The specifics of the product features were left to marketing, engineering, and manufacturing. The actual laptop design evolved from the business unit differentiation strategy.

Marketing Strategy

Customer Requirements. Extensive market research of the personal computer market in the U.S. was conducted through the joint efforts of a U.S. consulting firm and Toshiba's marketing and engineering personnel. Personal computer dealers were the prime target for focused interviews concerning the new product. The results showed a distinct need for a significantly smaller personal computer that performed exactly the same as a desktop model.

Product Concept. The product that was conceived based on customer requirements was a desktop computer that could be used on any desk and perform exactly as a conventional desktop computer would.

Product Design. The design of the product was to match the needs of the end users, who liked the look and convenience of flat displays. This concept led to the laptop style.

Target Pricing. To set the sales price, Toshiba carefully determined what sales price would be accepted by consumers for significant market penetration. A target sales price was established and, after deducting a desirable profit, a target cost was introduced. Prior to the T-3100, Toshiba

Case Study, concluded

made pricing decisions on a cost-plus basis. The target costing method set the standard for joint product and process development and, subsequently, for the evolution of the Toshiba manufacturing strategy.

Manufacturing Strategy

Integration Choices. Having established a target cost, Toshiba worked on an integrated approach to the development of detailed product design and production processes. Key elements of the manufacturing strategy were the use of value analysis and computer-aided design. It was clear to the Toshiba team that since the product concept was so new, they would have to introduce innovative parts and manufacturing technologies. Toshiba chose both internal and external integration strategies (see Figure 3-3).

Since no currently available hard disk drive was small enough to install in the T-3100, an entirely new component had to be developed. The printed circuit boards also had to be redesigned. To minimize the size and thickness of the printed circuit boards, the Toshiba development team worked with the company's LSI (Large Scale Integration) Division to produce a super-integrated LSI chip. Simultaneously, the Toshiba Manufacturing Engineering Laboratory was involved in the development of a special surface-mount technology that would ensure that the LSI chips were soldered flatly on the boards. A special soldering machine was engineered for this process.

Along with internal integration, supplier integration efforts with Matsushita Electronics Devices were initiated for the production of the plasma display. Toshiba asked Matsushita to develop and introduce specially designed plasma displays. Toshiba also advised Matshushita on the introduction of a robotic manufacturing system for the special plasma displays that would be required for the T-3100.

Structural and Infrastructural Choices. To produce the T-3100, Toshiba employed JIT, TQC teams, advanced manufacturing technology including CIM, and robotics. These manufacturing strategy choices were useful for continuous improvement of product and processes and provided the necessary capacity to meet demand. Toshiba's business strategy dictated a global playing field. Consequently, new production facilities were established in the U.S. and West Germany.

As depicted in Figure 3-3, Toshiba's manufacturing strategy had been engineered to use all the company's capabilities as a comprehensiveelectronics manufacturer. It required extensive use of boundary management to enable the product design to meet the size, weight, and performance attributes required by customers and to enable the production processes to meet specifications at relatively low cost. Toshiba's manufacturing processes supported its corporate strategy, making the company a global market leader in laptop computers.

This case was prepared by A.V. Roth as a demonstration of the application of manufacturing strategy choices. Mr. T. Katsuta translated the source materials: A. Iwabuchi, *The Resistance of Toshiba*, Kodanshai, 1990 and the intercompany "Toshiba News."

FIGURE 3-3 Toshiba's Manufacturing Strategy for Production of T-3100

Business Unit Strategy
Penetration of World Market for Desktop Computers

Manufacturing Critical Success Factors
Cost/Performance

Manufacturing Strategy
Pattern of Resource Allocation

Structural Choices

Processes	Robotics/New Processes
Materials	JIT
Facilities	Global Networks (Japan, U.S., and West Germany)

Infrastructural Choices

Information	CIM
Quality	TQC teams

Integration Choices

Suppliers	Support with New Manufacturing Process
Customers	Target Pricing/Market Research
Other Functions	Value Analysis/CAD/Value Engineering
Business Unit	Cross-functional Teams

Source: A.V. Roth, "Boundary Management: Manufacturing's New Imperative," Duke University, Working Paper, 1990.

compromising each element of the production system. They are a radical departure from the traditional general purpose, do-it-all plants that satisfy no strategy, no market, and no task.[8]

John Deere & Company's Horicon Works brings together the people who work on and assemble riding mowers and lawn tractors using the focused factory concept. As a mini-factory within the complex, welders and other primary function workers who prepare parts for the assembly station work cells work in proximity to those stations so they can see how their performance affects the final assembly process. Workers focus on the objective of building a high-quality product efficiently to hold the line on production costs. The assembly workers are the customers for the products completed by the welders. The people who make individual components work to meet the needs of the people who assemble them into finished products; all are part of a team.

Best manufacturing practices, such as JIT and TQM, are greatly facilitated by focus. In focused factories, simplified coordination of production and shipping schedules is possible and, in some cases, shop floor scheduling systems have been eliminated because workers can tell suppliers what parts are needed and when. All operations are closely linked; downtime or quality problems in one area quickly affect all others. Operators learn to function as a team and to serve internal customers. Further discussion of focused manufacturing organizations appears in Chapter 8.

Revamping Strategic Directions and Priorities: U.S. on the Defensive

Manufacturing organizations frame their policies and actions around strategic business directions, which, in turn, stem from underlying market dynamics and industry conditions. They are the linchpin between the business unit strategy and the competitive capabilities required of manufacturing. Strategic business unit directions take two primary forms:

- Strategic market directions which describe the products and marketing thrusts of the business.
- Strategic directions pertaining to the scope of the business.

Realigning Products and Markets

The product-market matrix, as depicted in Figure 3-4, is a way of gauging the strategic market directions of manufacturing companies.

- Market maintenance denotes a *hold* position, in which maintaining market share with existing products is perceived to be the most critical thrust.
- A market development strategic direction coincides with the firm's desire to enter new markets with existing products.

FIGURE 3-4 Strategic Market Directions: Product-Market Matrix

Market Focus	**Product Focus** **Current Products**	 **New Products**
Current Markets	Market Maintenance	Product Development
New Markets	Market Development	Repositioning

Source: A.V. Roth, "Matching Business Directions and Manufacturing Capabilities," Duke University Working Paper, 1990.

- The product development strategic direction emphasizes the replacement of new products with old to build share within existing markets.
- The repositioning strategy is distinct; the objective is to increase share by entering new markets with new products.

How are the global players staking out their territories? The International Manufacturing Futures Project (MFP) provides insights. The MFP asked respondents a series of questions relative to their broad plans for products and markets over the next five years. Presented here are the key findings.[9]

- In each global region in 1986, the primary strategic market directions pertained to product development and market development. Leading U.S., Japanese, and European manufacturers emphasized the development of new products for existing markets and the increase of market share in existing markets.
- Intense competition forced notable changes in the strategic market directions of global players. By 1987, some reshaping of the competitive turf was occurring (see Figure 3-5). A defensive market maintenance posture was the top priority in the five-year plan of American manufacturers. The primary strategic directions of both the Japanese and the Europeans focused on product development, the same basic position observed in the previous year. Product development took second place in importance for the Americans. It is no secret that Toyota and Honda have put General Motors, Ford, and Chrysler on the defensive.
- Examination of the second most important Japanese and European strategic market direction was revealing. The Europeans' second course of action was directed to a defensive position — that is, market maintenance. As the Japanese enter the luxury car market, European automakers, including BMW, Mercedes-Benz, and Volkswagen, are becoming more defensive.
- At the same time, the Japanese are adopting a more aggressive strategy, namely, market repositioning. The Japanese emphasis on the development of new products for new markets ranked higher than any other geographic region. Illustrative of Japanese repositioning are Honda's transition from motorcycles to cars and Canon's expansion from cameras to copiers.

FIGURE 3-5 The Global Arena 1987-1992: Product-Market Matrix

Market Focus	Product Focus	
	Current Products	**New Products**
Current Markets	U.S. (1)/Europe (2)	Japan (1)/Europe (2) /U.S. (2)
New Markets	----	Japan (2)

Numbers in parentheses (n) represent the relative rank order importance of the strategic direction by senior manufacturing executives in the global region.

Source: A.V. Roth, "Matching Business Directions and Manufacturing Capabilities," Duke University Working Paper, 1990.

Rearranging the Territory

As manufacturing firms change their scope, broad-based restructuring occurs. Restructuring takes many forms:

- **Backward/Forward Integration.** The Japanese place relatively high emphasis on backward integration strategies, gaining control over suppliers of raw materials and components. For U.S. and European manufacturers, relatively little importance is given to backward integration as a strategic direction. The Japanese also attach more importance to forward integration over the channels of distribution than either the Americans or the Europeans. The rationales behind Japan's posture are related to improvements in both quality and lead time. The Japanese are exploiting vertical integration opportunities to strategically position themselves for global competition and, thereby, eliminate trade barriers in other countries.

- **Growth by Acquisition.** Both U.S. and European companies place much more emphasis on growth by acquisition than do Japanese companies. For Japan, growth by acquisition is the second lowest priority; only complete withdrawal from certain businesses and markets ranks lower.
- **Assault on Foreign Markets.** The assault on foreign markets is a continuing Japanese strategic direction with respect to both markets and structure. This strategic direction indicates continued growth through the penetration of new markets with new products. Vertical integration is being pursued to position the Japanese favorably in global marketplaces. U.S. and European manufacturers are more conservative in their approaches, relying more on their strength in existing markets.

The American style of restructuring is a weak and near-sighted response to competition and, typically does not provide long-term sustainable value to the business. U.S. manufacturers are employing market maintenance positions as a first course of defensive action. Growth for U.S. manufacturing firms emphasizes acquisition rather than vertical integration or entry into new markets with current manufacturing capabilities.[10]

International Manufacturing Networks

Leading Japanese companies are focusing on vertical integration strategies that position them as global players. These strategies create value for the business, particularly in terms of cost and fast response time. A 1990 update for American manufacturing suggests that, on average, American firms are not preparing for the global marketplace and are ignoring its importance for their future success. As shown in Figure 3-6, American manufacturers see relatively little value in offshore production and only slightly more than average value in overseas sales.

This picture, however, does not apply to American multinationals. The volume of world trade in manufactured goods is on the rise, and multinationals have been setting up foreign manufacturing facilities aggressively over the past 10 years. Yet, there is no theory of international trade or capital movement that offers a satisfactory explanation of managerial decisions to produce abroad. Common reasons for locating plants internationally are grouped into five categories:

- **Low Cost of Production Factors.** Exploit lower labor, material, energy and/or capital costs.

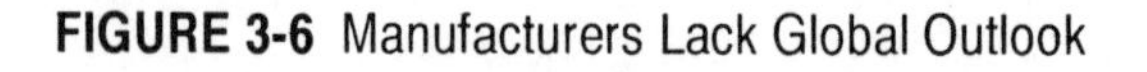

FIGURE 3-6 Manufacturers Lack Global Outlook

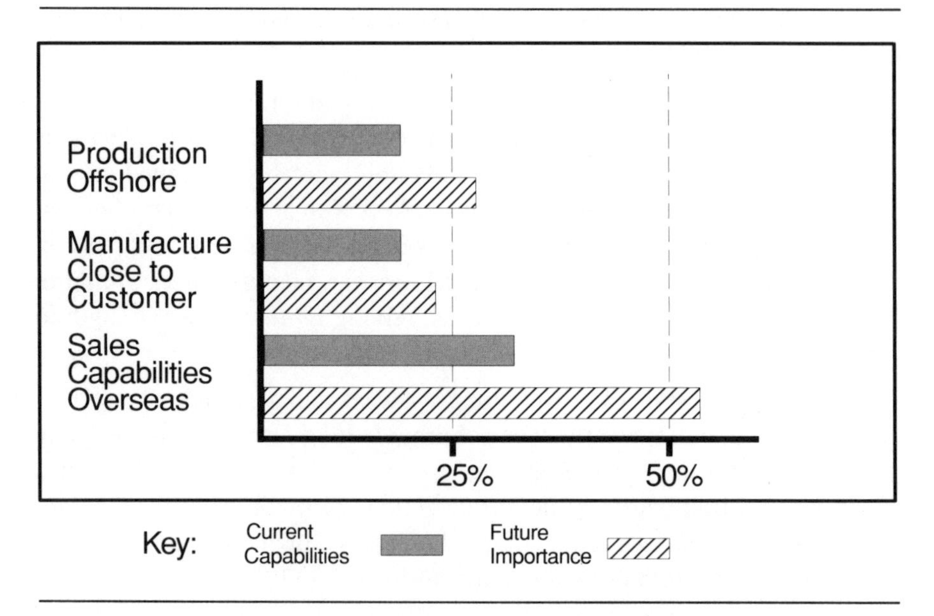

Source: C. Giffi and A.V. Roth, "Making the Grade in the 1990s," Deloitte & Touche Third Annual Survey of North American Manufacturing Technology, 1989.

- **Closeness to Customer.** Offer better customer service and/or reduce financial and trade restriction risks.
- **Tap Local Technological Resources.** Locate close to source of up-to-date technology—e.g., universities, competitors, customers, research centers.
- **Control and Amortisation of Technological Assets.** Earn higher returns on technology than would be possible through licensing.
- **Pre-empt Competition.** Develop local experience, customer loyalty, and barriers to competition.[11]

Ferdows presents a simple framework in Figure 3-7 to differentiate strategic roles that production facilities can serve in an international network. The framework provides a new perspective for tracking the patterns of change in the strategic role of offshore factories, for developing appropriate management and technology, and for organizing global capabilities.

Both *offshore* and *source* factories capitalize on low-cost production factors. Additionally, source factories develop and produce products or components for worldwide distribution. *Servers* are designated to specific national or regional markets with minimum investment. *Contributor* factories also furnish products to a specific geographic region, but their strategic role extends to the development of technological capabilities and know-how. The primary role of *outpost* factories is gathering intelligence information by locating close to technologically superior suppliers, customers, research laboratories, or competitors. *Lead* factories build strategic capabilities that tap into local technology sources.

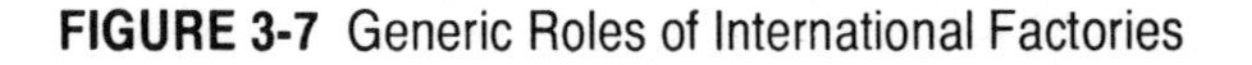

FIGURE 3-7 Generic Roles of International Factories

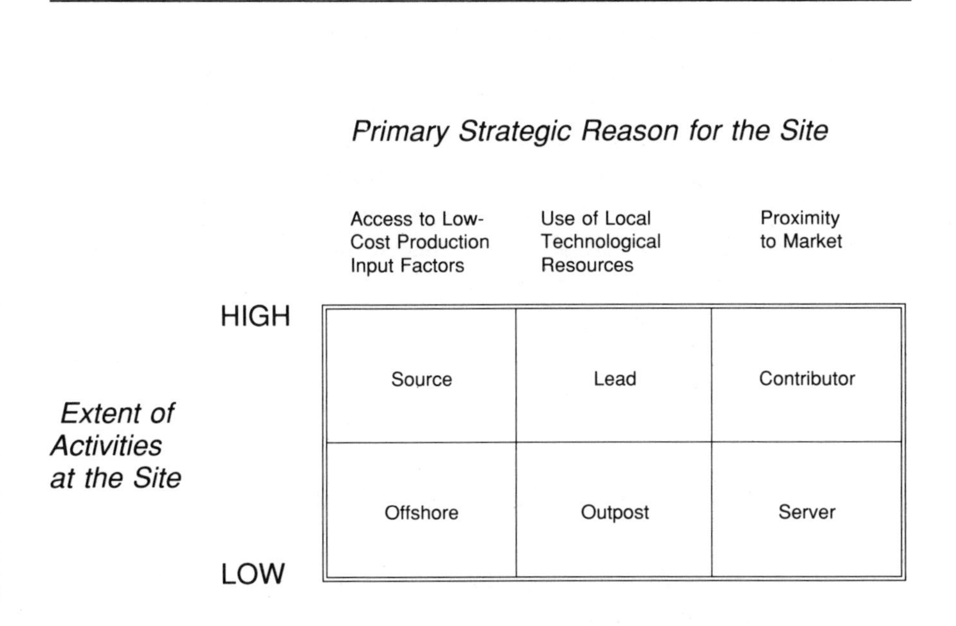

Source: K. Ferdows, "Mapping International Factory Networks," *Managing International Manufacturing* (K. Ferdows, ed.)(Amsterdam: Elsevier Science Publishers, 1989), pp. 3-11.

Maneuvering for New Advantages

The importance of acquiring competitive strength through manufacturing cannot be overstated. Directly or indirectly, the factors critical to success will be redefined by competition, technology, and the marketplace. What capabilities are perceived as critical for success today? For tomorrow?

Global players are differentiated by capabilities. In the 1987 International Manufacturing Futures Project, European and U.S. manufacturers ranked conformance quality, high product performance, and delivery schedule performance as their top competitive priorities; low prices, fast deliveries, and the ability to compete through rapid design changes (flexibility) ranked below all of these as competitive weapons.

The Japanese had an entirely different profile. Low price is the top competitive weapon in Japan. This has been true each year the survey has been conducted in Japan, since 1983. Ranked second to price as a competitive weapon by the Japanese is the ability to compete based on rapid design changes (flexibility). High quality, universally considered to be the most important competitive weapon, ranked only third for the Japanese.

The most common explanation for these rankings is that the Japanese have realized that most competitors will be able to produce high conformance quality products and that price will become the deciding factor in the future. Heavy pressure has also been placed on price due to the world valuation of the yen. It is clear, however, that the Japanese have achieved high levels of perceptible quality through more than two decades of experience. The Japanese are positioning themselves to exploit quality and cost advantages simultaneously.

Building Strategic Bills of Materials

Competitive advantage remains the hottest buzzword in business strategy. Yet, current strategic planning systems provide managers with models that not only obscure new business opportunities but hamper their ability to prepare for the future. Many businesses view the competitive game at the strategic business unit level and neglect the major competitive strengths that the functions provide. Since the concepts of strategy are those designed to beat competitors, the notion of developing functional capabilities for competitive advantage is drawing more attention. The identification of manufacturing capabilities for competitive advantage has emerged as an important factor over the past decade.

If a business neglects to plan properly for the maintenance and development of manufacturing capabilities, significant competitive advantage is lost. But what are the capabilities that are critical to success? How can they be planned? The 1989 North American Manufacturing Technology Survey of 759 American manufacturing firms found that American manufacturers continue to view quality as the key to competition in the 1990s (see Figure 3-8). After quality, come customer service, price/cost, marketing, technological leadership, and flexibility.

Within each broad-based capability are a number of key attributes of success. The critical success factors in each category reported by 759 senior manufacturing executives are shown in Figure 3-9.

The key attributes of success are as follows:

- **Quality:** Manufacturing executives rate the customer's perception of quality (perceived quality) as the single most important success

FIGURE 3-8 1990s Require Improved Capabilities

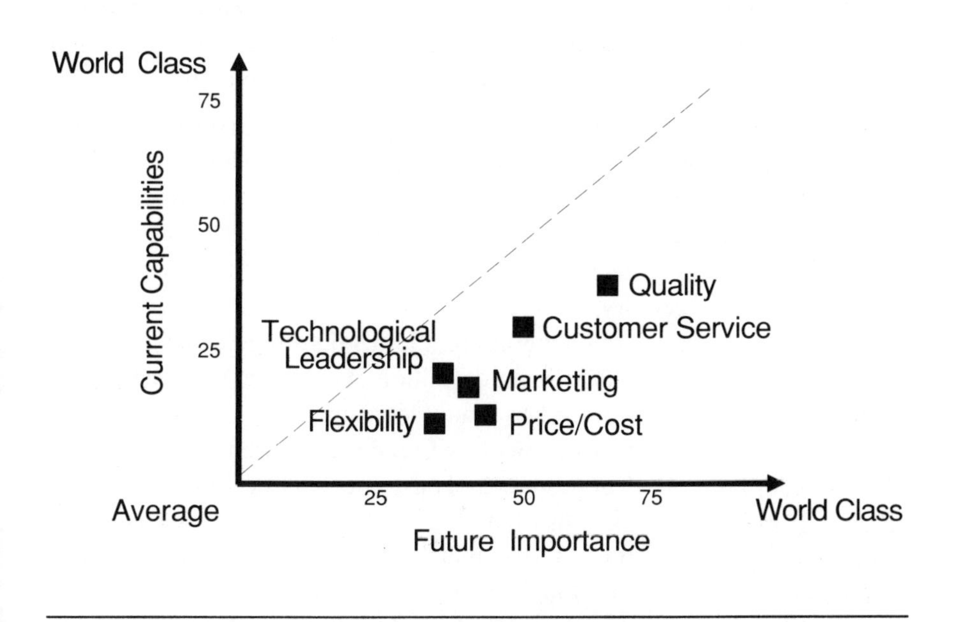

Source: C. Giffi and A.V. Roth, "Making the Grade in the 1990s," Deloitte & Touche Third Annual Survey of North American Manufacturing Technology, 1989.

factor overall. Conformance, product reliability, and performance reflect top-rated quality attributes.

- **Customer Service:** Delivery time reliability heads the list of service priorities. It is followed by promptness in handling customer complaints and the ability to confirm customer order delivery dates rapidly.
- **Price/Cost:** American manufacturing executives emphasize higher value products and the ability to meet competitors' prices. They are less inclined to manufacture at lower costs than competitors. Least appealing to American executives is the ability to offer lower-priced products than competitors.
- **Marketing-Manufacturing Linkages:** Critical marketing activities are seen to be important to manufacturing managers. At the top of the list is the ability to expand customer relationships and develop the technical knowledge of the sales force.
- **Technological Leadership:** American executives place more importance on developing core product technologies than on developing unique manufacturing process capabilities.
- **Flexibility:** Delivery speed, speed to market, and customization lead the list of responsiveness factors. Offshore and geographic proximity to customers are at the bottom of the overall list.

Over the past decade, American manufacturers have made great strides in improving the overall quality of their products. Those who have achieved high levels of perceived quality have become market leaders. Second only to product quality is quality of customer service. As mentioned earlier, customer service may determine the outcome of competition for business in the twenty-first century.

While the findings are encouraging, the results suggest that American manufacturers still have a narrower definition of customer service than that given in Chapter 2 for world-class manufacturing.

Manufacturers must move toward a broader definition of product, one that includes both physical product and customer service. Successful manufacturers build "strategic bills of materials" that include service as well as product components. Figure 3-10 depicts a typical strategic bill of materials in rank order of their overall importance.

Note that time-based critical success factors head the list of service-related components. Customer demand is substantially affected by the expanded choices and increased responsiveness associated with better service.

FIGURE 3-9 Factors Critical to Success on Six Capabilities (In Rank Order of Importance)

Overall Product Quality

Perceived Quality
Conformance
Product Reliability
High Product Performance
Durability
Aesthetics
Product Serviceability
Enhanced Product Features

Overall Flexibility/Responsiveness

Delivery Speed
Speed of New Product Introductions
Handle Custom Orders
Broad Product Mix with Same Facilities
Reduce Product Changeover Time
Change Production Volumes
Change Product Mix
Modify Methods for Material Quality Variances
Increase Number of New Product Introductions
Geographically Close to Customer
Global Manufacturing

Technological Leadership

Innovative Products
Unique Mfg Process Capability
High R&D Content Products

Price/Cost

High Value Products
Meet Competitor's Price
Lower Mfg Cost than Competition
Lower Priced Products

Customer Service

Reliability of Delivery Time
Promptness in Handling Customer Complaints
Confirm Customer Order Delivery Dates
After-sales Support
Pre-sales Service
After-sales Repair and Maintenance Service

Marketing-Manufacturing Linkages

Expansion of Customer Relationships
Technical Knowledge of Sales Force
Strong Sales Capabilities
Broad Product Lines
Broad Distribution Channels
Sales Capabilities Overseas

Source: A.V. Roth and C. Giffi, "Changing the Basis of Competition," Duke University Working Paper, 1990.

FIGURE 3-10 Strategic Bill of Materials for the 1990s

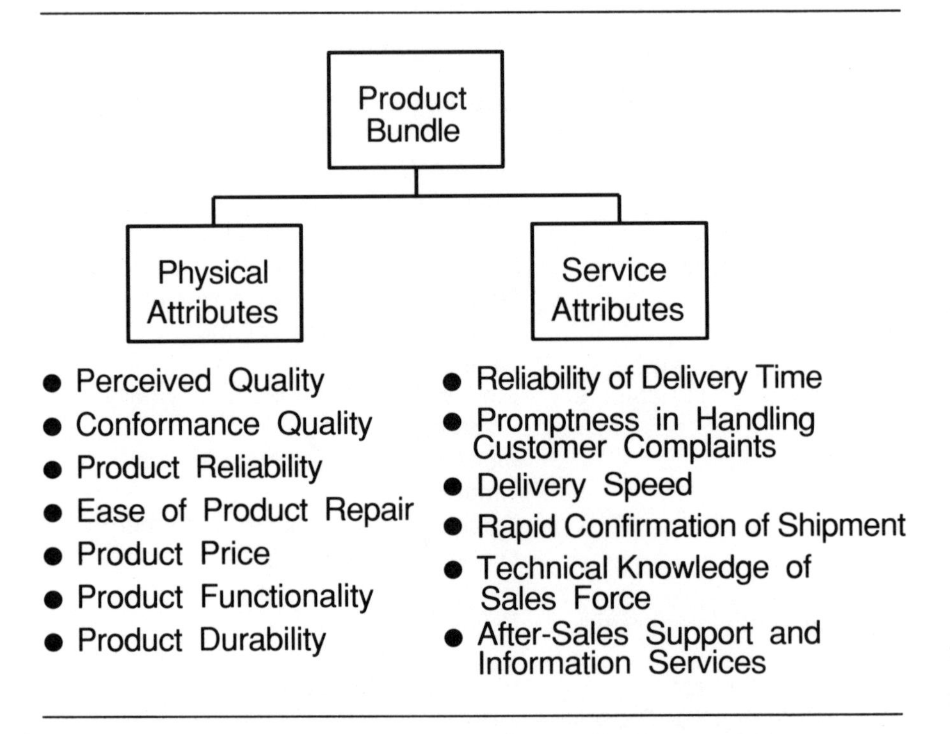

Source: C. Giffi and A.V. Roth, "Making the Grade in the 1990s," Deloitte & Touche Third Annual Survey of North American Manufacturing Technology, 1989.

Research shows that companies that can respond quickly to customer expectations and choices dominate the most profitable segments of the market. Thomasville Furniture, for example, responds 75 percent faster than its competitors and enjoys a fourfold profit advantage; it is the leading performer in its industry.[12]

Service is more than remedial after-sales repair and maintenance. It is something that can add value to the physical product and build relationships with customers. American manufacturers, such as Xerox, have learned that unreliable products, regardless of the strength of their after-sales repair service, fail to hold market share.

Closing the Competitive Gaps

By plotting present manufacturing capabilities against those required to compete in the future, a competitive gap graph can be developed. When this was done by the researchers conducting the 1987 International Manufacturing Futures Project, they found that the Japanese perceive their largest gaps in the areas they identified as most important for future competition—flexibility and price. In contrast, the U.S. and Europe portrayed a more comfortable picture, one in which the most important competitive requirements are those perceived to be easiest to reach. Both U.S. and European executives reported their largest gap to be flexibility—the ability to make rapid design changes and/or introduce new products quickly. These same companies, however, reported flexibility as among their least important priorities.

While quality remains the top-rated capability, American manufacturers believe that requirements for world-class competition in the 1990s exceed their current capabilities on flexibility and price. The 1989 North American Manufacturing Technology Survey confirmed this belief. Flexibility and price capabilities are the Achilles heel of the Americans; they are the two primary capabilities which the Japanese are attacking. Given the time it took American manufacturers to focus on and to improve quality in the 1980s, the across-the-board increase in competitive capabilities seen as required in the twenty-first century represents a formidable challenge.

Within the U.S., on an industry-by-industry basis, significant improvements have been noted in key manufacturing capabilities. Most industries have reported improvements in conformance quality and delivery schedule performance. Only modest gains have been made in flexibility, speed, and price/cost (see Figure 3-11).

The Emerging Battleground: Time-based Capabilities

Speed kills competition! Speed to market, manufacturing delivery speed, and speed in new product development are elements of the next competitive battleground. Hewlett-Packard Chief Executive John Young, a leading proponent of speed, explains, "Doing it fast forces you to do it right the first time."[13] Young reported that speed has yielded both quality and cost advantages for the company. Hewlett-Packard planned to produce computer terminals at a lower cost than those produced by the master clonemakers. Thinking about time forced attention on all activities. Young caused Hewlett-Packard to focus on "breakeven time," trying to cut in half

FIGURE 3-11 Percent Improvement in Capabilities by Industry: 1985-1988

	(Percent)					
	Total	Consumer	Industrial	Basic	Machinery	Electronic
Conformance	19	13	16	20	16	29
Price/Cost	8	6	8	4	4	14
Flexibility	7	3	0	2	13	17
Delivery	13	12	7	12	10	24
Speed	8	7	7	4	10	13

Source: A.V. Roth and J.G. Miller, "Manufacturing Futures Factbook," 1988 North American Manufacturing Futures Survey, Boston University Manufacturing Roundtable Monograph Series, 1988.

the time between conception of a new product and profitability across the organization. Hewlett-Packard is one of manufacturing's new "speedsters." Figure 3-12 lists other manufacturing speedsters.

Time-based competitive advantage can be defined in several ways, including those shown in Figure 3-13. Despite the proven payoffs to those who have successfully developed time-based capabilities, there exist for the typical American manufacturer significant gaps between current abilities and future requirements. The steps, shown in Figure 3-13, that American manufacturers must climb in the 1990s are steep. Delivery time reliability is the most important requirement and among the highest steps. Speed in customer service, manufacturing lead times, product innovations, and customization lag in perceived importance.

The ability to create, produce, and distribute products more quickly; make design changes to existing products faster; and establish customer service responsiveness are critical components of time-based competition. What are the barriers? One of the greatest is the tendency to "expedite," rather than modify and simplify current processes.

When GE began producing a newly designed refrigerator compressor in 1986, the manufacturer thought it had a revolutionary product that would allow it to leapfrog its Japanese rivals. GE staked its $2 billion refrigerator business on the rotary compressor to prove that "America could still be a world leader in manufacturing." But the company experienced its own worst nightmare when a significant design flaw led to early failure of the

FIGURE 3-12 Speedster Innovators and Producers

Speedster Innovators	Product	Development Time	
		Old	*New*
Honda	Cars	5 yrs	3 yrs
AT&T	Phones	2 yrs	1 yr
Navistar	Trucks	5 yrs	2.5 yrs
Hewlett-Packard	Computer printers	4.5 yrs	22 mos

Speedster Producers	Product	Order-to-finished-goods Time	
		Old	*New*
GE	Circuit breaker boxes	3 wks	3 days
Motorola	Pagers	3 wks	2 hrs
Hewlett-Packard	Electronic testing equipment	4 wks	5 days
Brunswick	Fishing reels	3 wks	1 wk
Matsushita	Washing machines	360 hrs	2 hrs
Harley-Davidson	Motorcycles	360 days	< 3 days

Adapted from "How Managers Can Succeed Through Speed," *Fortune*, February 13, 1989 and G. Stalk, Jr. and T. M. Hout, *Competing Against Time* (New York: The Free Press, 1990).

compressor. The new rotary compressor flopped so badly that GE had to take a $450 million pretax charge in 1988. Since then, GE has voluntarily replaced nearly 1.1 million defective compressors.[14]

GE management did not appreciate the magical balance between getting it right and getting it fast. The firm pushed the compressor development too fast and failed to test the design adequately. The engineers made some bad assumptions and failed to ask the right questions. Managers eager to cut costs forced inadequate "life testing" and rushed the product into production. The saga worsened as GE committed $120 million to a new factory capable of turning out rotary compressors at a rate of one every six seconds, instead of the 65 minutes required to make a conventional GE compressor. Its Italian and Japanese rivals' production cycle was 25 minutes. Despite the disastrous introduction of its rotary compressor, GE's production cycle time was an amazing feat, given that the new design

FIGURE 3-13 Will North American Manufacturers Be Ready "in Time"?

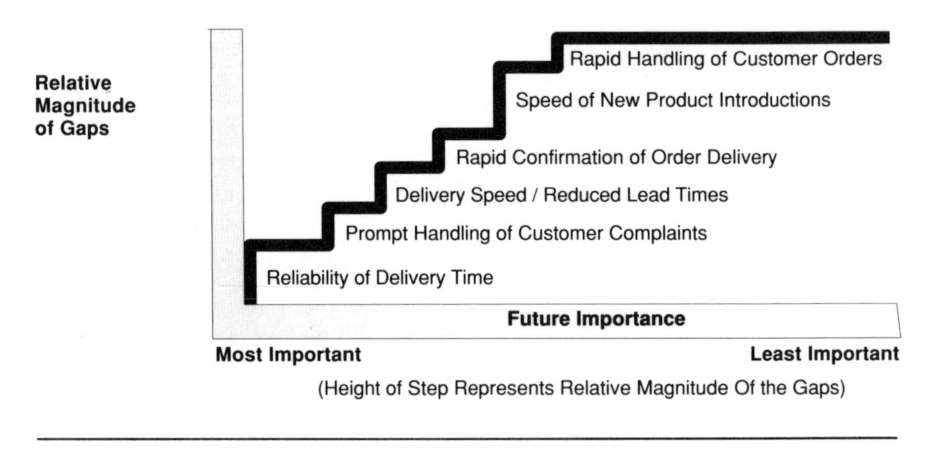

Source: C. Giffi and A.V. Roth, "Making the Grade in the 1990s," Deloitte & Touche Third Annual Survey of North American Manufacturing Technology, 1989.

required key parts to work together with a tolerance smaller than 1/100 the width of a human hair—a production success, but an engineering failure.

Proven tactics used by masters of time-based competition include:

- **Start from scratch**: Don't expedite current operations.
- **Wipe out approval**: Cut away layers of management.
- **Form multifunctional teams**: Increase communication.
- **Develop clear but honest deadlines**: Benchmark capabilities.
- **Improve distribution**: Close the sales-engineering-factory-warehouse loop.
- **Develop "time-based" culture**: Change performance metrics.

Delivery Time and Service as Bases for Competition

American manufacturers are beginning to understand the strategic advantage created by time-based capabilities. Time-based competitive capabilities have been increasing in importance. This theme has been echoed by numerous researchers over the last several years. While conformance and performance quality remain the primary focus of U.S. manufacturers, time and service continue to increase in importance.

The trend toward faster deliveries and better after-sales service was echoed by the executives interviewed as part of the National Center for Manufacturing Sciences study. Almost unanimously, executives in all industries indicated that delivery speed and after-sales service were significantly increasing in importance, although their current focus was on quality and delivery schedule performance.

The Japanese ranked delivery speed and after-sales service last on their list of competitive priorities, a similar position held in the U.S. and European companies. Unfortunately, no data from the International Manufacturing Futures Project indicated the relative change in importance of these two competitive weapons over time in Japan. There may be an opportunity for American firms to capitalize on both delivery speed and after-sales service as competitive weapons if quality and dependability capabilities are brought into line.

FIGURE 3-14 Time-based Capabilities Are Becoming More Important

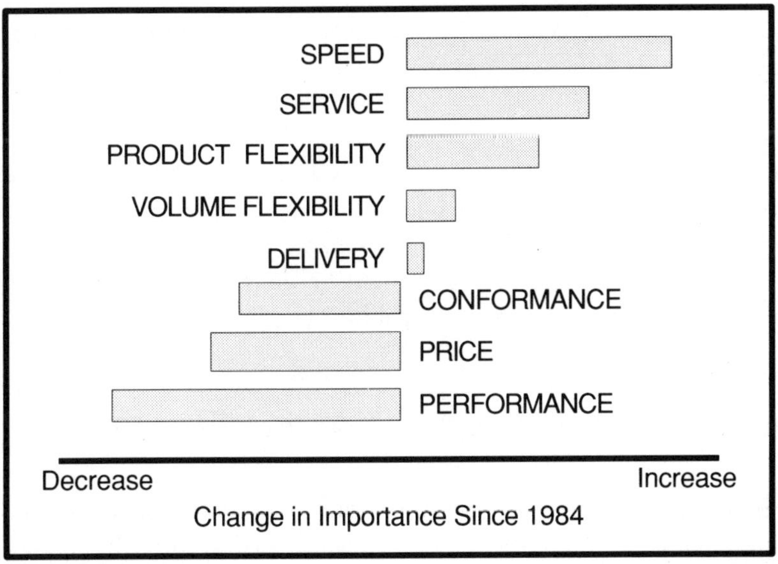

SPEED
SERVICE
PRODUCT FLEXIBILITY
VOLUME FLEXIBILITY
DELIVERY
CONFORMANCE
PRICE
PERFORMANCE
Decrease
Increase
Change in Importance Since 1984

Source: J.G. Miller and A.V. Roth, "Manufacturing Strategies: Executive Summary of the 1988 North American Manufacturing Futures Survey," Boston University Manufacturing Roundtable, 1988.

Flexibility as a Basis for Competition

Flexibility is a critical component of time-based competitive capabilities and includes both product and volume flexibility.

- Product flexibility is the ability to introduce new products and make design changes quickly.
- Volume flexibility is the ability to produce a wide variety of products and quantities with the same manufacturing process.

These capabilities can give any competitor a significant advantage, regardless of market.

Consider the automobile industry. Japanese automakers have long been able to introduce new automobile designs faster than their global competitors can. Taiichi Ohno, designer of the Toyota production system, said his system was "born out of the need to make many types of automobiles, in small quantities, with the same manufacturing process."[15]

In the late 1970s, Japanese companies exploited flexible manufacturing to the point of creating *variety wars.* The classic variety war erupted between Honda and Yamaha over motorcycles in 1981. Over a two-year period, Honda introduced 113 new models—more than three times the number of new Yamaha models. Not only were the Honda designs new, they also were innovative. Yamaha was not capable of such technological innovations.

Many industries have been affected. The International Manufacturing Futures Project indicates that the battle is just beginning. While quality continues to be the focus of efforts in the U.S. and Europe, product flexibility is increasingly the focus of efforts in Japan. World-class American and European manufacturers also appear to be gearing up to make critical improvements in their flexibility capabilities.

Future Directions and Key Action Programs

When senior manufacturing executives were asked in 1987 about key action programs, tools, and activities to be emphasized in the future, they identified these key factors:

Japan

- Manufacturing lead-time reductions
- Computer-aided Design (CAD)

- Value analysis
- Developing new processes for new products
- Integration of systems

U.S.

- Vendor quality
- Statistical process control
- Integration of systems
- Manufacturing lead-time reductions
- Worker safety

Europe

- Zero defects
- New product introduction
- Production planning and inventory control systems
- Vendor quality
- Integration of suppliers[16]

It appears that the Japanese are focusing their attention on programs that will improve flexibility and cost effectiveness. They are aggressively pushing forward in multiple areas simultaneously.

The drive toward flexibility, innovation, and low cost on the part of Japanese companies was also confirmed by a study, entitled *Japan 2001,* undertaken at Waseda University in Japan. In a presentation at the International Conference, Boston University Manufacturing Roundtable, held at Boston University on October 21, 1988, Jinichiro Nakane identified four key alternatives for Japanese firms in the future:

- Emphasis on R&D with the goal of becoming leaders in innovation and new product introductions.
- Emphasis on becoming the low-cost producers by automating manufacturing processes and employing offshore production.
- Emphasis on the development of flexible manufacturing systems to deal with volume changes, product mix changes, and the speedy introduction of new products.
- Development of effective multinational organizations to compete on a global basis.

In 1987, U.S. companies overall were still focused on overcoming their quality problems. They were emphasizing tools and procedures for managing the direct as opposed to the indirect manufacturing functions—namely, materials, processes, and human assets.

What is the prognosis for the early 1990s in American manufacturing? Manufacturers are indicating a much higher commitment to human assets; 6 of the top 10 programs for the early 1990s deal with improving human resources. American manufacturing executives appear to be genuinely committed to improving their infrastructures through improved labor/management relations and workforce improvements. In 1990, the emphasis remains on managing the direct manufacturing activities, but the balance in key action programs has clearly shifted to developing human assets (see Figure 3-15). The shift in American manufacturing to continuous improvement appears to be taking hold. Americans are developing people who are now being asked "to bring their brains to work"!

FIGURE 3-15 Management Focus on Human Resources, 1990-1993

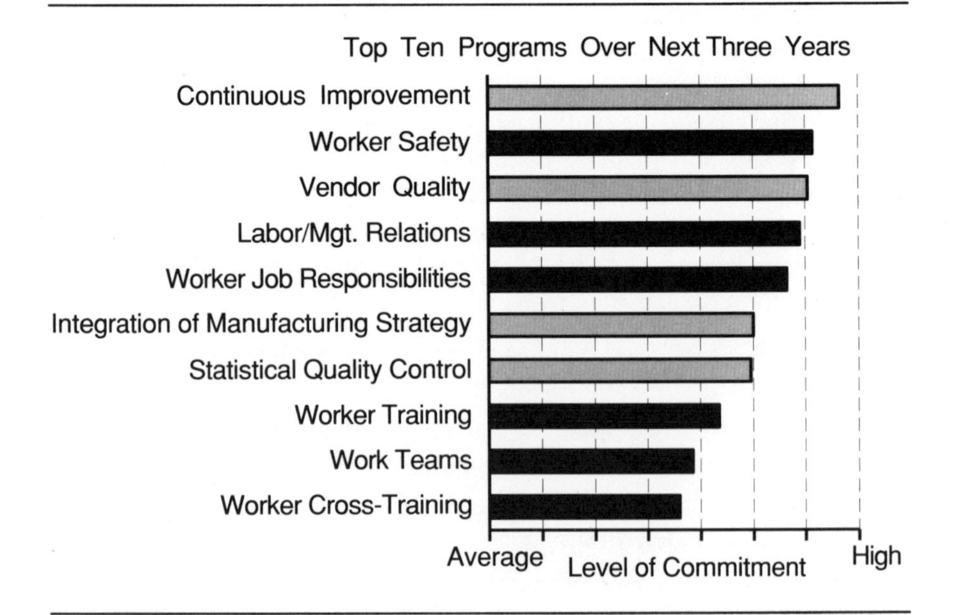

Source: C. Giffi and A.V. Roth, "Making the Grade in the 1990s," Deloitte & Touche Third Annual Survey of North American Manufacturing Technology, 1989.

The European companies, taken as a whole, also appear to be placing emphasis on quality and linking quality to their competitive priorities and action programs.

Strategy Development Process

In the previous sections, manufacturing strategies were operationally defined. If the manufacturing function is committed to being competitive tomorrow and beyond, it must embrace strategic manufacturing planning. The purpose of this section is to introduce the processes by which manufacturing strategies are developed.

Having a plan to follow and communicate to the organization appears to be a key ingredient of successful manufacturing. A written strategy, however, does not necessarily equate to implementation; nor does it equate to the living, dynamic process that is required for world-class manufacturing. The plan and, more important, the implementation process must be flexible and adaptable to changing requirements.

The 1988 North American Manufacturing Futures Project revealed that the development of a manufacturing strategy was one of the most useful management techniques cited by executives. The 1989 North American Manufacturing Technology Survey showed that a manufacturing strategy, fully integrated with the business strategy, ranked among the top 10 key action programs of leading manufacturers in 1990. This section briefly describes some of the trends and issues in manufacturing strategy formulation.

Competitive Approaches

A wide variety of approaches to strategy formulation and communication exists. In some companies, strategies are documented in writing; they identify goals, objectives, action plans, and responsibilities over both the short and the long term. Other companies run smoothly without any clearly expressed or documented strategies.

While there may be no single state-of-the-art approach to manufacturing strategy formulation, most approaches contain the same basic steps. For the purposes of this study, we have adopted the top-down analysis espoused by Wickham Skinner, a recognized expert in the field of manufacturing strategy formulation.

Skinner states that executives should think of manufacturing strategy formulation as an orderly process or sequence of steps. He defines six key steps in developing a manufacturing strategy:

1. **Perform a competitive analysis.** Develop an analysis of how rival companies are competing in terms of product, markets, policies, and channels of distribution.
2. **Perform an internal assessment.** Develop a realistic appraisal of your company's skills, resources, present facilities, and manufacturing approaches.
3. **Develop a company strategy.** Define how your company will compete; combine your strengths with market opportunities and define niches in markets in which a competitive advantage can be gained.
4. **Assess the implications of the strategy on specific manufacturing tasks.** Based upon the competitive strategy chosen, explicitly and precisely define the implications for your company's manufacturing organization and capabilities.
5. **Study constraints and limitations.** Determine the limitations imposed by the economics and technology of your industry—factors which are usually common to all competitors. Explicit recognition of these limitations is a prerequisite to an understanding of manufacturing problems and opportunities.
6. **Synthesize facts into a manufacturing strategy.** Define the supporting manufacturing choices required, given the facts regarding current capabilities, economic and technological limitations, and the corporate strategy for future competition.[17]

Skinner also points out the critical components of the planning cycle. They include:

- Development of the implementation requirements and action plans to execute the manufacturing strategy.
- Definition and development of critical systems to support the implementation of the strategy. These include:
  - Management information systems
  - Production planning and control systems
  - Reward systems
  - Performance measurement systems

- Development of the required manufacturing controls for cost, quality, material flows, inventory, and time.
- Selection of specific manufacturing operation ingredients critical to success — labor skills, material yields, utilization levels, for example.
- Constant review and monitoring of the strategy and its execution to make modifications as required, on a timely basis.

In practice, it is likely that the process of developing a manufacturing strategy will be iterative. Use these additional questions to guide discussion as your firm progresses through the six steps.

- Who are our current and future customers?
- What capabilities must we possess to win and maintain orders?
- What are the greatest challenges? The greatest opportunities?
- Where are the weaknesses of the competition?
- How will we measure performance?
- What is the desired economic outcome in the long run?

Common Development and Implementation Problems

Some of the most common problems in developing and implementing a manufacturing strategy were identified by Electronic Data Systems (EDS) in a recent survey of 40 Fortune 500 companies. These problems included:

- Failure to link the corporate competitive strategy to the manufacturing strategy because of emphasis on techniques and technologies (i.e., CIM, FMS, and JIT), rather than on the priority of manufacturing tasks required for successful competition, such as:
  - Quality/reliability
  - Dependability/delivery
  - Flexibility/innovation
  - Efficiency/cost
- Vague descriptions of performance requirements such as *low-cost producer*, *high-quality producer*, or *reliable-delivery producer*, as simultaneous unquantified objectives.
- Independent strategy development for manufacturing tasks, rather than integrated strategies for the entire company. Individual managers attempt to optimize their functional areas, while executive management attempts to optimize the company as a whole.

- Lack of definition of success drivers or required outputs from each functional area in the organization.[18]

The EDS survey team found that integrating manufacturing plans into the company's actual infrastructure of people, machines, and computers generates "more fireworks and disagreement" than virtually any other issue. This phenomenon was perceived to be largely the result of fragmented decision-making authority within the manufacturing organization; in some cases, however, it may be due to normal resistance to change. In any event, the issue of integration must be handled.

One of the most common problems mentioned during discussions with executive manufacturing management was the tremendous amount of resistance to planning because of the time-consuming, never-ending nature of the task. *Nevertheless, the relentless process of continuous improvement to meet changing customer requirements requires dynamic planning capabilities.* Planning demands responsibility and commitment and reveals inadequacies, inefficiencies, and the need for change.

Solutions offered to improve the planning process include:

- Relying on small incremental steps which makes the path clear, rather than large strategic leaps in which the next step cannot be seen by anyone but top management. This facilitates involvement and understanding at lower levels of the organization, where the plan must be executed.
- Using a bottom-up entrepreneurial planning vs. top-down staff-dominated planning process. This reduces the size, scope, and responsibility of the corporate planning staff and shifts the emphasis to the manufacturing unit managers.[19]

Conclusion

Implementing less than a totally customer-driven strategy (Phase 5) may have potentially dangerous side effects; too tight a coupling between the business strategy and the manufacturing strategy may restrict the means by which business goals and objectives can be achieved. It may force distortions in commitments that will inevitably influence patterns of manufacturing choices for resource allocation and skill deployment and may cause underinvestment in developing the critical competencies required by the business. By focusing on the customer and customer requirements, the company is forced to explore new opportunities even if it doesn't have the correct resources in hand.

Research has shown that the companies that have risen to global leadership over the past two decades have done so using initiatives that were out of proportion with their capabilities and resources. This obsession with winning has been termed *strategic intent.*[20] The basics of strategic intent encompass an active management process which can guide manufacturing strategy in the 1990s. Strategic intent should be established and communicated to:

- Capture the essence of winning.
- Lengthen the organization's attention span and stability over time.
- Set targets that foster personal effort and commitment by company chairperson and employees alike.

Strategic intent implies that the organization will be constantly stretched and challenged so that:

- Individuals will focus on short- to medium-term ends.
- A sense of urgency is created to move the organization forward.
- A competitor focus is generated at every level of the organization through the use of benchmarks against "best-in-class."
- Employees are given the skills they need to work knowledgeably.
- The organization has time to synthesize each challenge before launching another.
- Clear milestones and review mechanisms to track progress and reward desired behavior are maintained.[21]

Remaking manufacturing strategy requires a radical shift in management's ability to consolidate technologies and production skills into competencies that empower the entire business to adapt to an ever-changing environment. Manufacturers should shift gears toward customer-driven strategies and boundary management to develop the competencies required to achieve their strategic intentions.

In Chapter 4, factors of manufacturing performance are presented. These factors bridge the gap between the definition of manufacturing strategy in this chapter and the discussion of the performance measurement system in Chapters 5 and 6.

CHAPTER 4

THE POWER OF MANUFACTURING

Being a world-class company means more than being the first to get your product to market. A world-class company is significantly better than its competitors on a number of fronts. Take Merck, Sharp and Dohme Pharmaceuticals, for example. Described by many as the miracle company, Merck not only has the R&D savvy to develop blockbuster drugs, but also knows how to get through FDA red tape, build customer relationships, and manufacture effectively. What's the Merck formula? A mixture of scientific acumen, good judgment, and devoting a lion's share of resources to an offensive position – creating future products. The secret ingredients in the Merck recipe are hiring and maintaining the best people and utilizing state-of-the-art technical equipment.

Manufacturing capabilities matter to Merck and to other leading firms. Manufacturing matters to the American economy as a whole. The belief that manufacturing is an important ingredient in corporate and national success has stimulated the need to increase our understanding of the manufacturing function and its relationship to business success.

Assessing performance in any industry poses significant managerial challenges. What is lacking is a structured approach to the strategy-performance linkages, so that typical relationships between manufacturing strategy and performance can be scrutinized. These relationships can be found by analyzing the experiences of many firms across a wide variety of settings.

In particular, it is important for today's manufacturing executives to understand the general types of impacts that their counterparts have experienced when developing action programs and establishing approaches for achieving competitive priorities. To understand the linkages of specific strategies and action programs on performance, performance must be specifically measured. Most of the measurement systems in place are deficient in their ability to capture the relationships between performance and manufacturing strategies. Specific measures and measurement systems are explored in Chapters 5 and 6.

A new route to competitive revitalization implies a new view of manufacturing strategy. In Chapter 3, manufacturing strategy was defined as a pattern of choices pertaining to resource allocation and management strategies. The central theme of this chapter is that behaviorally defined manufacturing strategies can be linked to manufacturing capabilities and performance and, ultimately, to business success.

This chapter does not provide formulas for resolving specific business issues. It does, however, illustrate the manner in which common patterns of relationships between manufacturing strategy and performance generally contribute to greater effectiveness. It describes how world-class firms are creating opportunities through manufacturing strength. By reviewing systematic behavior patterns of sufficiently large numbers of business units, manufacturing managers can obtain a rich foundation for the more specific situation analysis required for good decision making. In this chapter, broad-based performance linkages are viewed from three perspectives:

- The linkages between critical competitive manufacturing capabilities and international manufacturing strategies.
- The link between managerial success and manufacturing strategy.
- The impact on the Strategic Business Unit (SBU) and/or manufacturing performance of several of the most commonly researched components of manufacturing strategy.

Staking Out the Global Competitive Agendas

How is manufacturing being used in the realization of business unit objectives? One premise is that manufacturing choices should mesh with business unit goals. While operating decisions may make sense individually, from a tactical perspective, it is the pattern of choices reflecting the manufacturing strategy that works cumulatively to reinforce business objectives.

Business unit objectives describe the distinctive competencies that set the company apart from its competitors. Often, these are expressed in terms of the primary manufacturing task or order-winning capabilities required of manufacturing:

- **Low price/cost** – The ability to sell similar products at a lower price than competitors due to low-cost production capabilities.
- **Quality** – The ability to produce products with significantly higher levels of conformance and performance than competitors.

- **Delivery**—The ability to meet all delivery commitments on time and to manufacture products more quickly than competitors.
- **Flexibility**—The ability to make significant changes in product design, introduce new products quickly, and be responsive to demand shifts in volume.

How do the manufacturing strategies stack up internationally? Are all strategies rationalized in a similar fashion when the basis of competition is the same? Data gathered by the 1986 International Manufacturing Futures Project shows that the linkages between competitive capabilities and manufacturing strategies are clearly different on a global, region-by-region basis.[1] This research forms the basis for the discussion that follows in this section.

Diverse global manufacturing strategies exist. Figures 4-1 through 4-4 provide insight into the way in which leading manufacturers worldwide organize for competitive advantage on low price, flexibility, quality, and delivery. Note that manufacturing strategies are characterized by the key action programs being pursued by large numbers of leading manufacturers in each region. These key action programs are considered the manufacturers' action program *portfolio*. They are the systematic aspects of regional manufacturing strategies.

This section does not present empirical evidence of the success of the action program portfolios selected. Instead, it illustrates the existence of specific linkages between the competitive priorities and action programs of leading manufacturers worldwide. It also summarizes the impact on performance or success of several critical components of manufacturing strategy. Note, also, that individual competitive capability dimensions are treated independently, so firms can compete on one or more capabilities at any given point. For benchmarking purposes, firms competing on multiple competitive capabilities can combine manufacturing strategies across dimensions.

Competing on Price

American manufacturers that place a priority on low price as an important competitive capability develop manufacturing strategies aimed at reducing costs through worker productivity improvements such as labor and management relationships and worker job enlargement programs, robotics, and other automation programs aimed at reducing direct labor content. Notice that Americans have systematically excluded product redesign and standardization from their cost-reduction plans.

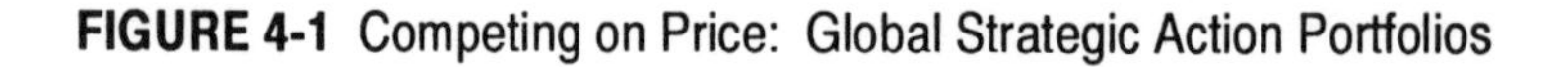

FIGURE 4-1 Competing on Price: Global Strategic Action Portfolios

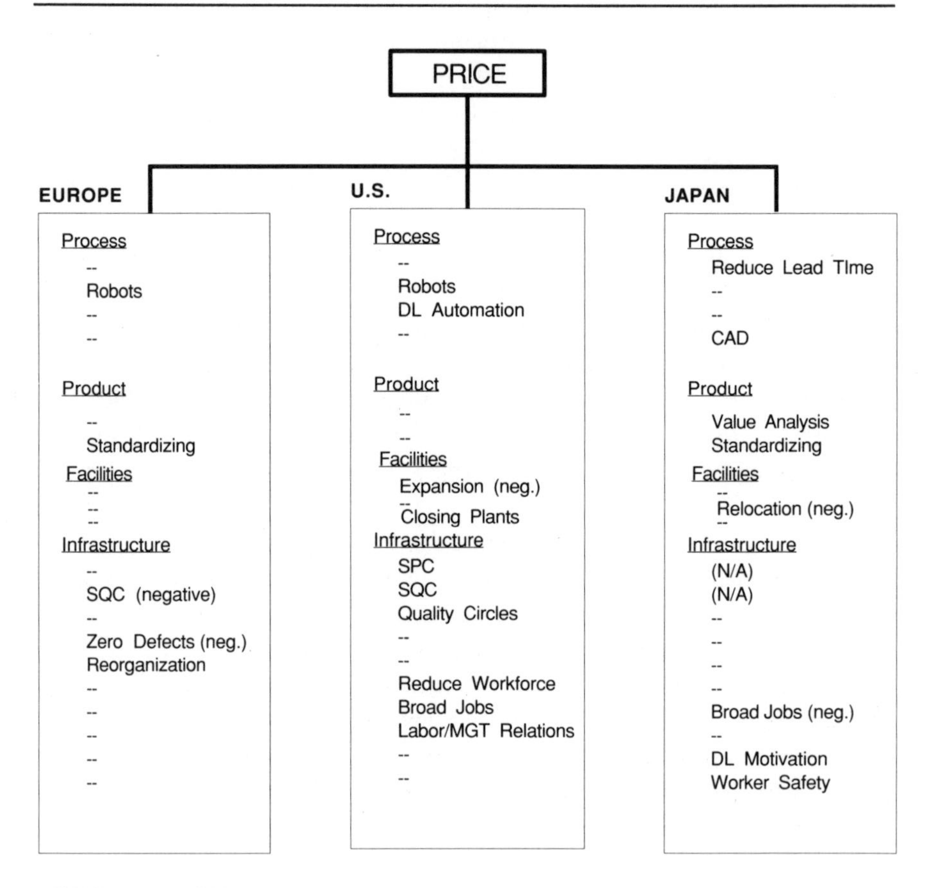

(N/A) Data not available

Source: A.V. Roth, A. DeMeyer, and A. Amano, "International Manufacturing Strategies: A Comparative Analysis," *Managing International Manufacturing* (K. Ferdows, ed.) (Amsterdam: Elsevier Science Publishers, 1989), pp. 187-211.

Strong emphasis is also placed on improving quality through the use of Statistical Process Control and quality circles. American manufacturers have adopted Crosby's concept that "quality is free" and the Deming approach for implementation. Not surprisingly, these same American manufacturers place significant emphasis on headcount reductions and plant closings. Apparently, low-price priorities drive American manufacturing strategies toward:

- Automation, which substitutes for labor and downsizing
- Conformance quality
- Increasing worker flexibility

By contrast, Japanese manufacturers pursuing low price as a competitive priority focus on enhancing the productivity of labor by process and product improvements, using value analysis, and standardizing products. Further, the Japanese improve their cost positions by continuing to pursue lead-time reduction programs and implementing selected technologies.

Unlike the Americans, the Japanese do not see any relationship between cost reduction and enlarging worker jobs. They respond in an opposite manner—employing job specialization in their action program portfolio aimed at reducing cost. Since quality programs are built into Japanese manufacturing processes, no specific data on the application of statistical process and statistical product control are available.

While there is overriding evidence that the Japanese recognize the relationship between quality programs and cost, it is likely that Japanese manufacturers pursuing low price as a competitive priority have already mastered quality. Japanese cost-reducing manufacturing strategies are instead focused on revitalization and renewal of the manufacturing function through:

- Product design
- Reducing complexity
- Reducing cycle times

The primary action programs of European manufacturers who have identified low price as a competitive priority include:

- Product standardization
- Robotics
- Reorganization

The Europeans place little emphasis on formal quality programs to achieve low cost capabilities.

Competing on Flexibility

Flexibility, or the ability to introduce new products and designs quickly, is one of the most difficult goals for a manufacturer to achieve. The rapid introduction of new products and designs was one of the most difficult challenges of the 1980s. The complexity of this manufacturing task ensures its continuation as a formidable manufacturing challenge in the 1990s as well. Yet, flexibility is the area in which the largest competitive gaps exist internationally and is a primary area of emphasis for the Japanese.

For American manufacturers, flexibility is linked to:

- Process flow improvements
- Communications
- Product and vendor quality programs
- Process technology innovations

To deal with the complexity and uncertainty of new product design introductions, leading American manufacturers place significant emphasis on structural, product, and infrastructural programs:

- Process technology such as CAD/CAM, group technology, and FMS (flexible manufacturing systems).
- Process flow improvements such as setup-time reduction and lead-time reduction programs.
- Quality management in areas relevant to design innovation, such as value analysis, supplier programs, and SPC.
- Communication improvements through better cross-functional information systems and the reduction of the size of manufacturing organizations.

Programs being linked to flexibility in Europe are more similar to the American programs than to those of the Japanese. They are more comprehensive than either the American or Japanese strategies. Europeans identifying flexibility as a competitive priority have the following structural, product, and infrastructural action program linkages:

- Implementation of advanced process technology, including group technology, robotics, flexible manufacturing systems, and CAD/CAM technology.

FIGURE 4-2 Competing on Flexibility: Global Strategic Action Portfolios

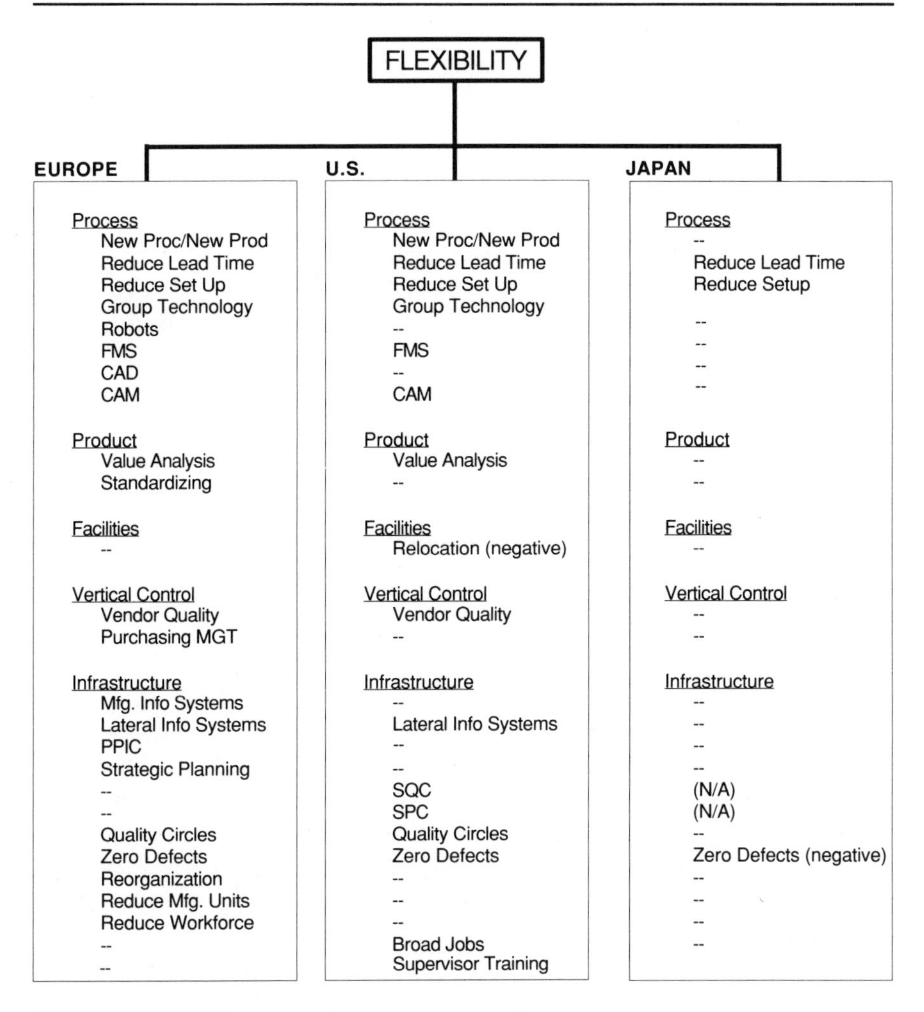

(N/A) Data Not Available

Source: A.V. Roth, A. DeMeyer, and A. Amano, "International Manufacturing Strategies: A Comparative Analysis," *Managing International Manufacturing* (K. Ferdows, ed.) (Amsterdam: Elsevier Science Publishers, 1989), pp. 187-211.

- Decrease of cycle times through lead-time and setup-time reductions.
- Modification of products through value analysis and standardization.
- Integration through vendor quality programs, strategic planning, and information systems.
- Adoption of quality circles and zero defect programs.
- Reorganization and downsizing.

Notice that process technology plays a significant role in the Europeans' flexibility-related programs. Technology is complemented with improved vertical control (supplier control), product design changes, streamlining of operations, and improved information systems.

The Japanese, by way of contrast, have a much shorter list of action programs linked to improvements in flexibility. They focus on cycle time process improvements, decreasing cycle times through lead-time and setup-time reductions.

Flexibility is also an integral part of quality and delivery for the Japanese and may explain their sparse action program list. Although the Japanese identify flexibility as a by-product of action programs aimed at other competitive capabilities—quality and delivery—flexibility was identified as an area with a large gap. The Americans fit somewhere in between the Europeans and Japanese, with greater emphasis on infrastructural programs than structural ones.

Competing on Quality

American manufacturers whose competitive priority is quality emphasize CAD/CAM technology. These programs are aimed at simultaneously enhancing both design and manufacturing operations; that is, design for manufacturability. Statistical control of product is also important for Americans. In addition, programs defined for *zero defects* are strongly emphasized along with worker job enlargement programs. Value analysis is included as a program for improving product design.

Interestingly, American manufacturers pursuing a quality goal displayed no consistent tendency to employ group technology, flexible manufacturing systems, or the standardization of product lines. Nor were they systematically involved in downsizing or streamlining activities.

The Europeans have a broader action program linked to quality than either the Americans or the Japanese. As with the Americans, the Europeans' top programs for pursuing quality are CAD/CAM. They exhibit a proclivity toward adopting advanced technology for improving product

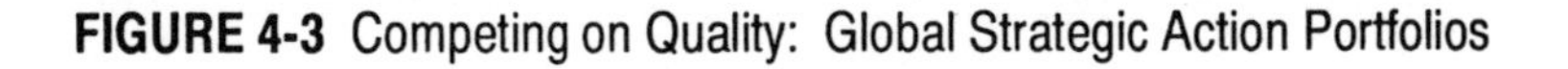

FIGURE 4-3 Competing on Quality: Global Strategic Action Portfolios

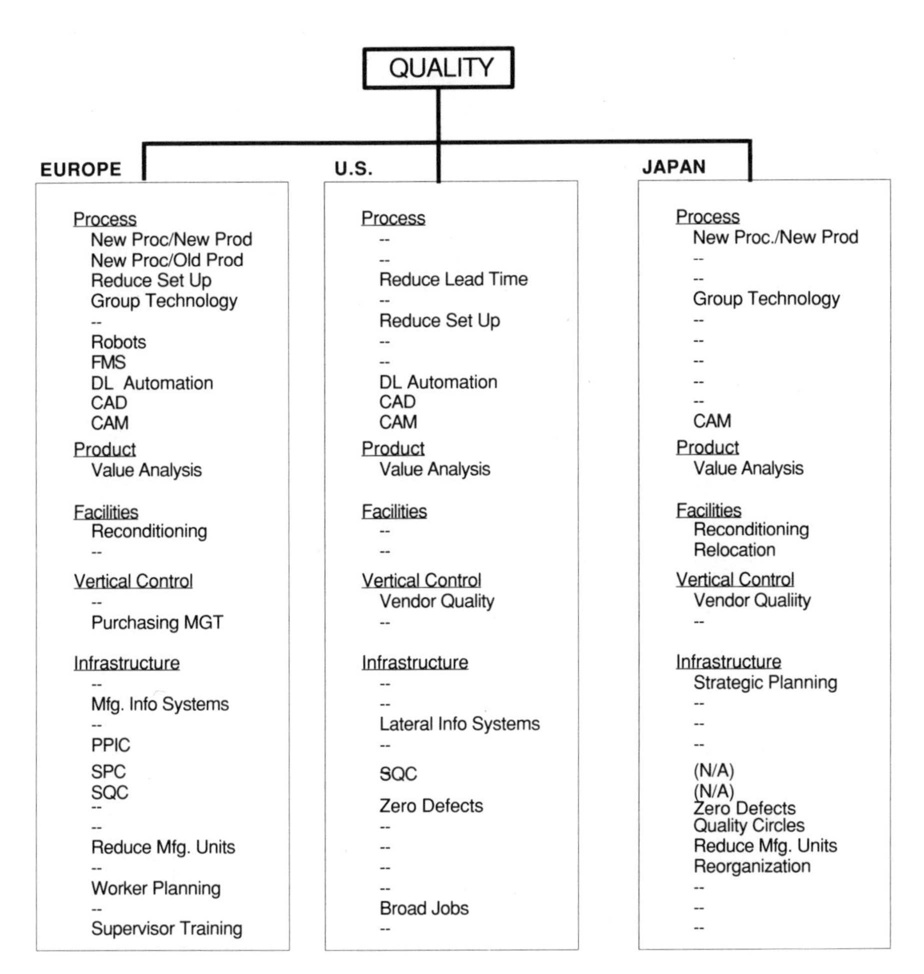

(N/A) Data Not Available

Source: A.V. Roth, A. DeMeyer, and A. Amano, "International Manufacturing Strategies: A Comparative Analysis," *Managing International Manufacturing* (K. Ferdows, ed.) (Amsterdam: Elsevier Science Publishers, 1989), pp. 187-211.

quality, including the use of robotics, flexible manufacturing systems, and direct labor automation equipment, in addition to CAD/CAM technology.

From an infrastructural perspective, the Europeans emphasize worker involvement in the planning process and increased supervisor training. The Europeans also show linkages between downsizing manufacturing and efforts to improve product quality.

For the Japanese, the primary linkages to quality are the infrastructural elements of strategy, including programs for zero defects, quality circles, and reorganizing and reducing the size of the manufacturing units. While the International Manufacturing Futures Project did not include SPC as a key action program on the Japanese survey, SPC continues to be a key linkage to quality for the Japanese.

From a structural perspective, the Japanese quality focus is on group technology, new processes for new products, CAM, value analysis, facilities, and vendor quality programs.

Competing on Delivery Time

Those American companies identifying delivery as a key competitive capability had limited portfolios of action programs in this area. The American time-based delivery manufacturing strategy included:

- Cycle-time reduction
- Advanced automation
- Facility relocation
- Labor relations

Corroborative data collected through a series of interviews with both NCMS member companies and nonmember companies, indicated a stronger awareness on the part of American executives of both on-time delivery performance requirements and delivery speed requirements. Though a number of these companies had both programs and systems to deal with improving delivery performance, few had clearly defined programs or activities aimed at delivery time improvements.

U.S. manufacturers were eyeing technology to evoke time-based capabilities. U.S. manufacturers did not attend to infrastructural components of manufacturing strategy in any consistent way. Improving labor and management relations was the notable exception. This finding, coupled with the delivery gaps noted in Chapter 3, suggests that the U.S. has not yet developed a systematic approach to its manufacturing strategy in this area.

FIGURE 4-4 Competing on Delivery: Global Strategic Action Portfolios

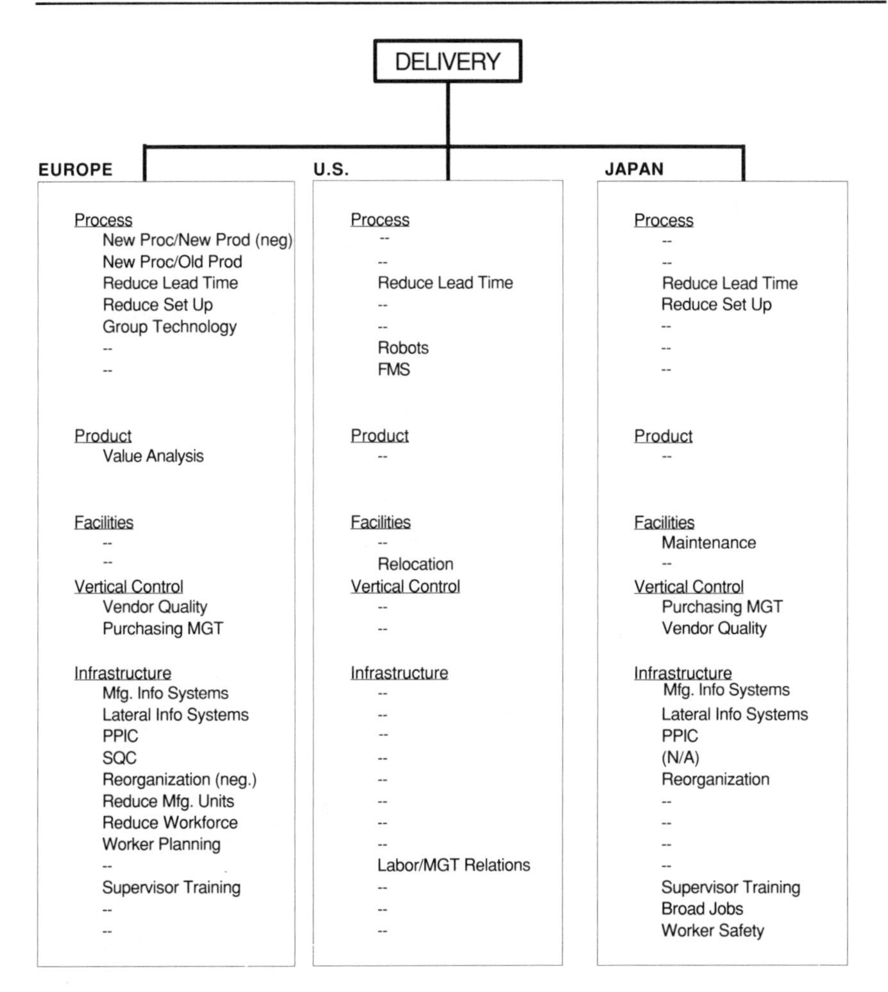

(N/A) Data Not Available

Source: A.V. Roth, A. DeMeyer, and A. Amano, "International Manufacturing Strategies: A Comparative Analysis," *Managing International Manufacturing* (K. Ferdows, ed.) (Amsterdam: Elsevier Science Publishers, 1989), pp. 187-211.

Both the Japanese and the European companies had much broader programs aimed at increasing their competitiveness in the area of delivery time. Both are attacking time-based priorities through significant structural and infrastructural choices.

The Japanese time-based manufacturing strategies are characterized by:

- Decreasing cycle time through lead-time and setup-time reduction programs.
- Improving maintenance of equipment and facilities.
- Integrating suppliers through vendor quality and purchasing management programs.
- Integrating systems by improving management information production planning and inventory control systems.
- Strengthening the workforce by giving workers broader jobs, improving supervisor training, and improving worker safety.
- Reorganizing the organizational structure.

The European manufacturing strategy for time-based competition involves:

- Decreasing cycle time through lead-time and setup-time reductions.
- Improving product design through value analysis.
- Implementing group technology and designing new processes for old products.
- Integrating systems by improving management information, production planning, and inventory control systems.
- Implementing statistical process control.
- Strengthening the workforce including giving workers more planning responsibilities and improving supervisor training.
- Streamlining operations through reduced workforces.

Implications for International Competition in the 1990s

From the examination of global strategic action portfolios, several managerial inferences can be drawn with respect to which strategies are most effective for world-class manufacturers.

- International competitors have different game plans and are systematically targeting different resource mixes to achieve similar competitive capabilities.

- Achieving strategic outcomes is not likely to result from tweaking the manufacturing system or making marginal improvements over competitors' tactics. Simple imitation of individual programs for JIT, SPC, or TEI will not reduce competitive risks in the long run; however, they may yield significant improvements over current operations.
- The basic architecture of manufacturing strategy must be designed for the competitive advantage. Managers cannot take on global competitors by playing the same manufacturing game better. They must intelligently make fundamental changes in the ways their game is played so that they put their competitors at a functional disadvantage. For example, Yamaha's entry into the grand piano market was built upon nontraditional assembly line production techniques, while Steinway was producing its grands in a job shop environment. Toyota's manufacturing strategy was radically different from those of American manufacturers.
- Japanese companies are more consistent in applying their action portfolios. More Japanese companies tend to "fit the profile" than either American or European companies. American companies are the least consistent in their application of manufacturing action plans for the pursuit of particular competitive priorities. Could it be that American manufacturers, in general, are still experimenting with their strategies or that American manufacturers are not using manufacturing solely as a competitive weapon? Most likely the answer is yes, in part, to both questions.

These findings do not explain the systematic impact of pursuing different action programs on performance. Since most measurement systems lack the rigor required to clearly track the impact on performance of different strategies, research in this area is also lacking. A handful of important studies have been performed, however, that attempt to more specifically define the impact of strategy on performance. The remainder of this chapter examines the impact of the most common action programs on Strategic Business Unit or manufacturing performance.

Winning through Manufacturing

The information available on the performance of individual companies is overwhelming. Tom Peters in *Thriving on Chaos* says that successful organizations have the following characteristics in common:

- Flatter organizational structures
- More autonomous work units
- Quality and service controls
- Capabilities to be responsive, innovative, and flexible
- Highly trained and skilled workers
- Leaders, not managers, at all levels

In the context of the global strategic action portfolios, the characteristics appear to mix means and ends. Despite the logical argument that means and ends ought to be related, the linkages are poorly understood. For manufacturing to matter, a clearer understanding of the relationships between manufacturing strategy, manufacturing capabilities, and economic performance is necessary to operate intelligently in the 1990s and beyond.

Understanding the Performance Linkages

Recent manufacturing research addresses these linkages using the framework shown in Figure 4-5. Within the framework, manufacturing strategy is related to three different types of performance measures:

- **Relative Manufacturing Success**—The relative strength of the manufacturing firm compared to its competitors on quality, flexibility, delivery, and price capabilities.
- **Relative SBU Managerial Success**—General management success in accomplishing strategic business unit goals.
- **Absolute Economic Outcomes**—Return on Assets (ROA) and profit margins.

A good manufacturing strategy does not guarantee manufacturing success. A company must be able to implement that strategy to yield relative superiority on key competitive measures. The framework in Figure 4-5 illustrates that two key factors in determining managerial success are:

- The relative strength that exists in all functional areas and is available to be exploited.

- The ability of the management team to do the exploitation (general management ability).

Absolute economic outcomes, such as ROA and profit margin, are dependent upon relative managerial success and on environment. For example, the chemical and paper industries have had a very favorable market environment in recent years in the U.S., while computer manufacturers have been faced with more difficult market conditions.

Manufacturing Superstars Support Business Success

The conceptual framework shown in Figure 4-5 was tested by using data from a sample of 193 executives of large American manufacturing firms in 1988. Key findings are shown in Figure 4-6. They include:

FIGURE 4-5 Conceptual Framework Linking Manufacturing Strategy and Performance

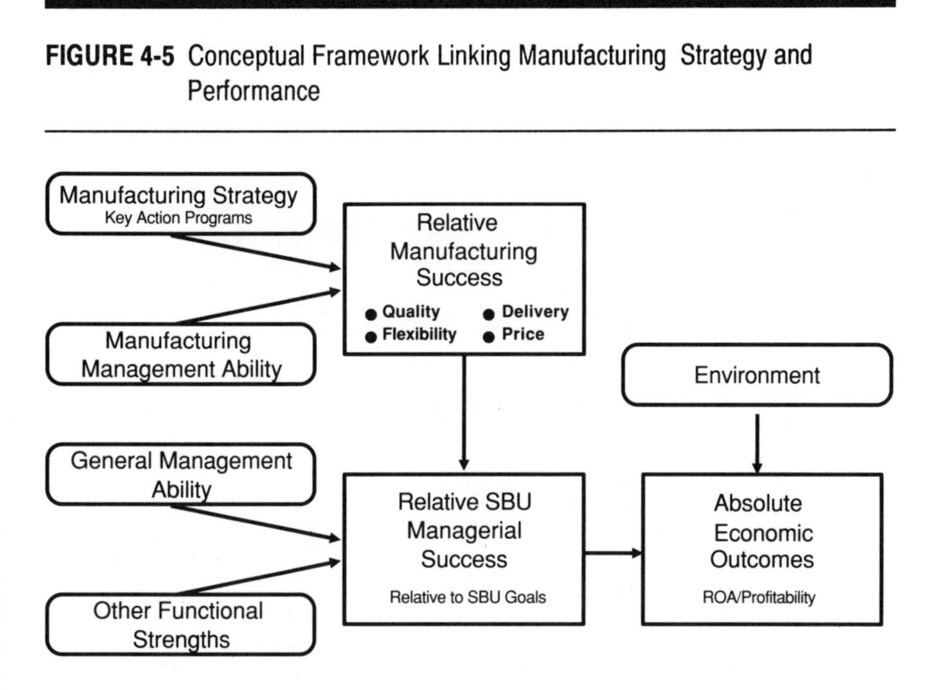

Adapted from: A.V. Roth and J.G. Miller, "Manufacturing Strategy, Manufacturing Strength, Managerial Success, and Economic Outcomes," in *Manufacturing Strategy* (J.E. Ettlie, M.C. Burstein, and A. Feigenbaum, editors) (Boston: Kluwer Academic Publishers, 1990), p. 98.

- Business unit management performance relative to goals was associated with economic outcomes after controlling for the size of the business. The managerial SBU leaders reported net pretax profits of 11.4 percent of sales and pretax returns on assets of 23.6 percent in contrast to managerial SBU laggers who had average profits of 7.8 percent and ROAs of 12.8 percent.
- Managerial SBU leaders, defined by the ability of their management to meet goals on qualitative performance measures, also displayed heightened manufacturing competitive strength. The SBU leaders outperformed SBU laggers in delivery, flexibility, price, and market scope capabilities.[2]
- Managerial SBU leaders demonstrated different degrees of strength on each manufacturing performance variable as well—that is, not all manufacturing capabilities were equal. Both managerial SBU

FIGURE 4-6 Relative Manufacturing Strength of Business Unit Leaders and Laggers

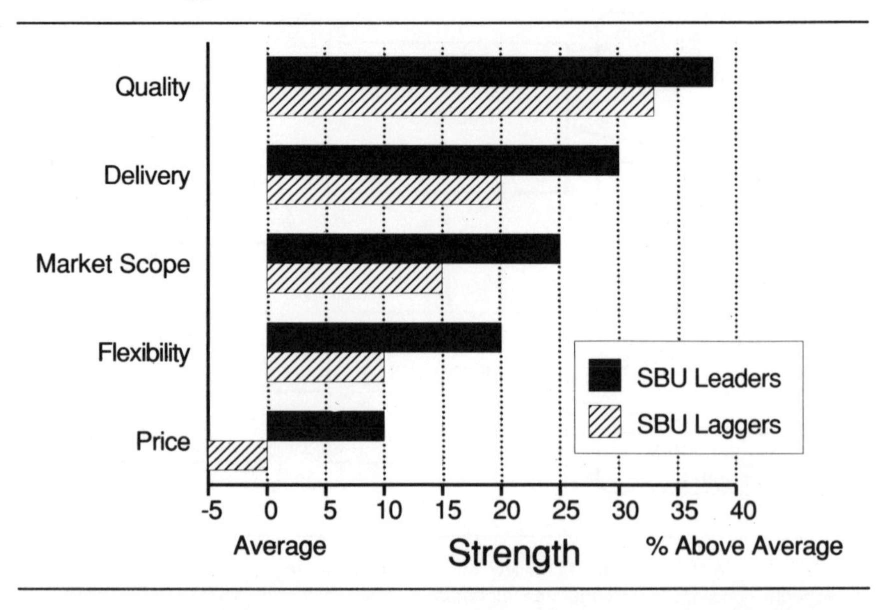

Adapted from: A.V. Roth and J.G. Miller, "Manufacturing Strategy, Manufacturing Strength, Managerial Success, and Economic Outcomes," in *Manufacturing Strategy* (J.E. Ettlie, M.C. Burstein, and A. Feigenbaum, editors) (Boston: Kluwer Academic Publishers, 1990), p. 102-103.

leaders and SBU laggers reported high levels of manufacturing strength on quality.

- Manufacturing weaklings, those with lower competitive strength, are more likely to be managerial SBU laggers. Manufacturing superstars, those with significant competitive manufacturing strength, were more likely to be managerial SBU leaders. Economic returns are highest, however, for those that score high as SBU managers and substantially lower for those with less business unit success. Further research is required to determine if manufacturing strength leads to business success or if managerially successful firms create better manufacturing.
- The manufacturing strategy choices are correlated with manufacturing competitive strength. Overall, a model describing the manufacturers' emphasis over several structural and infrastructural choices displayed a predictive power 30 percent better than chance with respect to identifying manufacturing superstars and weaklings.
- Manufacturing strategy content areas found to be *positively* correlated with building manufacturing strength are:
  - Resource improvements: The degree of attention to care and nurturing of human assets and physical resources, including tri-level training of employees (workers, supervisors, and managers), job enrichment and safety, and preventive maintenance of equipment.
  - *Quality programs:* The degree of emphasis on total quality management of processes and suppliers.
  - *Advanced process technology:* Application of group technology, flexible manufacturing systems, computer-aided design, and robotics.
- Manufacturing strategy content areas found to be *negatively* correlated with building manufacturing strength are:
  - *Restructuring:* The degree of emphasis on downsizing, reorganization, and relocation.
  - *Information and systems:* The degree of emphasis on systems for integration within the function and across functions.

The explanation for the inverse relationship between manufacturing strength and restructuring may be that manufacturing superstars have already restructured or have no need to restructure; whereas, weaklings perceive the need for radical surgery.

There is compelling evidence that a number of currently successful companies that underwent restructuring in the early to mid-1980s achieved impressive results. However, not all restructuring has been successful. Chief Executive Donald Kelly of BCI Holdings Corporation reported that "the trick is to distinguish true restructuring from the me-too variety... There's a lot of scrambling around, and many companies are doing it just so their top executives can go to cocktail parties and say they restructured. Many companies are doing it in a protective mode, and some are doing it poorly."[3]

The negative correlation between information systems and manufacturing strength may be because superstars have reduced the need for such systems because they have succeeded in reducing the inherent complexity of their production processes. The manufacturing weaklings may view integration of information as a way of reducing uncertainty or alter their information systems to coincide with facilities and organizational changes.

Laying a Solid Foundation for the 1990s

The managerial implications for the 1990s are:

- Manufacturing excellence alone is not sufficient for business success, but attaining world-class manufacturing status increases the odds of success.
- Everybody is in the quality game; product should not leave the factory without it.
- The primary building blocks of manufacturing excellence are total resource improvements, quality management, and application of appropriate advanced manufacturing technologies.
- World-class manufacturing enterprises require dynamic, ever-improving manufacturing capabilities and savvy business planning.[4]

Linking Perceived Quality and Performance

In the previous section, the broad-based linkages between manufacturing strategy and manufacturing strength were summarized from the perspective of a large number of manufacturing executives from leading firms. Product quality capabilities—conformance and performance—and their relationship to manufacturing strength were described.

An important question still remains with respect to quality and performance. Is the importance of product quality as a differentiator waning? A growing body of literature suggests that American manufacturers have made significant strides in improving the quality of products over the past decade; the perceived gap between quality capabilities among manufacturers is closing. Even so, *there remains a serious question about the ability of many manufacturing executives to judge their real competition.* For example, at the Ford Wixom plant, signs visible on the shop floor read "BEAT GM." There is no mention of foreign competition. The American automobile makers and Ford, in particular, have made impressive strides in quality over the past decade, but the Japanese have not been waiting for them to catch up.

The strategic implication of these findings is that product quality is a *minimum daily requirement*; it is more than the price of admission to the competitive arena. As suggested in Chapter 2, quality is a moving target. The standards are constantly being challenged. World-class manufacturing in the twenty-first century will reflect those firms that have been unrelenting in their quest to deliver superior quality and build customer trust and loyalty. Additionally, quality capabilities are basic to achieving competitive manufacturing strength in other areas, such as delivery, flexibility, and cost. A small gap in quality may reflect gaps in other capabilities. Therefore, the strategic importance of product quality cannot be dismissed.

Manufacturers in the 1990s must also realize that achieving a high level of product quality does not immediately translate into increased market share. How long will it take for the satisfied owner of a Honda Accord to recognize quality in and be willing to purchase a domestic car? *There is a lag between a change in product quality and its recognition by customers.* Customers purchase goods based upon the relative level of quality they perceive. This subsection describes the relationship between perceived relative quality and business performance based on two studies:

- The Profit Impact of Market Strategy (PIMS).
- The 1988 American Society for Quality Control (ASQC) Gallup Survey.

PIMS Findings

The most rigorous analysis of the impact of perceived relative quality on business performance has been done by the PIMS researchers.[5] Over the last 16 years, PIMS researchers have developed a database of over 450 companies and 3000 business units for the purpose of analyzing business strategy and performance impacts.

The PIMS program, which is housed at the Strategic Planning Institute (SPI), was initiated to determine how key dimensions of business strategy affect profitability and growth. The PIMS database includes North American and European companies of various sizes with many different products and markets including machinery, consumer products, high technology products, heavy industrial goods, basic industries (raw materials), and services.

Historically, the PIMS research has found:

- Relative perceived quality, above all other factors, drives market share.
- Quality is whatever the customer says it is, and the quality of a particular product or service is whatever the customer perceives it to be.
- Relative perceived quality and conformance quality are not the same. Relative perceived quality is viewed from an external, or customer, perspective. Conformance quality is viewed from an internal, or quality assurance, perspective.
- Achieving either superior perceived quality or superior conformance quality can result in a competitive advantage. However, since the two concepts are not mutually exclusive, achieving both is the best way to be the winner.
- Relative perceived quality and market share are correlated with regard to their impact on profitability (see Figure 4-7), however:
  - Relative quality affects relative price, but separate from quality, market share has little impact on price.
  - Conversely, market share affects relative direct cost, but perceived quality has little impact on cost.
- Achieving superior conformance quality ultimately results in both lower costs and superior perceived quality.
- As manufacturers move from inferior to superior quality positions, they typically trade the cost reductions of improved conformance quality (i.e., scrap and rework) for the increased costs of improving product or service performance on key attributes that result in the customer's purchase decision (i.e., perceived quality).

Therefore, *superior perceived quality typically leads to greater profitability as the result of higher relative prices but comparable direct costs.*

FIGURE 4-7 Quality and Share Both Drive Profitability

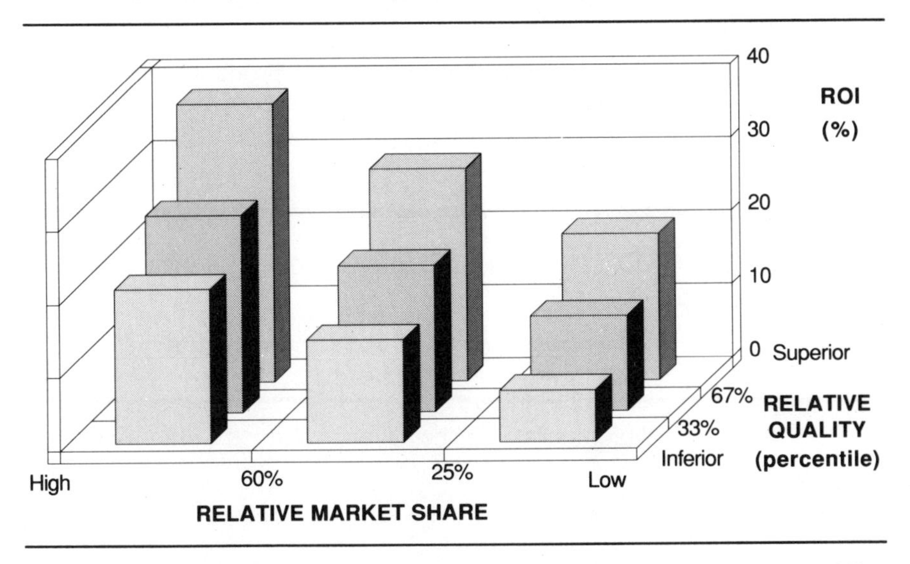

Reprinted with permission of The Free Press, a division of Macmillan, Inc. from *The PIMS Principles: Linking Strategy to Performance* by Robert D. Buzzell and Bradley T. Gale. © 1987 by The Free Press.

1988 Gallup Quality Survey

The 1988 Gallup Survey, conducted for the American Society for Quality Control, found that Americans are willing to pay substantially more for perceived higher quality. The Gallup Survey found that:

- The premium Americans are willing to pay for quality ranges from 21 percent more for a car to 40 percent more for a television set to 72 percent more for a sofa (see Figure 4-8).
- American consumers who have a preference for the quality of foreign products are significantly less resistant to paying a premium for high ticket items, such as cars.
- About 50 percent of the Gallup Survey respondents gave American products high marks for quality vs. foreign products. This echoes the 1985 survey results.
- Two-thirds of the adult population feel that American-made products more closely satisfy their standards of quality than foreign

products. However, 40 percent of the upper income individuals (household income more than $40,000 annually), with higher disposable income, believe that foreign products more closely satisfy their quality needs.

- The strongest support for American-made products comes from older, poorer, and less-educated groups.

FIGURE 4-8 1988 ASQC/Gallup Survey - Price Willing to Pay for Better Quality Product

	Automobile	Dishwashers	Shoes	TV	Sofa
Baseline cost	$12,000	$400	$30	$300	$500
Avg. over baseline cost	$2,518	$167	$20	$201	$362
% who would pay nothing extra	18%	6%	6%	7%	4%

Source: Reprinted with the permission of ASQC.

The 1988 Gallup Survey also found that in defining quality American consumers placed a great deal of emphasis on performance and durability factors and little emphasis on appearance, brand name, and price.

The linkage between perceived quality and specific companies was revealed in a question asking respondents which companies, globally, are most closely associated with high quality (see Figure 4-9). GE and General Motors, the top two companies in 1985, were again the top two companies mentioned in 1988. Four foreign companies improved their rankings within the top 20 in 1988; seven American companies dropped out of the top 20 between 1985 and 1988.

FIGURE 4-9 1988 ASQC/Gallup Survey - Companies Associated with High Quality

Top 20 Rankings

Rank	Company
1	General Electric
2	General Motors
3	Ford
4	IBM
5	Sears
6	AT&T
7	Sony
8	Chrysler
9	RCA
10	Procter & Gamble
11	Toyota
12	Zenith
13	Honda
14	Whirlpool
15	Panasonic
16	Mercedes Benz
17	Maytag
18	Del Monte
19	General Mills
20	Westinghouse

Source: Reprinted with the permission of ASQC.

While quality has been demonstrated to have a significant impact on business unit performance and customer perceptions in the 1980s, *time-based competition appears to be the primary differentiator for the 1990s.* Throughput-time reductions have been shown to have a significantly favorable impact on financial performance.

Time-based Competition and Performance

Throughput-time and delivery cycle-time reductions have been increasing in importance as competitive weapons. Time-based competitive capabilities, as previously described, are reshaping global markets and are on the basic agenda for world-class competitors in the 1990s. In their comprehensive account of time-based competition, *Competing Against Time*, Stalk and Hout report that time-based competitors outperform others in their industries, and that the strategic implications of compressing time are significant:

- Productivity increases of 20 to 70 percent, coinciding with significant work-in-progress turns.
- Premium prices can be charged—prices from 20 to 100 percent higher are typical.
- Risks are reduced as the costs of over- and under-forecasting demand are driven down.
- Market share rises, as companies meet deliveries more reliably and bring new products to market faster.[6]

Perhaps world-class manufacturers should no longer track and report labor efficiency and machine utilization but, instead, focus on tracking and reporting measures of throughput.[7] Time is a more effective metric than cost. Time is simultaneously a key cost driver and a revenue generator. When Caterpillar threatened Komatsu in 1970, Komatsu responded by first improving quality, then driving cost down and accelerating new product development. Komatsu transformed itself in four years, developing a strategy to make its manufacturing as flexible as possible.[8]

Nowhere have the lessons of throughput-time reduction been more closely observed than in the auto industry. Many studies have examined the success of the Japanese. Cusumano's five-year study of the Japanese automobile industry, which focused on Nissan and Toyota, found that:

- The performance of Japanese auto producers has depended not on the employment of Japanese workers but on Japanese innovations in management techniques such as JIT and the intelligent application of manufacturing process technology.
- While a casual review of the data makes it appear as though Japanese workers are more productive because of higher investment levels, the study found that not to be correct. Based upon an analysis of Japanese productivity and capital investment in the auto industry,

> Cusumano found that capital productivity was not higher in the U.S. than in Japan; rather throughput per worker per year was twice as high in Japan.[9]

Toyota, continues Cusumano, through the leadership of Taiichi Ohno, departed from traditional U.S. manufacturing methods after World War II and became the leading example of the impact of Japanese manufacturing techniques on performance. At that time, there was significant pressure on the Japanese auto industry to produce in very small lots at low cost to be competitive.

Further, U.S.-based techniques including high levels of worker and equipment specialization, extensive automation, long production runs on large machines requiring long setup times, large in-process buffer inventories, and the push concept of production control became inappropriate. Toyota developed a manufacturing strategy based on flexible manufacturing concepts that gave it a substantial operating advantage over its American competitors. Manufacturing practices associated with cycle-time reductions are addressed in Chapter 6.

The results Toyota achieved were superior to the results of the other Japanese manufacturers:

- Vehicles manufactured per worker year tripled at Toyota between 1955 and 1957 and then rose another 60 percent by 1964.
- Adjusted productivity measures (for vertical integration, capability utilization, and labor hour differences) indicate that Toyota had passed General Motors, Ford, and Chrysler by 1965 and had reached a level of productivity 2.7 times that of the Big Three by 1979.
- Inventory turnover rates for Toyotas went from three in 1950 to eight in 1955, but were still worse than those of U.S. auto manufacturers. With better synchronization and the use of kanban and pull inventory systems, as well as the extension of the kanban system to suppliers, inventory turnovers at Toyota reached 21 times by 1974, twice as high as U.S. manufacturers, and reached 38 times by 1985 and 1986.

The automobile industry is not the only industry that has benefitted from applying the concept of throughput-time reduction. The effect of throughput-time reductions and/or the application of JIT to achieve throughput-time reductions were analyzed in depth by Schmenner. His research, conducted both in the U.S. and internationally, involved 291 manufacturing facilities in diverse industries in the U.S. and 128 plants in 30 countries internationally.

Schmenner focused on differentiating the impact on performance of a wide variety of manufacturing technologies. Schmenner's findings included:

- Of the many potential means of improving productivity, only the JIT-related ones were statistically shown to be consistently effective.
- Plant location (domestic or international), plant age, union/non-union labor, industry, size, and manufacturing processes affected the results.
- Throughput-time reductions of 50 percent were shown to improve productivity by two to three percentage points.
- Only throughput-time reduction was shown to statistically and consistently impact plant productivity, quality, inventories, and overhead costs positively.
- Throughput-time reductions affected measures of performance other than productivity, but these impacts were less direct, although all measurements were positive and linked directly to the throughput-time reduction and the resulting productivity gains.[10]

Many U.S. and European companies excel in throughput- and cycle-time reductions. IBM, Westinghouse, and Motorola are among those that have instituted programs that focus on throughput-time reductions. Typically, these reductions are best accommodated and implemented in lean manufacturing facilities.

The Impact of a Lean Operation on Manufacturing Performance

Operationally *lean* organizations have long been thought of as efficient and effective. Kratick studied more than 50 automobile assembly plants in Japan, the U.S., and Europe under the auspices of the Massachusetts Institute of Technology's (MIT) International Motor Vehicle Program.[11] The focus of his study was to determine what characteristics distinguish high-performance plants from average plants. Key findings of the study included:

- Productivity performance was strongly correlated only with the leanness of the plant and the diversity of the product mix at the plant; a leaner and less diverse product mix correlated with higher productivity.

- Lean plants are more capable of simultaneously achieving high levels of productivity, quality, and mix complexity, than are buffered plants, i.e., plants with inventory to buffer production flow problems.
- On average, Japanese plants are both leaner and more productive than plants in the U.S., although the best U.S. plants are as good as the best Japanese plants.
- A high correlation was found between productivity and quality in lean plants, particularly those of Japanese parentage, but little correlation was found between productivity and quality in traditional U.S. and European plants, i.e., even plants with high productivity did not necessarily have high quality.
- The critical production system characteristics that differentiate superior performers and the direction of the characteristic are:

- Flexibility	*High*
- Span of Control	*Moderate*
- Inventories	*Small*
- Buffers	*Small*
- Repair Areas	*Very Small*
- Teamwork	*High*

A lean organization is important to other types of manufacturers as well as auto makers. Hundreds of plants and businesses have been shut down or sold off in the last 10 years. Of those that have survived, whole tiers of employees and managers have been laid off in the process of getting lean.[12] The new shape of the manufacturing organization is discussed in Chapter 8.

The Impact of Engineering Change Orders on Performance

Engineering change orders negatively affect productivity. In an in-depth study of 12 manufacturing plants located worldwide, Hayes, Clark and Lorenz found that ECOs specifying a change in either materials or manufacturing process had a clear, short-term, negative impact on productivity.[13]

Additionally, they found that ECOs had a long-term impact, with negative effects persisting for up to a year. This was true whether the change responded to customer specifications or was introduced to improve the product or production process.

Not only was the number of ECOs introduced on a month-to-month basis detrimental, but the variation in quantity introduced had a negative impact. Those plants that experienced higher levels of ECOs on average were less affected by variation than those that usually experienced low levels of ECOs. Clearly, some form of learning curve exists.

Finally, they found that the way the ECOs were introduced had an effect on performance. Major ECOs actually had a beneficial effect on performance, while ECOs considered to be minor were the most detrimental. Upon further investigation, they found that the major ECOs were given significant attention by managers, who feared possible disruption. This attention included careful pre-planning and early warning and training to all workers. Minor ECOs, on the other hand, were dumped onto the shop floor with no special attention.

The 1989 North American Manufacturing Technology Survey found that measures of manufacturing complexity, typically the drivers of overhead costs, were among the indicators that worsened between 1987 and 1989. Engineering change orders added to the complexity and, apparently, to the cost of the typical manufacturing environment (see Figure 4-10).

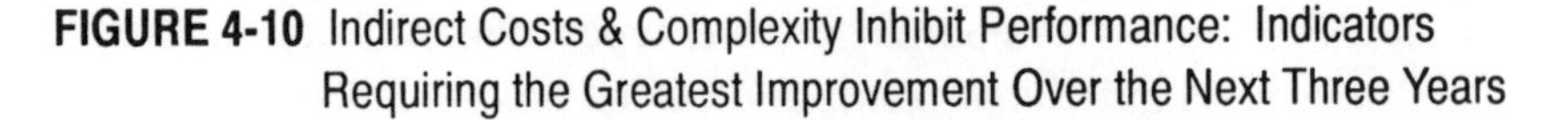

FIGURE 4-10 Indirect Costs & Complexity Inhibit Performance: Indicators Requiring the Greatest Improvement Over the Next Three Years

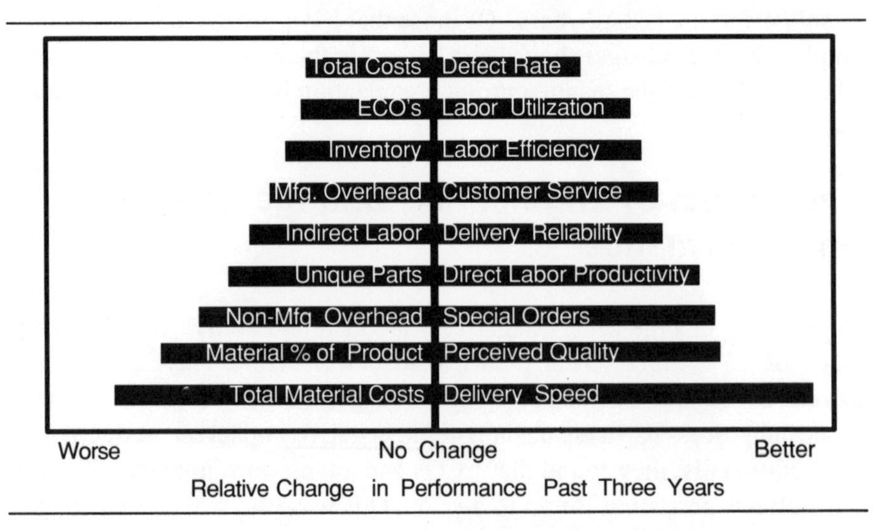

Source: C. Giffi and A.V. Roth, "Making the Grade in the 1990s," Deloitte & Touche Third Annual Survey of North American Manufacturing Technology, 1989, p. 7.

The Impact of Holistic Manufacturing Strategies on Performance

Performance improvements are measured along singular dimensions such as cost, quality, customer service, and flexibility. Little research, however, has been done in the area of defining the relationship between multiple performance improvement programs on overall manufacturing processes and those that contribute to manufacturing strength. The direct measures of manufacturing performance are under the control of the manufacturing function and should be directly affected by a manufacturing strategy.

The seven measures directly affected by a manufacturing strategy and the variables from which they have derived are:

- **Manufacturing Cost Improvement**—The relative degree of improvement in unit costs, unit labor costs, unit material costs, and total manufacturing overhead costs.
- **Inventory Turnover/Material Flow**—The relative improvement in work in process (WIP) turnover, raw materials, and finished goods inventory turnovers, manufacturing lead times, and changeover/setup time.
- **Internal Environmental Stability**—The relative improvement in reducing manufacturing complexity, including number of suppliers, number of parts/components, headcounts, and absenteeism.
- **Total Factor Productivity**—Productivity of direct and indirect labor, machine utilization, and materials yield.
- **Quality of Customer Interface**—Improvements in factors perceptible to the customer, including outgoing quality, percent of on-time deliveries, accuracy of inventory information, and master schedule performance.
- **Supplier Quality (Internal/External)**—Improvement in factors incoming to the production process, including the number of engineering changes and quality of materials.
- **Forecast Dependability**—Improvements in accuracy of sales forecasts and the percent of new products introduced on time.

As was evident in preceding sections, the manufacturing levers that are being pulled overlap to a considerable extent. This overlap makes it difficult to understand either the effect of a single program on a single performance measure or the impact of all programs on all measures. Of course, the impact of all programs on all measures may be best defined by total financial performance over time. Figure 4-11 illustrates the connections between manufacturing strategy contents and manufacturing performance indicators.

FIGURE 4-11 Mapping Manufacturing Strategy and Performance

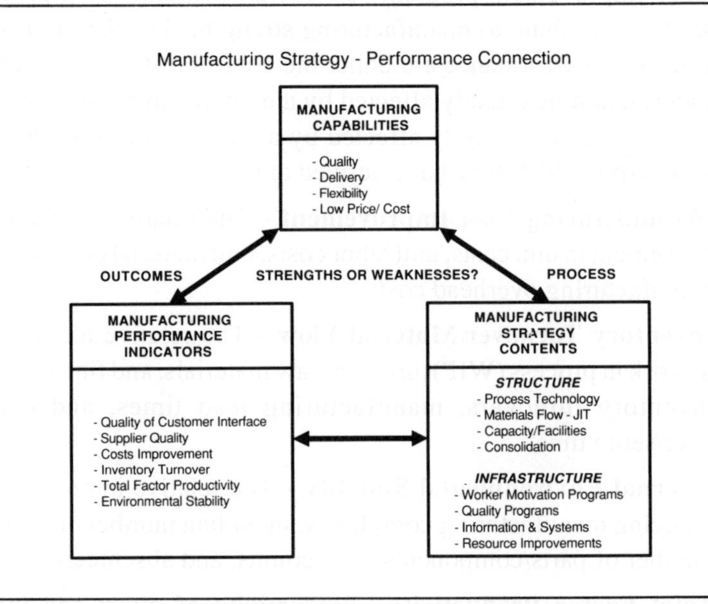

Source: A.V. Roth, "Linking Manufacturing Performance and Strategy, An Empirical Investigation," Duke University, 1990.

The correlation between strategy and performance suggests that those manufacturers that practice a holistic or balanced manufacturing strategy have superior performance overall relative to those manufacturers practicing very focused strategies (see Figures 4-12 and 4-13). Each manufacturing strategy choice specified in Chapter 3 must interrelate with and reinforce the others. Structural, infrastructural, and integration choices must fit together into an integrated whole.

Manufacturers who placed balanced emphasis on a number of elements of strategy, such as process technology, material flow improvements, capacity management, facilities consolidation, worker motivation programs, quality programs, and systems improvements, clearly had superior overall performance. Each of these decision areas dovetailed with the others, so that process and facility choices coincided with quality, human resources, and systems.

The implications for management are:

- Management performance is correlated with total strategic business unit performance; that is, manufacturing matters.
- Technology alone is not a panacea. It is only one element of overall manufacturing strategy; it must be balanced with appropriate infrastructure choices and boundary management activities.
- A holistic strategy offers both internal and external manufacturing performance.
- Superior performance can be developed by working incrementally on a large number of structural and infrastructural improvements simultaneously, as opposed to "betting the farm" on one or two areas.

These findings are of significant importance in understanding the power of the manufacturing function and its strategic choices. Performance improvements are closely related to balanced programs.

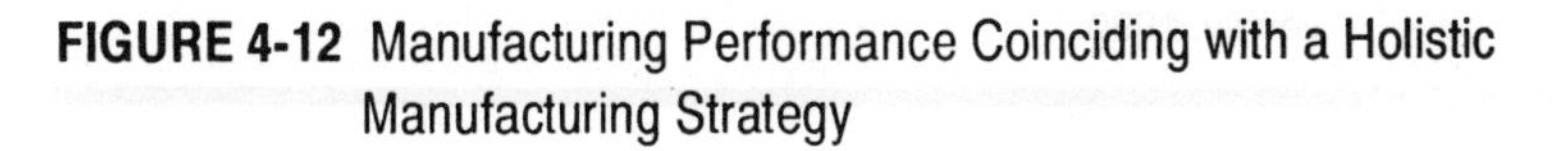

FIGURE 4-12 Manufacturing Performance Coinciding with a Holistic Manufacturing Strategy

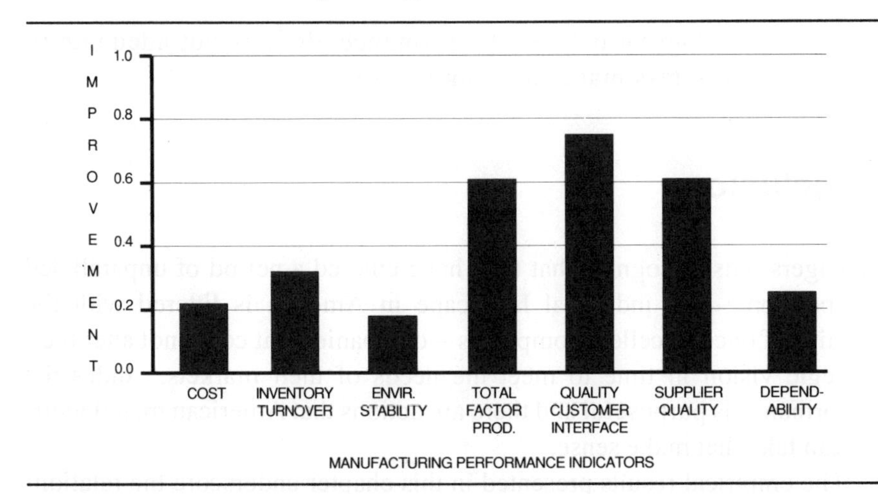

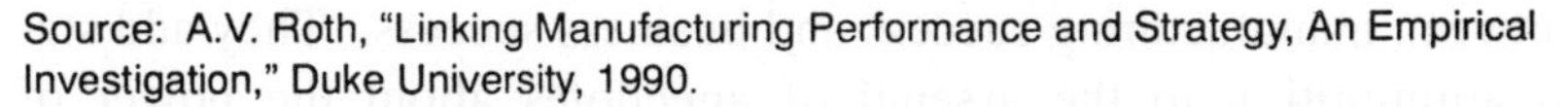

Source: A.V. Roth, "Linking Manufacturing Performance and Strategy, An Empirical Investigation," Duke University, 1990.

FIGURE 4-13 Manufacturing Performance Coinciding with a Disjunctive Manufacturing Strategy

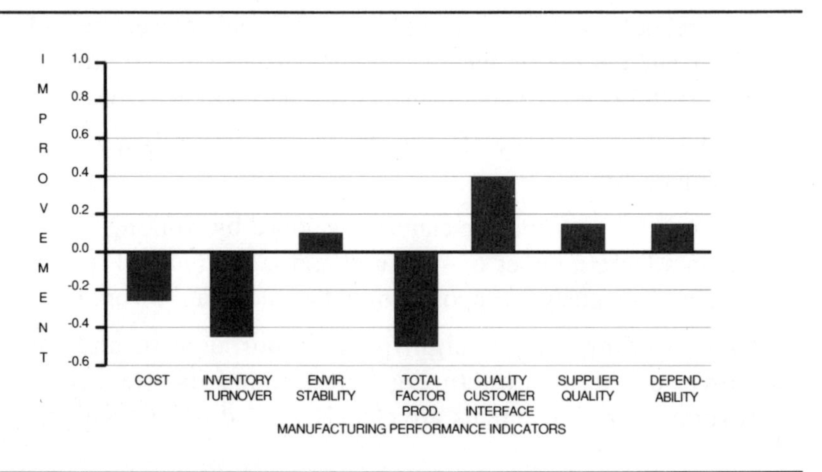

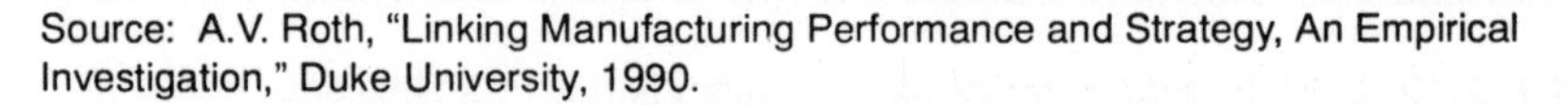

Source: A.V. Roth, "Linking Manufacturing Performance and Strategy, An Empirical Investigation," Duke University, 1990.

Quality, technology, or delivery performance alone is not adequate to achieve world-class manufacturing stature.

Conclusion

Managers must recognize that they have entered a period of unparalleled competition. The industrial landscape in America is littered with the remains of once-excellent companies—companies that could not alter their strategic vision in time to meet the needs of their markets. Industrial performance is purposeful and there are actions that American manufacturers can take that make sense.

The empirical results presented in this chapter underscore the relationship between manufacturing success and business success. They add statistical ammunition to the arsenal of anecdotes about the power of manufacturing to make a difference.

In Chapter 2, total quality management was emphasized as the underpinning of world-class manufacturing. In Chapter 3, an operational definition of manufacturing strategy was presented. This chapter showed the systematic linkages between manufacturing strategies and performance linkages as they occur in large samples of businesses. Important messages for developing world-class performance metrics follow in Chapters 5 and 6.

CHAPTER 5

RECONCILING ACCOUNTING AND MANUFACTURING

Traditional manufacturing performance measures have focused on cost: cost of labor, cost of materials, and cost of overhead. This focus induces *cost-myopic* behaviors that ultimately lead to poor business performance. Excessive attention to the expense side of the ledger can hamper employees' abilities to perform their jobs, stifle necessary renewals in processes, and create an unstable work environment.

Most leading manufacturers today, as well as public accounting firms, recognize that cost is only one element of performance measurement. In fact, using cost as a measure of performance at all is being seriously questioned. This is due in part to the belief that cost is closely related to other measures such as productivity and quality. Furthermore, product cost is not the only factor directly tied to customer requirements.

Experience shows that cost reduction, for decades nearly the universal and often the sole objective of manufacturing performance, has not produced concomitant quality and delivery-time improvements; however, paradoxically, improvements in quality and delivery time have resulted in lower cost. In the 1970s, most American manufacturers believed that they could make tradeoffs between quality and cost or between cost and delivery. Today's conventional wisdom is that the tradeoff model no longer works. Programs designed primarily to reduce costs have actually weakened product quality, reduced manufacturing capabilities, and damaged morale.[1] Much of this impact has been attributed to a poor understanding of the factors that drive manufacturing performance.

World-class manufacturers are tailoring their performance measurement systems to drive manufacturing excellence and to include their marketing and business strategies. As they reshape performance measurement, they must address three fundamental questions:

- How should manufacturing performance measurement be gauged?

- What are the current practices and trends in performance measurement?
- How can customer requirements be applied to manufacturing performance measurement?

Defining Manufacturing Excellence

The goal of manufacturing performance measures is to encourage manufacturing excellence—not just functional excellence, but excellence perceived by customers. Above all, performance measures should be viewed from the customer's perspective. The ultimate goal of manufacturing performance, through its integration with other functional business areas, should be totally customer-driven. The pursuit of excellence entails superior performance on more than one product attribute—cost, quality, delivery, and service, for example—so that customers are provided with products that surpass competing products in value and reliability. To encourage manufacturing's contribution to the total business success, performance measures must focus on competitive variables and be aligned with the critical success factors of the business.

Clearly, given this broadened definition, achieving manufacturing excellence requires more than becoming a low-cost producer. Customers see price, not cost. But price is rarely a direct function of manufacturing costs. Manufacturing excellence can simultaneously lower manufacturing costs and increase prices, as exemplified by Honda's Acura Division. Acura commands a premium price for its Legend luxury sedan, while enjoying per unit cost advantages *vis a vis* Detroit.

What are Good Performance Measurement Systems?

World-class manufacturing performance measurement systems must be capable of driving the highest levels of performance possible. A litmus test of a good performance measurement system is the degree to which it:

- Tracks attributes related to customer-driven critical success factors, including conformance quality, on-time delivery, and delivery speed.
- Fosters intrafunctional integration and communication to optimize total business performance. For instance, replacing departmental

cost reports with team-based performance reports and providing data on team and overall manufacturing unit performance.

- Provides feedback data on the gaps between best-in-class performance and the manufacturing unit's own performance over time, so that trends in the competitive environment can be observed and processes improved.
- Promotes learning and continuous improvement so that opportunities for improvement are highlighted. For example, measures capture the outcomes of processes.
- Is appropriate for the level of the organization at which it is being used. For instance, real-time detailed information about specific manufacturing processes is provided at the operational level; broad-based aggregate data is captured at the strategic level.[2]

In each case, the number of measures captured at any one time should not be limited.

Performance Measures Must Match Manufacturing Objectives

Performance measurement systems link manufacturing capabilities and manufacturing strategy. This linkage is a natural one. World-class performance measurement systems are tailored to effectively monitor the outcomes associated with a given set of manufacturing capabilities and provide feedback on the ability of the manufacturing unit to implement its manufacturing strategy.

Defining manufacturing strategy as a pattern of manufacturing choices in support of the business strategy, preferred performance measures must be congruent with current manufacturing objectives. Furthermore, performance measurements should be dynamic and modified as strategy evolves. Long-term success requires applying new knowledge to improve performance.

World-class manufacturers have broadened their outlook. Their performance measurement systems encompass manufacturing direction and rate of improvement on competitive measures benchmarked against best-in-class. Motorola, Xerox, Ford, GE, and other manufacturing winners use best-in-class performance on key product and process attributes to gauge goal levels to be achieved. The very act of benchmarking signifies a company commitment to nothing less than excellence.

Performance Measures Must Encompass More than Cost

State-of-the-art thinking regarding manufacturing performance measures encompasses far more than cost. World-class performance must be predicated upon both the financial and the nonfinancial measures that drive total business success and on the systems that capture them. A good performance measurement system targets measures of business investment and balance sheet measures, as well as cost and revenue measures relating to the income statement. Nonfinancial performance will be discussed in Chapter 6.

How Do Current Financial Measurement Systems Measure Up?

The financial measures of performance used by today's manufacturers are artifacts of outdated cost accounting systems still being used. These conventional systems focus upon the processes and procedures necessary to summarize the costs of an activity to report and/or book results to the financial statements. The primary purpose of traditional cost accounting systems is to perform inventory valuation, journal vouchering, and, in the case of standard cost systems, variance reporting.

Current Practices Are Stifling Performance. Many companies' accounting systems have not kept pace with the changes in their organizations and the technology used in production processes. In many ways, current accounting systems stifle manufacturing progress. For example, the traditional cost accounting system accumulates all overhead into large cost pools. The overhead and indirect expenses include manufacturing, engineering, and information systems; test engineering; capitalized engineering expenses; depreciation on machinery and facilities; general employee benefits; and manufacturing support, site, and financial services. They lump overhead and indirect expenses together into large aggregate cost pools and allocate them to production—and, subsequently, to products—on some subjective basis.[3]

As reflected in Figure 5-1, traditional cost accounting systems cannot keep pace with the profound changes occurring in manufacturing practice; they are incapable of providing manufacturing management with the decision-making information required for competing in the twenty-first century.

FIGURE 5-1 The Development Gap Between Manufacturing and Cost Accounting

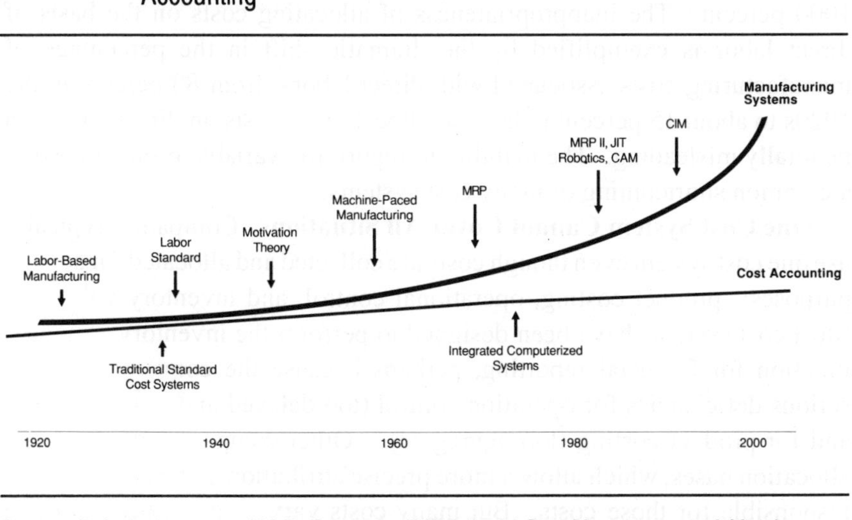

Source: D.A. Anderson and M.R. Ostrenga, "MRP II and Cost Management: A Match Made in Theory?" from *CIM Review* (New York: Auerbach Publishers). © 1987, Warren, Gorham & Lamont Inc. Used with permission.

Current Cost Systems Rely on Subjective Judgments. The issues for manufacturing performance under these traditional cost accounting systems are:

- Where will the allocation be placed?
- To what extent will it be applied?

Models of cost accounting assume that costs are drivers of behavior. These models require the judgment of people to bridge relationships. If this is the basis for judging an activity, then the performance will be biased due to the allocation process. Manufacturers need a completely objective, unbiased costing system that does not rely upon subjective allocation schemes. Such a system provides opportunities for more meaningful action.

Current Cost Systems Cannot Handle Variable Costs. The majority of current cost systems have characteristics that whitewash any potential for unbiased measurements. Most employ a two-stage cost allocation system. Costs are assigned to cost pools, commonly called *cost centers*, in the first stage. In the second stage, the costs are allocated from pools to products. The method used to allocate costs from planned overhead accounts to cost centers varies widely in the first stage. However, the majority of companies use direct labor hours in the second stage. The problem is intensified, because direct labor hours are typically used even when the production process is so highly automated that burden rates often exceed 1000 percent. The inappropriateness of allocating costs on the basis of direct labor is exemplified by the dramatic shift in the percentage of manufacturing costs associated with direct labor—from 80 percent in the 1920s to about 15 percent today. So allocation of costs on direct labor can be totally misleading. The inability to report true variable product costs is a common shortcoming of many cost systems.

One Cost System Cannot Cover All Situations. Companies typically use one cost system even though costs are collected and allocated for several purposes: product costing, operational control, and inventory valuation. Most cost systems have been designed to perform the inventory valuation function for financial reporting, perhaps because the company has had serious deficiencies for operation control (too delayed and too aggregate) and for product costing (too aggregate). Other companies use multiple allocation bases, which allow a more precise attribution of costs to products responsible for those costs. But many costs vary with the diversity and complexity of the product, rather than with the amount of direct labor input or the number of units produced.[4]

Cost Systems Are Biased. Peter Drucker reported than one serious limitation of the traditional cost accounting system is that it "ignores the costs of nonproducing, whether they result from machine downtime or from quality defects that require scrapping or reworking a product or part . . . And nonproducing time costs as much as producing time does—in wages, heat, lighting, interest, salaries, and even raw material," none of which are captured in traditional systems.[5]

Cost Systems Prevent Investment in Technology. One of the most dysfunctional impacts that these traditional cost accounting systems and their irrelevant allocation methods have is on manufacturers' investments in advanced manufacturing technology. Manufacturers cannot find the justification for technology investments; often because their cost accounting systems are unable to support their analysis. Traditional systems of measuring cost are seriously disadvantaged in highly automated environments, so much so that many manufacturers are discouraged from investing. Other manufacturers delay investing in advanced manufacturing technology, because their traditional approach to measuring its costs and benefits suggests that it will not offer as high a return on investment as they are used to.

Traditional costing systems do not take into account the profound strategic benefits that may result from the application of new technology.

Competitive product cost structures, flexibility, quality, and new relationships with customers are among the many market-driven benefits of advanced manufacturing technology. Moreover, traditional accounting systems fail to reflect the benefits of learning and system synergy that result from integrating new technology with current processes.

Cost Systems Do Not Support Automated Manufacturing. The international forum of Computer-Aided Manufacturing International, Inc. (CAM-I), a non-profit consortium of progressive industrial organizations, is attempting to define the role of cost management in the new industrial era. CAM-I has studied the changing nature of cost behaviors and concluded that existing cost accounting systems do not adequately support the objective of automated manufacturing.

The reasons for this lack of support are many and varied. For the most part, traditional cost accounting systems exhibit the following problems, which diminish their value in highly automated environments:

- Inaccurate overhead rates, distorting product cost structures, due to high overhead cost structure, and inadequately traced costs, which are common in highly automated environments.
- Failure to isolate the cost of nonvalue-added activities from the activities that increase the customer's perception of the product's value.
- Failure to penalize overproduction, so the application of overhead absorption, through the allocation of overhead cost based on production volumes, encourages excess inventories.
- Failure to adequately identify and report the cost of quality deficiencies in current products and processes.
- Focus on controlling the production process, while significant costs, determined at the design and development phases of a product's life cycle, are inadequately addressed.
- Failure to assess and monitor the potential benefits of the application of advanced technology, including the synergy and learning that occurs when the technology is implemented.
- Employment of performance measurements that conflict with strategic manufacturing objectives.
- Inadequate evaluation of nonfinancial measures such as quality, throughput, and flexibility.

- U.S. hurdle rates are significantly greater than those of major competitors (see Figure 5-2).

These problems are of critical importance to companies that have invested in advanced manufacturing technology or are planning such investments. While General Motors learned many lessons about the right and wrong ways to implement advanced manufacturing technology in the 1980s, it also learned that traditional cost accounting systems do not provide meaningful information about the performance of the new environments. Companies like General Motors are leading the way toward the new generation of cost systems discussed later in this chapter. Manufacturers that invest significantly in technology will find revamping of their cost systems high on their list of required action steps.

FIGURE 5-2 Cost of Capital

Country	Real After-Tax Cost of Funds	Equipment & Machinery	Physical Plant*	R&D**
U.S.	6.0%	11.2%	10.2%	20.3%
Japan	2.2	7.2	5.0	8.7
U.K.	5.0	9.2	7.9	23.7
W. Germany	1.8	7.0	5.4	14.8

* Factory with physical life of 40 years.
** R&D project with 10-year payoff.

Source: Robert N. McCauley and Steven A. Zimmer, "Explaining International Differences in the Cost of Capital, *FRBNY Quarterly Review*, Summer 1989, p. 7.

Japanese Firms Link Performance Measures and Strategic Goals

How have the Japanese approached these problems? Are their cost accounting systems more in tune with their strategic objectives? Do they automate because of or in spite of the information they receive from their cost accounting systems?

In general, Japanese companies do not have the same problems with their cost systems that American and European companies do. The Japanese do not let the accounting procedure determine how they measure and control their organizational activities. There are relatively fewer accounting and finance personnel in Japanese firms compared to similar U.S. companies. Japanese companies tend to use their management control systems to support and reinforce their manufacturing strategies. A more direct link, therefore, exists between strategic goals and performance measures in Japanese firms.

The Japanese are not concerned about using direct labor as a basis for allocating overhead, even though it does not reflect the manufacturing process in the factory's automated environment. They are committed to aggressive automation to promote long-term competitiveness. They believe that allocating overhead based on direct labor creates a strong pro-automation incentive throughout the organization. Accounting systems influence rather than inform.[6] One simple difference between the Japanese and Western industrials is that the Japanese treat direct labor as a fixed asset.

While this orientation permits a poorly designed accounting system to guide the organization to increasingly higher levels of automation, it says little about the accuracy of the information presented for competitive purposes. The Japanese have learned to motivate behavior by using the system to focus on elements they feel are important to their manufacturing success. Whether the costing information is accurate or not is of secondary importance. The Japanese are more concerned with motivation and problem solving.

By overhauling their cost accounting systems, U.S. manufacturers have the opportunity not only to motivate company behavior as the Japanese have done, but to create competitive advantage out of accurate and timely information.

Figure 5-3 compares the 1987 performance measurements of the U.S., Japan, and Europe. Where the U.S. and Europe are most concerned with quality and cost, Japan considers lead time and productivity most important. Costs do not even appear on the Japanese list, nor is the cost of capital an issue in Japan.

What Stands in Our Way?

A broader set of financial measures to guide management decision making is needed, but the definition of such measures is only in the discussion phase at this time. Why are we talking and not doing?

FIGURE 5-3 International Comparison of Performance Measures, 1987

U.S.	Japan	Europe
Outgoing Quality	Mfg. Lead Times	Outgoing Quality.
Unit Mfg. Costs	DL Productivity	Unit Mfg. Cost
Percent On-time Delivery	WIP Turnover	Unit Materials Cost
Overhead Costs	Incoming Quality	Overhead Cost
Unit Material Cost	Vendor Lead Times	Percent On-time Delivery
Inventory Accuracy	Indirect Labor Productivity	Incoming Quality
Incoming Quality	Materials Yield	DL Productivity
DL Productivity	Finished Goods Inventory Turnover	Materials Yield
Unit Labor Costs	Inventory Accuracy	Forecast Accuracy
Mfg. Lead Times	Absenteeism	Unit Labor Costs

Source: J.G. Miller, A. Amano, A. De Meyer, K. Ferdows, J. Nakane, and A.V. Roth, "Closing the Competitive Gaps—The 1987 International Report Manufacturing Futures Project," in *Managing International Manufacturing* (K. Ferdows, ed.),1989, p. 164.

- The requirements for financial reporting and the demands from Wall Street dictate that traditional measures of business unit performance be used. Performance systems are related to the way the goals are stated.
- The competitive environment for most manufacturers is changing so rapidly that the development of appropriate new measures is akin to hitting a moving target.
- The increasing use of automation has created a condition in which direct labor now represents only a small fraction of corporate costs, while expenses covering factory support operations, marketing, distribution, engineering, and other overhead functions have exploded.
- A way to account for the risk of *not* making an investment is not included in current systems.

Therefore, any new measurement system requires a complete break with the systems of the past, a move that is seen by most manufacturers as both costly and risky. World-class manufacturers, however, understand that systems designed mainly to value inventory for financial and tax statements are incapable of giving managers the accurate and timely information they need to meet the competition. These companies are implementing new methods of performance measurement and breaking the bonds that have tied them to traditional cost accounting systems.

Toward New Measures of Financial Performance

In cost control systems, operating costs are reported too late and are too aggregated to benefit production supervisors. These systems direct management's attention toward "product cost estimates that focus upon the least important cost component—direct labor—and ignore expenses involved in designing, marketing, distributing, and servicing goods."[7] Managers in companies that sell multiple products, therefore, are making important decisions about pricing, product mix, and product technology based upon distorted cost information. Even worse, alternative information rarely exists to alert these managers that product costs are badly flawed. Most companies detect the problem only after their competitiveness and profitability have deteriorated significantly.

As American businesses race to improve their competitive edge and close the performance gaps, measurement systems remain essentially unchanged. Most manufacturers are buried in traditional ways of reporting and measuring productivity. To provide cost information for more informed operational and strategic decisions in today's competitive arena, new methods of accounting for and measuring productivity are required.

Cost Management Concerns the Entire Product Life Cycle. Enlightened manufacturers have recognized the need for a more encompassing group of measures of financial performance. *Cost management*, the name given to this new way of measuring performance, deals with *the management of costs through the elimination of nonvalue-added activities*, whether or not these costs have direct impact on inventory or financial statements. Cost management is broader in scope than cost accounting and is consistent with the Just-in-Time philosophy of a companywide commitment to the continual reduction of costs regardless of whether or not those costs can be measured in purely financial terms.

World-class manufacturers are paying more attention to *total cost* in the product cycle—from order entry through after-sales service. Cost management also supports the JIT philosophy of total cost avoidance versus cost control or reporting. World-class manufacturers know that both product design and the amount of effort expended on trying to market a given product or product line are influenced by the anticipated costs and profitability of the product. Intensified global competition and radically new production technologies have increased the urgency of the need to develop more accurate product cost information. Cost information is crucial to competitive success.

Technology Has Dramatically Altered the Manufacturing Environment. In manufacturing today, not only does change take place faster, but its impact ripples through the national and global economy more quickly than ever before.

World-class companies know how complicated the interplay is between reject levels and factory throughput time, between purchasing and equipment maintenance and work in process. The idea of isolating "these interrelated costs into tidy packages on a monthly or evenly quarterly basis 'is ludicrous.' It would increase transactions and the total product cost. Moreover, it might induce companies to pursue a 'mirage of increased profitability' (as depicted in their financial reports, prepared according to GAAP) by avoiding actions and investments that are essential to long-term survival."[8]

Accounting Reports and Reporting Systems Are Being Ignored. World-class manufacturers have recognized the flaws in their existing financial measurement systems and have resorted to decision making by gut feel or through analysis of off-line management systems. Managers in these companies often relegate accounting reports to the circular file. In fact, an estimated 50 percent of all reports generated today are either not used or are so repetitive in content they should be eliminated.[9] The production managers' information is generated from their own personal computers. Certainly, world-class manufacturing in the 1990s warrants a comprehensive value analysis of reports and reporting systems.

Total Performance Management Systems Must Be Adopted. Advanced management systems must be broader in scope than cost management. While depending upon user requirements for operational, control, and strategic information, they must be simple, user friendly, and accurate, and provide appropriate detail for the function they serve. Advanced management systems will replace (or supplement) traditional cost accounting systems and play a number of different roles in total performance management in the 1990s and beyond.

Companies striving to win orders have recognized that the cumulative effect of their decisions on product design, introduction, support, discontinuance, and pricing, as well as on manufacturing processes, defines a firm's strategy. The new math of excellence over the next decade will be written by companies that know the fundamental differences between cost accounting and *total performance management* and that can apply those rules to the operation of their businesses. Figure 5-4 details the elements of a Total Performance Management System.

Leading manufacturers are investigating new financial and nonfinancial measures of performance. Measures that reflect their market strategies and assist them in applying scarce resources will be most effective. In the next section, the current trends in financial reporting systems development are addressed.

Today's Trends, Tomorrow's Practices

Manufacturers' responses to the irrelevance of existing financial reporting systems in guiding their decisions are varied. Many see the need to change their practices but are unclear about the directions to pursue. They are experimenting with a number of new systems that they believe may be more satisfactory.

Assign Overhead Costs Differently. Some are attempting to assign overhead costs differently. New assignments are directed toward capturing data on various factors that drive costs, including machine hours, number of customer orders or jobs, schedule, factory throughput times (or the amount of work-in-process inventory), and number of parts produced.

Exploit Manufacturing Technology. Others are attempting to overcome the deficiencies of current systems by exploiting the capabilities of new computerized manufacturing technologies to capture detailed data about a given job or process on a real-time basis. This capability permits individual operations or transactions to be assigned to specific support activities.

Develop New Performance Measures. Still others are developing entirely new measures and the systems to capture them. They hope to capture data on factors associated with winning orders such as customer perceptions of quality or service, time to launch new products, and total cycle time.[10]

Multiple Systems Are the Current Answer

Because of the varied nature of the information required for operational and strategic control and for maintaining financial information consistent with the current reporting requirements, the current solution is to maintain multiple systems of financial performance. The current economies of distributed computing for information collection, processing, and reporting have made it possible for decentralized data to remain available for opera-

FIGURE 5-4 Elements of a Total Performance Management System

Inventory valuation:	Allocate periodic production costs between goods sold and goods in stock for financial and tax statements.
Operational costing:	Periodic management feedback on the resources consumed (labor, materials, energy, overhead).
Activity control of cost drivers:	Information on nonvalue-added activities in the total organization.
Product costs:	Total product costs, including direct and indirect, as well as true contribution margins.
Performance measurement:	Take stock of the level of current capabilities and rate of change in critical success factors over time.
Reward systems:	Feedback on performance measurements and reward activities to human assets.
Manufacturing intelligence systems:	Adaptive systems that proactively support continuous improvements throughout the organization; for example, the use of real-time expert systems to predict machine capabilities so that appropriate preventative maintenance or retooling can be applied on the shop floor or to market response functions at the strategic level.
Market intelligence systems:	Provide periodic feedback on market expectations, the abilities of the best-in-class performers, and the competition's current competitive capabilities.

Source: A.V. Roth and H. Schneider, "Customer Driven Strategy: A Paradigm Shift," Duke University, 1990.

tional control and product costing. As managers gain experience with these new financial reporting systems, they may be able to develop the integrating mechanisms required for total performance management.

The pursuit of improved financial measures to expand manufacturers' decision-making capabilities is occurring in four major areas:

- Activity-based Costing
- Target Pricing
- Investment Justification
- Competitive Benchmarking

Activity-based Costing

Progressive manufacturers look beyond the transaction-based cost systems of today to determine if today's decisions will drive tomorrow's earnings engine. Activity-based costing systems are a relatively new approach to management accounting which recognizes that the way to achieve profitability is to manage activities. Activity-based information is collected to describe the work (or activities) with respect to the resources consumed and the value delivered to the business.

Because activities are a common denominator in cost accounting, nonfinancial performance measurement, and investment management, they provide a logical framework for integrating these critical areas. In today's manufacturing environment, these three measurement systems often remain independent. This independence causes each to seek to optimize its own objectives, thereby resulting in less than optimal performance for the company as a whole.[11] When managers attempt to achieve profits by managing costs, as they have for decades, they implicitly use cost to measure activities indirectly. Activity-based systems are based upon the principle that activities consume resources and products consume activities. Activity-based information focuses managerial attention on the underlying causes (or drivers) of cost and profitability.

What Are Cost Drivers?

Cost drivers and activities that add cost are not the same thing. Drivers link to activities in a cause and effect relationship. The typical, and often exclusive, driver of a traditional cost accounting system is direct labor. In an activity-based system, however, direct labor is only one of many potential drivers. Others might include the number of orders placed, the number of

CASE STUDY

GE: Activity Analysis

GE has successfully orchestrated one of the most impressive comebacks accomplished by any American corporation during the last decade. Sales per employee have doubled from $67K to $134K during the period 1981 to 1987. The company has had significant breakthroughs and market successes in dishwashers, medical imaging technology, and other product lines. It has managed downsizing gracefully and coped reasonably well with such controversy as the faltering of the automated systems supply business and problems in consumer electronics.

Although many factors, including outstanding project management techniques, simultaneous engineering, management development programs, employee involvement, and effective alliance management, could account for this impressive decade of performance, GE's use of *activity analysis* to enhance performance is the focus here.

GE does not mandate change by fiat from the corporate offices. The strategy for transition involves a complex set of managed evolutions that are firmly based in development programs, corporate staff support, and aggressive leadership in the divisions.

The "right" activities are those that satisfy stakeholders. That is, there is a hierarchy of activities that accounts for costs and benefits to the business unit. Often, when activity analysis is done, one finds that people are doing work that contributes to more than one activity and that it can be measured by one or more critical success factors—productivity, for example. Ultimately, that activity analysis changes decision alternatives taken. It helps allocate resources, like people, to the most important activities and may influence make-buy decisions by contracting for less important activities. It reveals the cause-effect pattern of the business.

GE has implemented activity analysis in at least two locations. The first location was staffed by about 30 people involved in electromechanical assembly. This assembly shop was low volume, producing about 10 units per week with high product value (i.e., thousands of dollars per unit). This shop was recovering from a sluggish implementation of JIT (Just-in-Time manufacturing). The unit used activity analysis as a solution, identifying the most important elements in planning, operation, and maintenance. The unit eventually called itself a "value-added activity center," which demonstrates the profound effect this analysis approach can have. Assembly and material accumulation were the most important activities identified with 10.7 and 3.9 equivalent people, respectively. In addition, material accumulation had high waste (no added value).

Internal drivers were identified for each waste activity. For material accumulation, this included stock layout and the assembly sequence.

Case Study, concluded

Overall, productivity was improved by 20 percent as a result of activity analysis.

At the second GE location, an entire business of several thousand people was involved. This location was engaged in heavy mechanical and electrical assembly with low volume (i.e., several thousand units per year) and high cost. Lead times could be as long as 18 months, and the market was declining overall. Lowering base costs became the objective. As a result of activity analysis, one layer of management was removed, and the program ultimately saved $20 million to make the business number one in the marketplace.

Activity analysis started with a team of six people that identified 250 activities in the business unit. Management and selling were the two most expensive activities. Secretarial support was third. As a relational database was used to map activities to people, the manipulation began to suggest change scenarios. Pareto, or 80/20, analysis techniques were used first to identify the most and least costly activities. Step Two was cause-and-effect analysis. Step Three was matching priorities with costs. An example of the latter is the finding that the general manager had established selling as the top priority and telephone inquiries as the lowest priority. Pricing and customer training turned out to be misplaced priorities. Not surprisingly, the mix and assignment of personnel were among the first changes that resulted from the analysis.

The final step in the analysis was a comparison between the amount of effort being applied to an activity and the amount being applied to similar businesses. Examples include percentage of payroll spent on employee training given the objectives of the business. Material handling is another area of comparison.

In identifying the drivers behind activities it was found that timekeeping accounted for $1.2 million, which came from the processing of 80,000 vouchers per week, driven by 11 wage payment instructions. Eventually, accounting procedures were modified to eliminate all but two of these instructions.

In summary, activity analysis at GE seems to boil down to three steps, regardless of the type of business or scope of activity: *identify activities, analyze activities, and improve performance by finding drivers for these activities*. This may not be the only secret of success at GE, but it certainly appears to be one of the critical factors.

This case, prepared by J.E. Ettlie, is based in part on Thomas O'Brien, "Improving Performance Through Activity Analysis," *Proceedings of the Third Annual Management Accounting Symposium*, San Diego, CA, March 1989, and on John E. Ettlie and Henry W. Stoll, *Managing the Design—Manufacturing Process* (New York, McGraw-Hill, 1990).

setups required, the number of changes to the master schedule, the number of customer complaints processed, etc. Figure 5-5 shows examples of typical activities and their associated drivers.

Driver identification requires a thorough analysis of the entire manufacturing process. This analysis is at a much higher level of detail than can be supported by an operational activity-based system. Driver identification requires a Pareto analysis of the causes and the activities they control to identify the 20 percent of the drivers that account for or cause 80 percent of

FIGURE 5-5 Examples of Activities and Drivers

Activity	Drivers
Enter Orders	Number of Orders
	Number of Line Items
	Standard vs. Custom Specifications
Set Up the Machine	Number of Setups
	Complexity (Tools and Fixtures Required, Setup People Required)
	Setup Time

the observed activity. Robin Cooper argues that three factors should be taken into account when selecting cost drivers:

- **The Cost of Measurement:** How much effort and expense will be required to obtain and manage the data required by the cost driver?
- **The Degree of Correlation:** How correlated are the driver and the activity? For example, is there a direct one-to-one relationship between the driver and the consumption of resources by the activity, or is the relationship only a surrogate of the actual conditions?
- **The Behavioral Effects:** How will individuals in the organization respond to the measurement of the driver? What behavior will be promoted? What will the effect be on the organization?[12]

Determine Nonvalue-added Activities

A nonvalue-added process is defined as any activity or procedure performed within a company that does not add value to a customer's perception of the product. By accumulating costs along lines of activities, management is in a better position to link activities with results. Scrutiny of these links determines which processes can be eliminated or are *nonvalue-added.* Each activity's potential benefits are weighed against its costs.

The key in identifying nonvalue-added costs is a solid reassessment of the factors, policies, and processes that drive costs in the plant. Both direct and indirect activities support the production and delivery of today's goods and services. Therefore, all of a company's costs must be considered product costs. Since most factory and corporate support costs can be traced directly to individual products or product families, this system supports the collection of cost by activity. Direct traceability of cost becomes the key to improved decision making for pricing, product line profitability analysis, make or buy decisions, and cost-reduction efforts.[13]

The concept of activity-based costing is often confused with improved product costing, which is a simpler approach. Why not just allocate overhead better to get more accurate product costs? *Many accountants, as well as manufacturing executives, fail to recognize that while improved product costing can be accomplished by a more rigorous allocation of traditional overhead costs, the improvement of product cost performance cannot be facilitated in this manner.* Activity-based costing and the understanding of the drivers of the activities are essential to responding to the "what do we do now?" question that is inevitably asked once product costs are accurately established.

Use Pilot Projects to Explore Your Options

Pilot projects are the method that manufacturers usually use to explore these options. Future trends in practice will result from pilot project experiences. For the most part, manufacturers have not overcome the resistance to change that is associated with the implementation of activity-based costing systems or the system design issues themselves.

The Two Types of Activity-based Information

Two types of activity-based information form the backbone of world-class management accounting. The first type produces nonfinancial information about sources of competitive value in a company's operating activities. World-class manufacturers will derive their own internal competitive capability data from a common database of activity information. In the near future, it may be possible to assess activities on a daily basis using personal computers.

The second type of activity-based information, strategic cost information, enables managers to assess the long-term profitability of a company's current mix of products and activities. Strategic cost information is an indicator of the cost-effectiveness of a company's activities compared to alternatives outside the company and the profitability of the mix of products management chooses to offer the marketplace.

The Sequence for Applying Activity-based Costing

According to Herb Schneider, manager, GE Corporate Business Development and Planning, the natural sequence in which activity-based costing models for world-class manufacturing should be applied in the next decade is:

1. **Get an accurate or better P&L for product lines.** Some lines are more profitable than others. Unless you understand the true contribution margins, allocations may lead to suboptimization.
2. **Use activity-based costing as a means for work groups, cells, or teams to focus on waste and nonvalue-added activities.** This "snapshot" permits self-analysis by the work group as a basis for improvement and produces significant value. Such snapshots can be periodically captured.
3. **Use dynamic models where activity-based inputs would be generated in real time as a basis for day-to-day decision making.** At present, few American manufacturers have dynamic activity data or the means to review it.

Schneider argues that the data managers use has traditionally been different from that used for financial reporting; he does not see that changing. In a world of continuous flow and shorter cycles, Schneider believes that performance measures need to be more transient. In the 1990s, we can expect a movement toward more semi-autonomous and ephemeral work groups, that will require measures to aligned with particular work tasks.

Affordable Versus Target Cost Strategy

Prices are always driven by the market. But estimated product costs and the competitive situation in the marketplace are important contributors to pricing strategies. When market prices are not readily available, as in the case of customized products produced in low volumes, product costs play a particularly important role in determining prices.[14] In this situation, approaches to setting prices are largely reactive.

In contrast, world-class manufacturers have learned to use price as a competitive weapon. When manufacturers use price proactively to stimulate market share, they choose either an affordable cost or a target cost strategy.

The Affordable Cost Concept

The affordable cost concept was popularized by IBM. It involves establishing a target sales price which management believes is consistent with its objective of increasing market share over time, while meeting a certain level of returns. Target prices are estimated based upon competitive factors, historical data, and forecasts of the desired market position two to three years into the future.

The affordable cost concept tries to answer questions like this one: "Our company holds a 40 percent market share for a particular product or product line today. What prices are necessary to maintain (or increase) this share over the next few years given a competitive environment and rapid technological advancement?" To answer that question, the affordable costs in each year are determined by the sales price required for the target market share, minus desired profits. *Once policies on the level of profitability and share are determined, affordable costs are pushed onto the company.* The company must then run its operation based upon affordable cost estimates. In some cases, this may call for immediate and radical changes in business and accounting practices.

The Target Cost Concept

In contrast, the Japanese are noted for their use of target costs. The primary objective is to grasp volume. The Japanese establish prices that stimulate market demand, so that volume goals are achieved. What happens to profitability is secondary—at least in the short run. Having achieved the requisite market share, they can exploit its benefits. Costs are lower because the company has learned how to make the product efficiently, and revenues are higher because it has established a large market share. This process results in a pricing strategy that has an incremental impact upon the firm's operation. Continuous improvement in process and product reduces cost.

Dahaitsu is representative of Japanese manufacturers who apply target cost policies. First, Dahaitsu establishes a selling price based upon what it anticipates the market will accept. While a target profit margin that reflects the company's strategic plans and financial projections may be specified, the target costs (difference between target sales price and target margin) are usually well below what is realistically achievable at the time. *The difference between the target sales price and the actual cost is the allowable cost.* The challenge is to incrementally tackle the current technologies and practices to diminish the gap between allowable cost and target costs.

The Activity Costing and Target Pricing Link

Manufacturers of the 1990s seeking to improve their competitive position through target pricing must have improved product cost information, so they can adequately address their policies. Thus, to make a target pricing program effective and to assess product costs and changes in product costs accurately, activity-based costing systems will become the foundation of target pricing programs. Activity-based costing enables management to make more enlightened decisions about product costs and, therefore, pricing.

Investment Justification

The justification of automation is severely hampered by a strangling web of accounting and financial procedures. Managers often feel strapped by unrealistic payback hurdles and inappropriate procedures and precedents. Failure to justify the implementation of advanced manufacturing technology has been found to be a key component of the overall failure to successfully implement strategic plans.[15]

What are the criteria that companies are using today to justify investments in advanced technology and the reasons they cite to reject such investments?

- Profitability, cost reduction, and the desire to utilize proven technology are the most frequently applied criteria.[16]
- Costs typically considered in the justification decision are associated with:
  - **Equipment:** Acquisition, redundancy, and utilization.
  - **Facilities:** Relocation and site preparation.
  - **Process:** Revision of plans, NC tapes, time standards, and inspection procedures.[17]
- Hidden costs often arise due to complexities of integration:
  - **Startup:** Time to master technologies, training.
  - **Systems:** Management, maintenance, software customization, development of parts library, and added transactions.[18]

What are the primary reasons for postponing commitments to advanced process technology acquisitions?

- Uncertainty concerning the benefits, including an inability to quantify the potential strategic impact of the new technology and the perception that the risk is too great.
- Concerns over the lack of skills necessary to achieve the potential benefits and the existence of labor and organizational barriers.
- Investment criteria not met, hurdle rates set too high.
- Inability to adequately quantify returns.
- Inability to properly identify all the costs that will be reduced, including indirect, intangible, and opportunity costs.
- The wrong people involved in the decision-making process.
- About one quarter of the experts surveyed felt that the postponement of automation was tied more to general economic factors than to uncertainties related to the technologies themselves.[19]

Traditional Justification Methods Overemphasize Expenses

The application of traditional justification methods imposes an inordinate emphasis on expenses in the justification decision. This phenomenon has contributed significantly to the postponement of advanced technology acquisition. All too often, manufacturers sacrifice their long-run competitive advantage because the traditional financial approaches used in planning are focused on cost and strongly biased toward the *status quo*.

CASE STUDY

Simmonds Precision Products: Does CAD Work?

Simmonds Precision Products designs and manufactures measurement, control, and display systems for industrial and aerospace customers. This is a summary of the Simmonds Instrument Systems Division program of justification and implementation of computer-aided design and computer-aided manufacturing technology. The division is a $100 million annual sales unit of Simmonds located in Vergennes, Vermont.

Like many other companies, Simmonds debated whether or not to use traditional economic justification and planning models to assess CAD/CAM technology. Capital equipment justification at Simmonds, like other companies, was dominated by conservative rules like using direct labor savings and a requirement that payback be obtained in two years or less. Available and budgeted dollars were compared before projects were approved. Capital purchases were reviewed quarterly.

In reviewing source material for CAD/CAM adoption, glowing reports of productivity gains of 10, 20, and 40 to 1 were typical. The detailed information needed to support a payback justification of two years was difficult to find. When details were available, such costs as service, were often left out of calculations.

Printed circuit (PC) design at Simmonds presented the greatest potential opportunity to make comparisons, because historical design data was available and the area had adopted computer-aided drafting several years earlier. Mechanical design also had potential, but comparisons were more difficult and experienced users were hard to find. A third area, manufacturing and graphics, was considered. It was found that a 40-month payback was likely instead of a 24-month payback required by current policy.

Intangibles were then considered, based on the reports of other users. These included promotion of standardization and enhancement of creativity, but the company elected not to include these benefits in its analysis, because it was concluded that hidden costs, such as "psychological factors," would offset these benefits. This is typical of the way people address risk and unanticipated consequences in these cases. These risks are real and should not be diminished. It is difficult to quantify them and fit them into these decisions, but it can be done.

Simmonds top management continued to support study of the technology and reinforced the strategy that was moving toward integration. Some new technology was going to be needed to facilitate this strategy implementation. A consultant was called in and helped with the capital proposal but was conservative in projecting benefits from adoption.

A CAD/CAM system was adopted. The implementation plan included training, a full-time, centralized design group, and diversion of high-pres-

Case Study, concluded

sure tasks out of the project. The system achieved nearly 100 percent reliability.

Post-audit information is available for PC design. On average, 67 hours and $75 in vendor fees were saved on every standard design made during the first year. PC design time was reduced on average from 119 to 50 labor hours. This data was collected from a computerized project management information system. In the time that it used to take to do a feasibility layout, the company could now do four or five layouts. Excluding intangibles, PC design savings were $154,000 during the first year. By the end of 20 months, a total saving of $498,000 had accrued along with a 100 percent reduction in cycle time for production of designs.

In mechanical design, results were not documented since the design process was not automated. A saving of $18,600 was projected in manufacturing as a result of reduction in numerical control programming (NC), but a loss resulted when compared to training costs. Graphics, however, were greatly enhanced.

Production time for graphics was reduced from 2 to 6 hours to 15 to 60 minutes. CAD/CAM had a significant impact on error reduction at Simmonds. A 24 percent reduction in errors per drawing (.26 to .20 errors on average) was documented. The cost to correct errors had been running about $14 per error and was reduced to $9.50 per error.

Direct labor reduction in engineering was reduced by 27 percent, permitting an equivalent increase in workload. Overall, after-tax savings for the first 20 months ran about 20 percent ahead of two-year payback expectations.

Overall, the centralized design staff experienced lower than average absenteeism but had to work 10-hour shifts due to staffing limitations. Further, spread of the user base has been a problem at Simmonds, in spite of this positive initial experience.

Compared with experiences in other settings, such as construction, engineering consulting, and architecture (see footnote), the Simmonds experience seems typical. Simmonds chose the area easiest to automate (PC) and found it was difficult to spread the application of this technology across the firm. The short-term payoff was high, but the long-term application of this and other technologies of integration is in doubt. However, they have a success to build on, and there is hope that it can be done.

This case prepared by J. E. Ettlie. Drawn in part from R.C. Van Nostrand, "CAD/CAM Justification and Follow-Up: Simmonds Precision Products Case Study," *CAD/CAM Management Studies,* New York, NY, Auerbach Publishers, Inc., 1987. Other material of interest is: Jan Forslin, et. al., "Computer-Aided Design: A Case Strategy in Implementing a New Technology," *IEEE Transactions on Engineering Management*, 36 (3), August 1989, pp. 191-201; J.E. Ettlie, "Innovation in Manufacturing," in *Technological Innovation: Strategies for a New Partnership*, Denis O. Gray, et. al. (eds), North Holland, Amsterdam, 1986, pp. 135-144, has documented the case of failure of a CAD system in an architectural firm.

Status quo does not necessarily mean technological stagnation. In their classic book, *Organizations*, March and Simon report there is a general tendency to change the level of acceptable standards of performance based upon the established norms of the relevant reference groups. This implies that in all organizations there will always be some changes in production technology, in the form of either replacements or upgrades of vintage equipment or frantic grabbing at quick fixes. Traditional approaches do nothing about the long-term competitive positioning of the company.

Consequently, technology acquisitions by U.S. firms during the 1970s and 1980s lacked any systematic connection with market requirements. The "patchwork quilt of technology acquisitions within plants has tended to contribute little or nothing so far as improvements in the manufacturing performance of these plants. Even where improved performance has occurred, this often has happened in areas with little or no significance for the contribution of the plant to the competitiveness of the firms."[20]

Traditional methods for justifying new technology often do not quantify the potential long-term market and process benefits. Traditional perspectives fail to take into consideration:

- The strategic opportunity cost of not investing in new technology due to lost sales to the competition.
- The impact on learning and the creation of a stock of knowledge that is necessary for the firm's future progress.
- The cost-effectiveness of the new technology's synergy with other operations or technologies.
- Improvements in quality; removal of nonvalue-added costs.[21]

World-class manufacturers are carefully rethinking their justification decisions. They are using multiple methods to aid decision making. Many are attempting to incorporate the intangible and soft benefits of new technology. They are developing business case scenarios for their firms regarding technological acquisitions.

More often than not, the technology decision becomes very subjective—based upon what feels right to management, given the competitive environment, customer requirements, and current business capabilities. One world-class firm's CEO made a $300,000,000 technology decision after he was led through a business case storyboard. The decision was justified by the market requirements for survival and not by reams of paper and financial analysis.

New Technology May Be Only Part of the Answer

The acquisition of new advanced process technology is not a panacea for competitive problems. In some instances, world-class manufacturers can derive significant benefits without spending heaps of money on new technology. Simple rearrangement of current equipment into group technology cells or work centers, often connected by conveyors or material handling equipment, may produce quantum payoffs. Furthermore, simplification of processes and product design should precede, or at least accompany, new technology acquisitions.

Operating in the 1990s and beyond requires altering our view of the justification process for advanced process technology. More focus should be placed upon meeting the customer's requirements and beating the competition than on short-term hurdles, which are often unduly high in the first place. The company's technology strategy can and should be a proactive competitive weapon. Technology as a proactive competitive weapon is described in Figure 5-6.

FIGURE 5-6 Investing in Technology as a Competitive Weapon

- Investment management extends beyond the traditional capital budgeting procedures and introduces factors associated with the long-term strategic impact of the technology.
- Multiple criteria, including subjective and market-related factors, such as customer response to shorter cycle times, quality, and flexibility, are used in the evaluation decision.
- Opportunity and other costs, relative risks of both acquiring and not acquiring technology, and potential long-term benefits are considered as integral elements of the investment strategy.
- Structured product-process analyses are undertaken in a systematic way so that every key performance criterion required of the product is linked to critical process capabilities.
- Investment decisions support the reduction or elimination of nonvalue-added activities and the achievement of target costs.

Source: Adapted from A.V. Roth, *Strategic Planning for the Optimal Acquisition of Flexible Manufacturing Systems*, Doctoral Dissertation, Ohio State University, 1986.

Competitive Benchmarking

In the late 1970s, Xerox found that it had lost its undisputed leadership position in the copier marketplace. Japanese competitors had captured the low end of the market and were making advances toward Xerox's home turf, the high-end, high-volume business copier. Xerox was watching the Japanese invade its markets in the same way it had invaded the motorcycle, automobile, and consumer electronics markets. However, unlike many U.S. companies, Xerox aggressively sought to find out how the Japanese were manufacturing such impressive high-quality products at such low costs. Xerox enlisted the support of its Japanese affiliate, Fuji-Xerox, and began looking closely at how the Japanese ran their businesses. Xerox then began comparing what it found to what it was doing in a structured manner and, in the process, established competitive benchmarking as a critical tool for manufacturers aspiring to achieve world-class capabilities. Xerox is not the only company that engages in competitive benchmarking, but it is considered by many to be the best.

Competitive benchmarking entails the comparison of one company's performance on critical measures of success with the performance of other companies. The other companies need not be competitors and, in fact, when searching for the best in class may not even be within the same industry. For example, Xerox looked to L.L. Bean for benchmarks concerning product distribution and to American Express as the best in class in billing operations.

Competitive benchmarking includes more than traditional financial measures of performance. Financial measures of performance, however, should not be ignored. Financial measures of competitive performance are based on the premise that a company cannot be *world class* unless it is consistently financially successful over time.

How do you establish a benchmark for measuring success? Two levels of benchmarking are typically required. The first pertains to aggregate measures of business unit performance relative to other competitors in an industry sector worldwide. These include:

- Growth
- Market share
- Profitability

The second level of benchmarking pertains to functional area performance of the manufacturing unit, including attributes of the manufacturing task:

- Quality

- Flexibility
- Delivery
- Price

The most prevalent financial measures of profitability are operating margin, asset turnover, and return on asset. The profitability of a business unit (measured by return on assets (ROA)) can be further broken down into these key determinants:

- The profit margin on sales
- The ratio of break-even sales to capacity
- The capacity utilization rate
- The ratio of sales to assets
- The investment in R&D

Breaking profitability down in this way enables one to scrutinize the financial impact of manufacturing on the business. Manufacturing decisions and capabilities principally affect costs, capacity, break-even sales, and asset intensity.

With comparative information, a company can gain insight into how different approaches to manufacturing may affect short- and long-term profitability. A world-class manufacturing company may sacrifice short-term profitability for long-term success. Japanese companies often accept low profits for a decade or more while they try to gain footholds in U.S. markets. For a number of years, in fact, Honda's management thought that a target of three percent ROA would be satisfactory. Yet, Honda has emerged as a world-class automobile manufacturer and has gained significant share during the 1980s.

A General Approach to Benchmarking

Benchmarking data for different businesses is available in public documents. This information can be successively broken down into industry-segment dynamics which form the basis of competitive benchmarking.

Figure 5-7 depicts the general approach to benchmarking aggregate business unit performance. Suppose that your firm is a machine tool manufacturer with *N* players. Competitive benchmarking should take into account analysis of the primary competitors within your industry.

- Be sure to consider global competitors within your own industry, as well as secondary firms sharing similar process technology capabilities who may become future competitors.

FIGURE 5-7 Competitive Benchmarking Analysis - Industry Segments

Competitors	Operating Margin	Asset Turnover	ROA	Annual Growth	Market Share	Overall
Company A	3	2	2	1	3	2
Company B	5	4	4	3	5	4
Company C	1	3	1	1	1	1
Company D	4	4	4	4	2	4
⋮	⋮	⋮	⋮	⋮	⋮	⋮
Company N	2	3	2	5	3	3
Your Company	4	4	4	3	2	3

SCALE: LOWEST=1 BELOW AVERAGE=2 AVERAGE=3 ABOVE AVERAGE=4 HIGHEST=5

- Rate the secondary competitors by their own industry averages. For each industry, collect publicly available industry information regarding growth, market share, asset turnover, and profitability.

Figures 5-8 and 5-9 show the results of a competitive benchmarking analysis performed using public data. The data represents the time period 1984 through 1987. Multiple industry segments were examined. Companies for which public data was either not available or inconsistent were excluded. The purpose of the analysis was to determine the advantages and limitations of this type of analysis as part of a company's overall evaluation of relative performance. The industries were selected to provide a broad perspective of competitive issues in the U.S.

FIGURE 5-8 Competitive Benchmarking Analysis—Selected Industry Segments: Leaders

Industry	Company	Performance Measures					Overall
		Operating Margin	Asset Turnover	ROA	Annual Growth	Relative Market Share	
Machine Tool	Ingersoll (Tool Div.)	5	4	4	3	5	4
	Cincinnati Milacron	5	3	4	2	4	4
	Toyoda Machine	2	3	2	5	3	3
Forging	Chrysler Forge	4	4	4	3	2	3
	Alcoa	4	3	3	4	5	4
	Ladish	4	3	3	3	4	3
Industrial Automation	Emerson	5	4	5	3	3	4
	Ametek	5	5	5	3	1	4
	Telemecamique	4	2	3	5	2	3
Micro Computer	Tandy	3	5	3	4	3	4
	Compaq	4	4	3	5	3	4
	Apple	3	3	3	3	3	3
Aerospace/Defense	Rockwell	4	4	3	2	4	3
	Raytheon	3	5	3	3	3	3
	TRW	3	4	3	2	3	3
Automotive	Ford	3	5	4	2	4	4
	Peugeot	3	5	3	2	2	3
	Honda	2	5	3	3	2	3

SCALE: LOWEST=1 BELOW AVERAGE=2 AVERAGE=3 ABOVE AVERAGE=4 HIGHEST=5

Which of the companies in Figures 5-8 and 5-9 are world-class performers? What other criteria are necessary to make this determination? Based upon our benchmarking analysis of the broad business unit measures of performance, the following conclusions can be reached.

Benchmark Results Must Be Analyzed Over Time. Not every leader may, in fact, be world class. A company may be riding temporary good fortune that will not be sustainable over time. Similarly, today's followers may include tomorrow's world-class manufacturers. Benchmark statistics must be gathered over time, so that important trends can be noted. For example, though General Motors did not fare well in this analysis based upon the snapshot of time reviewed, it remains relatively profitable and its significant investments in technology, new products, and new organizations may make it the world-class manufacturer of automobiles for the 1990s. In performing this type of analysis, great care

FIGURE 5-9 Competitive Benchmarking Analysis—Selected Industry Segments: Followers

Industry	Company	Performance Measures					Overall
		Operating Margin	Asset Turnover	ROA	Annual Growth	Relative Market Share	
Machine Tool	Acme-Cleveland	3	2	2	1	3	2
	Monarch	2	2	1	1	2	2
	Wean United	1	4	1	1	2	2
Forging	Sumitomo	1	1	1	4	4	2
	Japan Steel	1	1	1	3	1	1
	Daido	2	1	1	4	2	2
Industrial Automation	Foxboro	1	3	1	1	1	1
	Gould	1	3	1	1	1	1
	General Signal	2	4	2	1	2	2
Micro Computer	NEC	1	2	1	5	2	2
	Fujitsu	1	2	1	5	1	2
	Toshiba	1	2	1	5	2	2
Aerospace/Defense	Boeing	2	3	2	2	5	3
	ITT	3	1	2	3	3	2
	Grumman	2	4	2	2	2	2
Automotive	Nissan	1	3	1	1	3	2
	Mazda	1	5	1	2	1	2
	General Motors	1	3	1	1	5	2

SCALE: LOWEST=1, BELOW AVERAGE=2, AVERAGE=3, ABOVE AVERAGE=4, HIGHEST=5

has to be exercised to not reach erroneous conclusions from the snapshot of data collected and analyzed.

Further Analysis of Operational Measures Is Needed. While competitive benchmarking of traditional measures of financial performance and market position is vital to developing an effective strategy, further analysis is required. *It must include operational, nonfinancial benchmarks such as parts-per-million defect rate, best-in-class performers, and customer satisfaction.* The nonfinancial operational measures of performance often more closely define success relative to the business strategy and the company's long-term success. Progressive manufacturers agree that the competitive assessment process resulting from the analysis of the financial measures of performance is merely a first step in gaining a thorough understanding of competitive performance. World-class manufacturers of the 1990s will

FIGURE 5-10 Measures Frequently Used in Benchmarking at Xerox

Cost and Cost-related Metrics

- Percent of cost of function to revenue
  - Sales
  - Service
  - Customer administration
  - Distribution
  - General and administrative
- Labor overhead rate (percent)
- Material overhead (percent)
- Manpower performance ratio
- Months of supply
- Cost per page of publication
- Cost per order
- Cost per engineering drawing
- Occupancy cost as a percent of revenue
- Return on assets

Quality

- Percent of parts meeting requirements
- Percent of finished machine quality improvement
- Number of problem-free machines
- Internal and external customer satisfaction rates
- Billing error rates

Service

- Work support ratio
- $/SCAT hour (Standard Call Activity Time)
- Service response time
- First time fix of service call problem
- Percent of supplies delivered next day or on time
- Percent of parts available for the technical representative

Source: *Competitive Benchmarking: What It Is and What It Can Do For You* (Stamford, Conn.: Xerox Corporate Quality Office, 1984), p. 17.

expand this process, as Xerox has done, to include dynamic competitive benchmarking of both financial and nonfinancial measures of performance.

World-Class Capabilities Will Result in Financial Success. World-class manufacturing cannot occur in a financially unsuccessful business over the long run. Obviously, this becomes complicated when looking at a particular business unit in a large multi-national manufacturer. However, there is a growing body of empirical evidence that suggests manufacturing success can and does impact business unit success. Over the long run, world-class manufacturers, who by definition strive passionately to exceed customer requirements, will experience financial success. On average, year in and year out, the financial performance of a world-class manufacturer is better than that of its competitors, despite industry cycles. In the final analysis, world-class performers are those that are successful over time.

Conclusion

In the future, world-class organizations will not have accounting systems in the traditional sense; instead, they will have cost management systems that provide meaningful and timely information to line managers. At the heart of these systems will be activity-based costing systems that will enable managers to identify the direct effect of their actions and decisions and support their analysis of value added vs. nonvalue-added activities. These same systems will enable managers to establish meaningful target costs and to manage the attainment of those targets. Competitive benchmarking will be used to establish both financial and operational targets and to provide insight into the steps necessary to reach those targets.

The point of all these new approaches to accounting for manufacturing, including activity-based accounting systems, is that if they work, decision making will change in the firm. For example, when the high-cost activities that truly impact performance with customers are identified, the decision as to how to allocate resources becomes easier. One assigns the best people or hires the right talent to concentrate on those activities that add value. Alternatively, one might decide to buy some service on the outside if it relates to activities that are not critical. If decisions are not altered in some way, then the performance management system is probably flawed.

Currently, manufacturers must contend with multiple measurement systems. These assume multiple goals, since much of measuring is not just for external reporting and problem solving but for evaluation of goals and problem solving. Ironically, publicly-held stock companies appear to be divided in their statements of goals. One major goal is to make money for the stockholders; another is for long-term survival of the business. Consequently, there is a dynamic tension between the performance measures valued on Wall Street (return on equity, expected cash flow, and return on total assets) and those valued by customers (quality, delivery, flexibility, and cost).

The measurement paradox, of course, is how to achieve both simultaneously. The trick is to recognize that customer-driven manufacturing performance is correlated with financial performance. *In the end, customers determine the financial outcomes.* It becomes obvious that manufacturing strategies that improve customer satisfaction, such as superior quality and reduced lead times, also reduce costs, increase share, and, ultimately, improve financial performance. Furthermore, the need for financial justification of investments will not disappear, but debates about the realization of potential payoff become moot.

Most certainly, the accounting system will be replaced and the cost management systems described in this chapter will be augmented with critical operational and strategic measures that result in a total performance management system, attuned to the critical market variables defining success and failure. In the next chapter, the critical operations measures being used by world-class companies to define total performance management system are explored in detail. These nonfinancial measures are the new weapons of world-class manufacturing.

CHAPTER 6

VALUE-ADDED PERFORMANCE MEASUREMENT

When Harley-Davidson found itself near extinction in the early 1980s, it turned to new methods of manufacturing and new measures of performance. Quality became the number one concern, and measures of quality the most important indicators of success. Inventory turns and throughput time also became paramount, as Harley-Davidson strived to drive products through its plant with as little investment in inventory as possible, thus generating the cash required to service the huge debt incurred during its leveraged buyout (LBO). Measures of productivity centered on bottleneck operation availability and availability of machines when they were needed. While financial performance was critical to Harley-Davidson, as it narrowly avoided bankruptcy, a new set of nonfinancial performance metrics guided the company back to its position as a competitive force in its industry. These new metrics were operational in nature and were directly tied to the company's business and manufacturing strategy.

The prevailing literature on nonfinancial performance measurement in manufacturing identifies productivity, time, and quality metrics as the nonfinancial *drivers* of manufacturing performance. Recent findings from both domestic and international surveys, as shown in Figure 6-1, confirm the continuing emphasis on productivity, time, and quality as the key measures of manufacturing excellence.

Executives interviewed in conjunction with this research indicated they are expanding their operational measures of performance beyond merely costs. The issues they are grappling with include:

- How should productivity, including labor and machinery performance, be measured?
- What are the impacts of measures of time-based competition, including delivery, throughput, and flexibility?

- How should quality, including the cost of quality, be measured?[1]

This chapter addresses each of these areas and identifies the performance measures that are most important for achieving and sustaining world-class capabilities.

FIGURE 6-1 Nonfinancial Performance Management

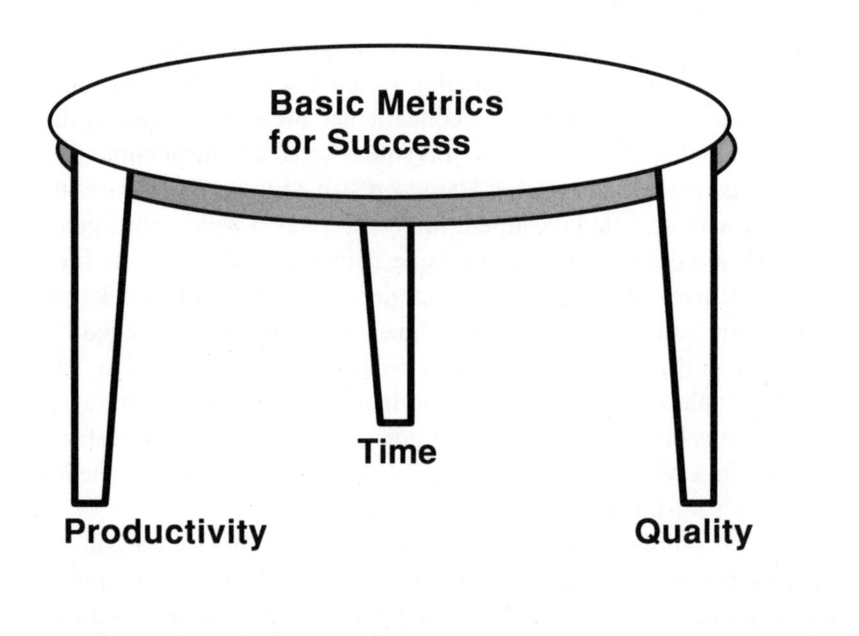

Productivity Measurement

Although productivity measurement is among the most traditional measures of manufacturing performance, it provides a great deal of information about what determines performance success. Executives interviewed were strongly influenced by Eli Goldratt's book, *The Goal*, which implies that a manufacturing decision is productive only if it:

- **Increases throughput** (the rate at which money is generated through sales).

- **Decreases inventory** (things we buy which we intend to sell).
- **Decreases operating expense** (money spent to convert inventory into throughput).

A decision is productive if all three manufacturing performance criteria are satisfied simultaneously and no tradeoffs have been permitted.

Managing Bottlenecks Is Critical. Most often referred to by executives in companies practicing JIT, Goldratt's perspective calls for manufacturers to consider the following factors in assessing productivity:

- A bottleneck is any resource the capacity of which is equal to or less than the demand placed on it.
- A nonbottleneck is any resource the capacity of which is greater than the demand placed on it.
- Every part has its own bottleneck if it is being produced to be sold. Bottlenecks do not exist for parts, such as inventory, that will not be sold.
- Plant capacity is increased only by increasing capacity at the bottlenecks.

According to Goldratt, the real manufacturing productivity problem is not operator or machine efficiencies but the backlogs of overdue orders that result from mismanagement of bottlenecks.

How You Measure Labor Productivity Makes a Difference. Today, many discrete parts manufacturers and researchers subscribe to Goldratt's basic tenets. The concept of direct labor efficiency is a measure that is leading manufacturers to overproduce. Labor utilization in a work cell is usually measured in one of two ways. The traditional measure of direct labor hours worked divided by hours available is sometimes adopted to reveal whether or not a work cell is overstaffed. Another, more relevant, measure being used is output divided by total headcount. World-class manufacturers are pooling both full-time equivalent (FTE) direct and indirect headcounts (staff, managers, janitors, information specialist, etc.) into the denominator of the labor productivity metric (see Figure 6-2).

This concept of total labor headcount, direct and indirect, is becoming popular because of the substantial reductions in direct *touch* labor that have occurred over the past decade, while indirect labor has increased. These reductions in direct labor diminish its effectiveness as a productivity metric. At best, direct labor efficiency provides too much precision. Coupled with significant transaction efforts to run the reporting system, direct labor is responsible for an increasingly smaller proportion of total cost. While direct

FIGURE 6-2 Labor Productivity

Old Measure	New Measure
Direct Labor Hours Worked divided by Direct Labor Hours Available	Total Output of Product divided by Total (FTE) Headcount

labor reporting thoroughly analyzes direct labor costs, a disproportionate amount of time is spent in this effort. Direct labor content is estimated to be 10 to 15 percent or less of total cost in many industries. Even a jump to 20 percent of the total cost would not yield significant profitability improvements and would not reverse the trends in industries damaged by imported goods.

When Harley-Davidson made its turnaround in the 1980s, it determined that monitoring direct labor hours by job or workcenter was both irrelevant and costly. As one of the many changes required to support the JIT environment, Harley-Davidson eliminated direct labor reporting.

Figure 6-3 reviews traditional performance measures, the performance incentives that justify them, and the results of using them.

Machine Utilization Measurements Are Inappropriate. Measurements of machine utilization have the effect of keeping a machine busy whether or not the parts it makes are needed. Machine utilization does not differentiate between bottleneck and nonbottleneck workcenters. Instead of machine utilization, many manufacturers are now keeping track of machine hours actually run. They divide this figure by the hours scheduled or *needed* to indicate how well the equipment performs when required.

Based upon Goldratt's definitions of bottleneck and nonbottleneck work centers, many companies are singling out those pieces of equipment most critical to plant capacity. If the equipment limits the ability of a plant to produce products that are to be sold (not inventoried), it should be used 100

FIGURE 6-3 Traditional Measures that Inhibit Performance

TRADITIONAL MEASUREMENT	PERFORMANCE INCENTIVE	RESULTS
Machine Utilization	Maximize machine utilization by overproducing.	Excess inventory buildup.
Direct Labor Efficiency, Utilization, and Productivity	Produce more standard direct labor hours.	Excess inventory buildup.
Purchase Price	Accept larger quantities of unneeded inventory to secure lower prices.	Lower-quality materials and potentially higher inventory levels.
Overhead Absorption	Produce enough standard hours to over-absorb the overhead cost budget.	Excess inventory buildup.
Variances	Direct management attention only to unfavorable variances.	A scrap allowance with no incentive to reduce the amount and cost of scrap.
Direct Labor Reporting	Thoroughly analyze direct labor costs.	Disproportionate time spent analyzing direct labor costs, which are normally 10% of total costs.
Composite Overhead	Summarize overhead costs.	Underemphasis of cost drivers and nonvalue-added overhead.
Quality	Measure quality costs against the quality control department budget.	Underemphasis of preventative measures, to reduce internal failure costs.

Source: D.A. Anderson and M.R. Ostrenga, "MRP II and Cost Management: A Match Made in Theory?" from *CIM Review* (New York: Auerbach Publishers) © 1987 Warren, Gorham & Lamont Inc. Used with permission.

percent of the time, excluding downtime for maintenance. All that is required of nonbottleneck workcenters, however, is to support production through bottlenecks.

Notice that the critical measure of equipment performance is equipment availability when it is needed. *Downtime when a machine is not required is not a relevant measure of performance.* Therefore, it should not be considered as either good or bad machine management. Labor and machine utilization, if they are based upon need rather than availability, more accurately reflect work cell performance and, in some sense, supervisor performance as well.

Traditional Machine Utilization Causes Overproduction. In most manufacturing environments, traditional cost accounting measures reflect continuous negative variances for work cells when machine utilization is not very high. Traditional methods of measuring machine utilization, overhead absorption, and direct labor efficiency result in inventory buildups due to overproduction. This conflict between traditional measures and the new metrics of world-class manufacturing must be resolved by accounting and manufacturing personnel so that the new metrics can replace the old measures—or at least coexist for different purposes.

Unfortunately, the prevailing norm is to report labor productivity in terms of direct labor efficiency and machine utilization. In a recent survey of over 1000 firms, manufacturers reported that the more sophisticated and potentially useful measures of performance are seldom used. These include:

- Measuring the productivity of the total production facility (24 percent).
- Using productivity data to determine compensation (19 to 23 percent).
- Using productivity indices to assess trends in performance (24 percent).
- Measuring productivity using constant dollar data (15 percent).
- Measuring productivity using value-added data (16 percent).
- Using total factor productivity measures (28 percent).[2]

Total Factor Productivity

Of the more sophisticated techniques, *total factor productivity* (TFP)[3] appears to be the most widely discussed and most promising measure of resource utilization. It has been defined as:

$$\text{TFP} = \frac{\text{Total Output Of Product}}{\Sigma \text{ Total Resource Inputs}}$$

Total factor productivity provides the means of measuring the relative efficiency of a manufacturing unit in transforming a variety of inputs into outputs. Most companies already collect the basic raw data required for computing total factor production but need new ways to process and report it. TFP can be used at a product level or at a total plant level. It has several advantages:

- TFP provides a normalized measure of performance by comparing changes to a base period. TFP should be measured on a constant dollar basis to eliminate inflationary effects.

- TFP has the ability to sort out the impact of a number of factors contributing to the change in productivity, including price changes, mix changes, capital productivity, labor productivity, and material productivity.
- TFP is a very comprehensive measure of performance. It takes into account all resource inputs and all plant outputs. It reflects total plant performance over time.

The primary disadvantage of TFP is that it does not take into consideration the value of the output to the customer. For example, what is the impact on performance of high TFP if the firm goes out of business? Total factor productivity is a measure of *efficiency*, not necessarily of *effectiveness*. Therefore, *manufacturers must also capture two additional nonfinancial indicators that are critical success factors for their market—time and quality*.

Removing Time Barriers

Quality attributes, particularly conformance quality, were clearly the competitive battleground of the 1980s. During the 1990s, the competitive war will be waged primarily on the basis of time. Globally, manufacturers are turning to time as a way of differentiating their products and services. Yet for most U.S. firms, time-based capabilities exhibit significant gaps between current abilities and future requirements.

When We Talk about Time Barriers, What Do We Mean?

American manufacturers view delivery time reliability as both the most important future requirement and one of the largest obstacles to overcome. The way in which manufacturers measure their own time-based capabilities, as well as their competitors,' is increasingly important. Unfortunately, the measurement of time-related capabilities is not a part of most American manufacturers' measurement systems. A 1987 survey by the Oliver Wight Companies showed that time-based competitive capabilities were measured by less than half of the companies surveyed. This percentage is expected to increase dramatically as the importance of time increases. The more traditional time-based measures and the percent of companies indicating use of them are:

- Reduction of delivery time (54 percent).

- Reduction of manufacturing lead times (60 percent).
- Reduction of vendor lead times (45 percent).
- Reduction of engineering lead times (25 percent).[4]

Ninety-four percent of the companies responding to this survey, however, reported they measured *on-time* deliveries. The information collected in interviews with executive manufacturing managers was very similar. Most companies measured on-time deliveries, but nowhere near as many companies measured other elements of time performance. Further, there remained questions as to exactly how *on-time* was defined. Is a delivery *on-time* if it meets the original manufacturing ship date? A revised ship date? The original customer-requested ship date? The receipt date required by the customer? Or the date the product is required for customer use? It is clear that no standard definition exists. *For purposes of world-class measurement, on-time delivery must be defined in terms of customer requirements—the right products must be delivered at the right time, at the right place, and ready for use.*

Time-based Competition Is Much Broader than On-time Delivery

There is growing recognition on the part of leading manufacturing executives of the need to measure time itself. Time is not just *manufacturing* time; it is the total order cycle time from the moment the order is placed until it is shipped to the customer. This requires a closer look at order entry processing and engineering time, which for many manufacturers is the largest component of total cycle time. One company visited directed the initial effort of a CIM plan toward reducing order entry time prior to looking at manufacturing lead times. This orientation toward reducing time requirements in areas other than the shop floor will become common as the 1990s unfold.

The issue of time reduction or *speed* affects not only the order delivery cycle but the entire product specification and design process. Research conducted by McKinsey & Co. has shown that high-technology products that come to market six months late but on budget earn 33 percent less profit over five years.[5] The same study found that when the product comes out on time but over budget by 50 percent, only a 4 percent reduction in profit is realized over five years.

The more dynamic the environment, the more important is speed to market. The U.S. auto industry has learned the painful lessons of the competitive advantage created by new product design and introduction speed from the Japanese.

While Ford's *Concept to Customer* program was put in place to address the need for improved product introduction speed and quality, it is estimated that the popular Ford Taurus will go a full 10 years before a major model change. Honda continues to introduce its new products at three-year intervals on average. Ford is not unique, however; General Motors and Chrysler have faced the same problems and, many would argue, have been less successful in responding to the challenge. The Chevrolet Camaro, which is slated for a new introduction in 1992, will be the first major change in a decade. Volvo, Saab-Scania, Jaguar PLC, and, to a lesser extent, Mercedes-Benz have all developed unique market positions in spite of new product introduction rates that are slow even by Detroit's standards.

Numerous success stories exist in the U.S. where progressive companies, forced to speed up or lose their markets, have responded to the call for speed.

- Motorola's Bandit Project resulted in reducing by almost three weeks the time from order to finished goods for their pagers. In their fully automated work area in Boynton Beach, Florida, they can now produce a pager every two hours.
- GE has had similar success in their circuit breaker business – a reduction from three weeks to three days for their order entry to product completion cycle.
- Hewlett-Packard cut from nearly five years to just under two years the time required to develop computer printers and has established a firm hold on the market.
- Likewise, AT&T cut in half the time required to develop new phones.

The Japanese have brought the requirement for speed to all industries in which they compete and have forced U.S. and European manufacturers to respond. This requirement will continue to shape the performance measurement systems of all manufacturers in the 1990s and beyond.

Measuring Delivery and Throughput Performance

When measuring delivery and throughput performance, the following indicators are most common:

- On-time delivery
- Lead time – order to shipment

- Waste time (nonvalue-added time = lead time minus process time)
- Setup time
- Percentage of orders delivered on time
- Average level of order fulfillment
- Vendor lead times
- Engineering cycle times
- Order processing cycle times
- Value-added time as a percent of total cycle time
- Percent of orders on schedule
- Average ECO execution time
- New product introduction lead time

Throughput is often stated in terms of inventory measures such as WIP turns, raw material turns, and total inventory turns. Measures of product flexibility include new product introduction/design time and machine or product line changeover time.

World-class manufacturers should break down time into relevant components of the total customer cycle time:

- Pre-order cycle time
- Order cycle time
- Manufacturing cycle time
- Delivery channel cycle time
- Post-delivery cycle time

Market-related activities, including time to configure custom products, design new products, and prepare customers for the sale, occur in the pre-order stage. The order cycle begins at the inception of an order by the customer and ends when the order is released to the shop floor; it is followed by the manufacturing cycle and, subsequently, by the time in distribution channels before receipt by the customer. Post-delivery time is the time required for some products to be prepared for ultimate customer usage. Post-delivery time may be attributed to installation at a customer site by a company or its agent, the time required by a customer to assemble the product, the time required for orientation and training, and/or the time required to respond to customer complaints and service needs.

The purchasing order cycle time occurs on a parallel track and, if not attended to properly, can delay the manufacturing cycle and, ultimately, the delivery of product to customers. With this broad competitive model in mind, key indicators, like those suggested above, can be developed for each component of total customer cycle time. Managing time requires a comprehensive understanding of the impact of total customer cycle time upon customer requirements and perfect execution of appropriate cycle time reduction (see Figure 6-4).

FIGURE 6-4 Total Customer Cycle Time

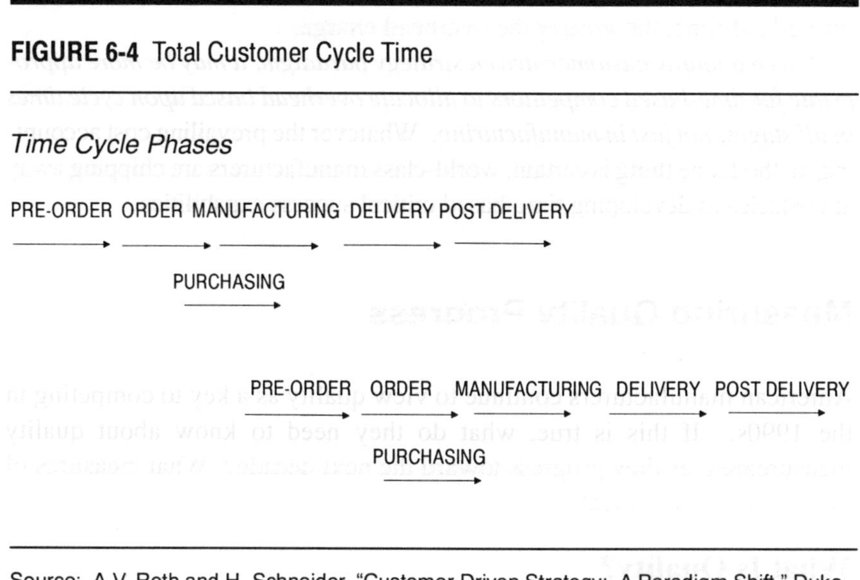

Source: A.V. Roth and H. Schneider, "Customer Driven Strategy: A Paradigm Shift," Duke University, 1990.

Manufacturing does not take place in a vacuum, however. There are situations where, in spite of precise planning, cycle times are lengthened because of outside pressures such as government testing requirements, e.g., FDA approval for pharmaceuticals, environmental issues, and other exogenous factors. In the twenty-first century, world-class manufacturers will prepare employees to tackle environmental issues as part of product plans to reduce time to market. The technology they develop—for example, biodegradable diapers or coolants that do not use fluorocarbons—will also eliminate environmental problems. Coming to grips with environmental, health, and government regulatory issues in the design phase of a production process will be the trait that makes world-class manufacturers leaders in their fields.

Selecting the Most Appropriate Time-based Measures

The selection of the most appropriate time-based performance measures is a function of a company's competitive priorities, the nature of the product, and the cycle time required in each phase. To marry time-based priorities with accounting systems, some researchers have suggested that overhead be allocated on the basis of time in the plant. This method of costing is termed *velocity costing* and is based on lead times; the longer the manufacturing lead time, the greater the overhead charge.

Given a totally customer-driven strategy paradigm, it may be more appropriate for time-based competitors to allocate overhead based upon cycle times in all stages, not just in manufacturing. Whatever the prevailing cost accounting method, one thing is certain: world-class manufacturers are chipping away at obstacles to developing time-based critical success capabilities.

Measuring Quality Progress

American manufacturers continue to view quality as a key to competing in the 1990s. If this is true, what do they need to know about quality measurement as they progress toward the next decade? What measures of quality are most critical?

What Is Quality?

Quality is what your customer says it is. Based upon this definition, and utilizing the findings of the 1988 ASQC Gallup survey of American consumers, the five factors consumers deem most important in judging product quality are:

- Product performance
- Durability
- Availability of service
- Ease of repair
- Ease of use

This survey shows that, while rated number one by consumers, performance quality has decreased in importance significantly since 1985. The importance of ease of use has increased over this same time period. The decrease in the importance of performance quality may be due to significant improvements in product performance attributes over the last decade. For example, overall quality is getting better through redesign and process improvements. Hence, many customers can no longer easily differentiate between products or vendors.

Most of the CEOs interviewed reported that their companies had some kind of a recognized quality program in place. These programs were usually strongly supported by top management and grew out of experimentation with a variety of approaches. The approaches and techniques most commonly applied today were described in Chapter 2.

From a measurement perspective, most of the methodologies are *standard.* Although manufacturing executives clearly believe that the customer defines quality, their companies lack an organized customer-focused quality measurement system with which to capture the "voice of the customer."

Measures of quality are both internal and external, or perceptible to customers. For example:

- **Internal Measures:** Yield rates, scrap rates, percent of products reworked, percent of total labor performing rework, incoming vendor defects in parts-per-million (ppm).
- **External Measures:** Outgoing product defects in parts-per-million, customer complaints, warranty claims, returns, and allowance percentages.

The trend in quality measurement systems for world-class manufacturing firms is extending into nonmanufacturing areas as well. Quality measures in most companies can be enhanced, particularly with regard to translating measures of quality from the customer perspective into manufacturing requirements.

As manufacturing managers begin to engage in boundary management, a major portion of their time will be spent on integration activities. There is a need to extract measures of *interface quality* to reflect the value of the services, information, and physical product transferred between manufacturing and nonmanufacturing areas. Interface quality measures pertain to the way in which manufacturing connects with engineering, customer service, marketing, and customers. For example, a measure of the quality of interface between manufacturing and engineering is the number of engineering change orders; between the supplier and the manufacturer, the rate of incoming "good" quality.

Texas Instruments has successfully developed an environment in which each employee defines and services his or her own customer; each customer-employee contact is treated as an opportunity to better serve the customer. This approach to developing interface quality, however, has not been widely adopted by U.S. manufacturers. The concept of interface quality may be difficult to grasp at first, because it reflects intangible coordination, communication, and synergistic activity between manufacturing and nonmanufacturing and because the idea is so new.

Execution of the Quality Program Is Important

The way companies measure quality is less important than the way the quality program is executed. Whatever quality measures are included must be consistent with program goals and support continuous improvement and organizational learning.

The total cost of quality remains elusive for most manufacturers. Comprehensive systems for measuring the cost of quality only add to overhead burden. Further, they are difficult for manufacturers to understand and accomplish. Many enlightened manufacturers, including Xerox, Baxter, and GE, build quality improvements into their systems and do not keep track of the costs in the typical Philip Crosby *Quality is Free* fashion. Keeping track of the cost of quality is viewed by many manufacturers as a nonvalue-added activity. This view is consistent with many Japanese applications of quality improvement programs.

Leading manufacturers are more likely to measure the outcomes of critical processes and gaps with respect to customer requirements. For example, Motorola's *Six Sigma* corporate initiative tracks parts-per-million defect rates on critical products and benchmarks these against best-in-class performance as is described in Chapter 2.

Recognizing Quality Organizations

Since 1951, Japan has annually awarded national Deming prizes to top manufacturers. The Japan Quality Control Society has awarded Japan Quality Control Medals since 1970 to promote the importance of quality. The U.S., however, did not have a national prize until 1988, when the Commerce Department's National Institute of Standards and Technology (NIST) made its first Malcolm Baldrige National Quality Improvement awards.

The competition, created by the Malcolm Baldrige National Quality Improvement Act of 1987, invited companies to compete for six quality awards. Companies were evaluated and recognized for making superior quality the keystone of their strategic planning process and for constantly improving quality in their operations.

Specifically, corporations and their senior executives received the Malcolm Baldrige National Quality Awards in recognition of their meeting of these award criteria:

- **Leadership:** "... in creating quality values, building the values into the way the company does business, and projecting the quality values outside of the company ... Evaluations are based upon the appropriateness, effectiveness, and extent of the executives' and of the company's involvement in relation to the size and type of business."
- **Information and Analysis:** " ... The scope, validity, and use of data to determine the adequacy of the data system to support total quality management." Information systems are used to support "management by fact" and analysis and use of data for decision making.
- **Strategic Quality Planning:** " ... The company's approach to quality leadership ... Competitive benchmark data is essential for planning quality leadership because it makes possible clear and objective quality comparisons."
- **Human Resource Utilization:** " ... The company's efforts to develop and involve the entire work force in total quality."
- **Quality Assurance of Products and Services:** " ... The consistency of execution of quality operations and incorporation of a sound prevention basis accompanied by continuous quality improvement activities."
- **Quality Results:** " ... The company's quality improvement and quality levels by themselves and in relation to those of competitors. Included are quality of products and services, internal operations, and suppliers. The number and type of measures depend upon factors such as the company's size, types of products and services, and the competitive environment. Evaluations take such factors into account and consider whether or not the measures are sufficient to support overall improvement and to establish clear quality levels and comparison."
- **Customer Satisfaction:** " ... The company's knowledge of customer requirements, service and responsiveness, and satisfaction results measured through a variety of indicators."

The quality measurement systems of recipients of Malcolm Baldrige National Quality Awards permeate their companies and extend the bounds of quality improvement for entire organizations.

The judging was administered by the American Society for Quality Control and the American Productivity and Quality Center. In the first competition, of the 10,000 application forms requested, only 66 were returned and only 13 were deemed appropriate by NIST for further investigation. Based upon further investigation of those 13 companies, only three companies – Motorola, Westinghouse Electric's Commercial Nuclear Fuel Division, and Globe Metallurgical – were granted awards in 1988. Of the six awards available, only three were handed out, due to a lack of qualified applicants. Two of the awards held back were intended for the service sector; one was intended for small manufacturers. In 1989, Xerox Business Products and Milliken & Co. took the prizes.

The Malcolm Baldrige National Quality Award has had a significant impact on American business in a very short time. Many leading manufacturers are using the award criteria to benchmark their own internal quality performance. While few companies applied in its first year, the Baldrige Award has picked up considerable momentum, and the battlecry, "Win the Baldrige," can now be heard in organizations large and small across the country.

The Baldrige Award has also caused U.S. companies to look differently upon the Deming Prize, which had previously been shunned by American companies. The Deming Prize has had nearly 40 years to mature and is different from the Baldrige award in a number of important ways, as shown in Figure 6-5.

The importance of the Deming Prize and the Baldrige Award lies not in an evaluation of their differences but, rather, in the way in which they influence behavior and actions. Both result in companies measuring performance in ways different from those used historically, focusing on the critical elements of quality and customer service. As a result of these visible changing criteria for successful performance, U.S. companies are beginning to change the way they behave along with the performance measures they use. *Moreover, improvement does not stop with the acceptance of an award.* The award winners, like Xerox and Motorola, are proceeding toward the goal of continuous improvement with as much zeal as ever.

FIGURE 6-5 Comparison of the Baldrige Award to the Deming Prize

Category	Deming Prize	Baldrige Award
Purpose	"Award prizes to those companies recognized as having successfully applied CWQC (Company Wide Quality Control) based on statistical quality control."	"Promote quality awareness, recognize quality achievements of U.S. companies, and publicize successful quality strategies."
Emphasis	Statistical Methods	Customer Satisfaction
Eligibility	Individuals, factories, and companies (global since 1984).	Companies only —confined to the U.S.
Number of Recipients	Any number of companies that meet a standard.	Maximum of: 2 manufacturing companies (plus divisions); 2 small businesses (less than 500 employees); and 2 service companies.
Award Psychology	Many winners—success breeds success.	Very few winners—many losers.
Judging Criteria	One page of guidelines—very vague and subjective.	20-plus pages of guidelines— more specific and objective.

Source: K. R. Bhote, "Motorola's Long March to the Malcolm Baldrige National Quality Award," *National Productivity Review*, 8 (4), Autumn 1989.

The New Quality Metrics

Excellent manufacturers in the 1990s and beyond will develop measures that reinforce quality behavior and practices, instill a commonality of purpose toward meeting customer requirements, and challenge and stretch the organization toward future competitive advantage, value-added functions, and profitability. New quality metrics will support planning for quality, organizing for quality, and monitoring progress toward quality (see Figure 6-6).

FIGURE 6-6 New Quality Metrics

Planning for Quality	**Examples**
Measurable Objectives	Improve product performance on each key parameter by 5 percent over next year.
Integration	Trends in number of cross-functional teams
Competitive Benchmarks	Who has the best service quality overall? Your Firm - Competitors A B C D E F
Suggestions	Trends in numbers submitted and accepted and by level of employee.
Organizing for Quality	**Examples**
Team Participation	Trends in number of teams, number of of enhancements.
Staff Development	Trends in training and development hours by type of personnel.
Product/Service Implementation	Lead roles of management, customers, marketing, engineering, manufacturing, from concept development to performance appraisal
Monitoring Quality	**Examples**
Trends in Quality Indicators by Function	
Engineering	Percent of new products successfully introduced.
Manufacturing	Throughput measures, setup times, queue times, cycle time, defect free, inventory, employee involvement, reduced process variability on critical processes.
Purchasing	Percent of suppliers providing fault-free materials and components.

Figure 6-6 concluded

Marketing	Percent of new accounts received based upon number bid.
Human Resources	Employee complaints, accident rate, absenteeism.
Customer Satisfaction Trends	Customer appraisals, customer complaints, customer retention, market share.
Internal Customer Satisfaction Trends	Employee assessments.
Product	Field deficiencies, warranty costs.
System	Total cost of business, reduced space, primary competitor benchmarks.

Source: Adapted from *The Baldrige Criteria*.

Conclusion

Nothing short of a major change in the way managers view manufacturing performance is required for success in the 1990s. Performance measurement systems of world-class competitors will be radically different from those used in the past in a number of key ways:

- **They will focus on the external drivers of change mandated by customers.** As business success is driven by market success, performance measures will reinforce and strengthen the company's competitive position and integrate functions.
- **There will not be a preoccupation with a single measure of performance such as unit cost or variance.** Multiple measures of performance will be used to guide action and promote continuous improvement. They will be the basis for improving organizational learning and promoting interfunctional synergy.
- **Nonfinancial measures will become significant barometers of success.** World-class manufacturers are moving toward a broader definition of "product" for the 1990s, one that includes both physical product attributes, customer service components, and market/distribution attributes. Nonfinancial measures will be developed to im-

prove attributes of the company's strategic bill of materials, as shown in Figure 6-7. In particular, measures will focus on monitoring critical manufacturing capabilities for future success.

FIGURE 6-7 Modified Strategic Bill of Materials for the Twenty-first Century

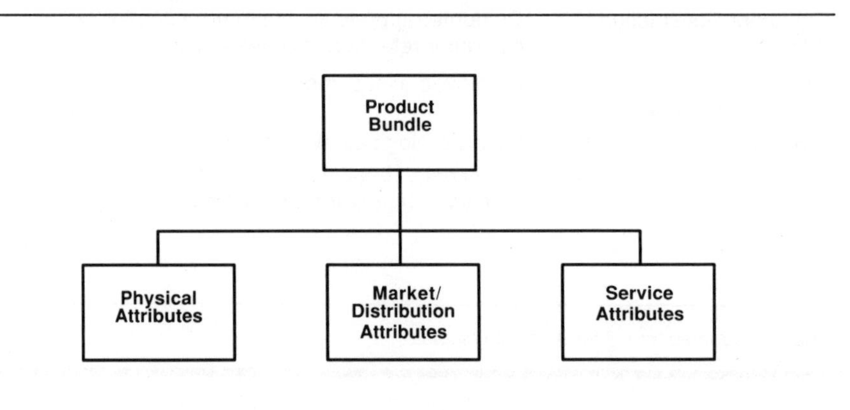

Adapted from: C. Giffi and A.V. Roth, "Making the Grade in the 1990s," Deloitte & Touche Manufacturing Research Reports, 1989.

- **Different measures of performance will be required at each organizational level, so that the measurement systems will mesh into a congruent framework for the total business.** At the highest levels in the organization, macro performance measures will be adequate. Moving down through the organizational hierarchy, performance measures will be tailored to employees' jobs in direct relation to the requirements of their immediate customers, who may be internal or external.

Regardless of the level of the organization that uses the output of the performance measurement system, the information must be timely. Performance information will be used increasingly by all levels of the organization to respond to changes in both market and plant conditions; old or dated information will not be useful. Simple, yet meaningful, measures will be adopted by world-class manufacturers. The measures will be constantly challenged regarding their relevance to current conditions. New measures will be integrated with old; irrelevant measures discarded. Competitive advantage is dynamic; so too must be the performance measurement systems of leading manufacturers.

CHAPTER 7

ADOPTING BEST MANUFACTURING PRACTICES

Only those enterprises that adopt an integrated approach combining new practices, new cultures, and new mindsets will be able to respond effectively to the global competitive challenges that lie ahead. Increased customer demand and competition are driving manufacturers to produce a greater variety of higher-quality products with reduced design and manufacturing lead times.

Ford Motor Company responded to these challenges in the 1980s. Unable to invest heavily in advanced manufacturing technology, due to limited financial resources, Ford set out to improve its operations, focusing on management and engineering practices. Results, drawn from Ford annual reports and information in the popular press, were impressive.

- In 1980, Ford averaged 6.7 defects per car. By 1989, Ford was averaging 1.5 defects per car.
- In 1980, Ford lost $1.5 billion. By 1987, Ford had posted a profit of $4.6 billion, making it more profitable than General Motors for the first time in history.
- Ford has increased its share of the U.S. automobile market in six of the past eight years, growing from 17 percent to 22 percent.
- Ford's recently introduced Explorer all-terrain vehicle is a first for a U.S. automaker; it will be sold by Mazda in Japan.
- Within the next two years, Ford will begin building minivans for Nissan.

Over the past decade Ford has epitomized a manufacturing giant evolving into a world-class company.

What did Ford do that made it different from its competitors? Ford made fundamental changes in manufacturing throughput and maintenance practices, adopting the *best manufacturing practices* available. What are best manufacturing practices, and will adopting them result in world-class competitive capabilities for every manufacturer?

"Best" Manufacturing Practices

Best manufacturing practice is a term used by the federal government to define any manufacturing technique or process that yields superior results. The U.S. government employs teams of reviewers to evaluate new processes and techniques that manufacturers submit as *best practices*. The surveys of these practices are available to the public on a subscription basis and through an on-line database.[1]

Which practices are world-class companies employing? Of the myriad options manufacturers have available to them, eight categories of best practices differentiate world-class competitors:

- **Quality management practices** that foster a long-term, total quality management perspective and a total dedication to customer needs. Techniques such as Total Quality Management (TQM) and Quality Function Deployment (QFD) are the foundation of any best manufacturing practice agenda. These practices are described in Chapter 2.
- **Innovative human asset management** that encourages excellence, teamwork, and accelerated learning, and rewards employees for their direct contributions.
- **Streamlined organizations** that are focusing their efforts on customer satisfaction and improving their manufacturing performance through such techniques as focused factories, cross-functional teams, and partnerships with other companies.
- **Restructured engineering processes** that use parallel approaches to product and process design, often referred to as *concurrent*, or *simultaneous*, engineering.
- **Just-in-Time (JIT) techniques** that focus on waste reduction, throughput improvement, and lead-time reduction.
- **Production planning and control** disciplines, such as Material Requirements Planning (MRP), that enable organizations to plan and execute production and material requirements effectively in complex environments.

- **Product and process simplification** efforts conducted on an ongoing basis using techniques that enhance value and maintain quality while eliminating cost, such as Value Analysis (VA) and Value Engineering (VE).
- **Progressive approaches to equipment and facilities maintenance** commonly described as Total Productive Maintenance (TPM).

The importance of quality in creating and maintaining competitive advantage is well-established. As we saw in Chapter 2, no single element is more important for developing world-class competitive capabilities than quality. The new approaches world-class companies are using for managing their human resources and organizing the manufacturing function are discussed in Chapters 8 and 9. In this chapter, the five remaining best practices are reviewed:

- Restructured engineering processes
- JIT techniques
- Production planning and control
- Product and process simplification
- Productive maintenance

Restructuring the Design Engineering Process

Significant improvements in the ability of manufacturing companies to compete in the global marketplace will largely be driven by their ability to rapidly specify, design, produce, and deliver quality products at low cost. Reducing the waste component of the product and process cost caused by engineering inaccuracies is a key strategy. The multiplier effect that engineering inaccuracies have on product cost, quality, and time to market are substantial. The opportunity for improvement is enormous. World-class companies are restructuring their engineering processes to eliminate these problems.[2]

The Effect of "Over the Wall" Engineering

Companies continue to operate in environments in which design and manufacturing organizations communicate infrequently. The traditional design/build process is error prone and time consuming; waste is built into the process. Product designs are tossed "over the wall" to manufacturing without any meaningful dialogue between the design and manufacturing departments. The result: designs are often rejected by the manufacturing organization, because they cannot be produced. When this happens, product engineering must change the design and release it again. The resulting engineering change orders delay product introduction, add cost, and frequently degrade performance and reliability.

What is the impact of these engineering deficiencies? It is commonly recognized that the first 5 percent of a product design effort determines 85 percent of the product cost. Studies conducted by McKinsey & Co. have shown that in a stable market, one without price erosion and with long product life cycles, delivering a product that has a cost 10 percent higher than planned will result in an almost 50 percent loss of profit over the life of the product. In a dynamic market, one with significant market growth, substantial price erosion, and short product life cycles, the same product cost overrun will account for only a 25 percent loss of total profit over the life of the product. In this dynamic environment, however, a six-month delay in the product introduction will account for an additional 30 percent to 35 percent loss of total profit over the product life, as compared to less than a 10 percent loss for the same delay in a stable market.

Early engineering inaccuracies and inefficiencies most affect product cost and product introduction time frames, resulting in significant cost overruns. Typically, it is too late to improve the outcome of the engineering process after even 5 percent or less of the design effort is complete.

The Root Causes of Design Engineering Problems

Behind these inaccuracies and inefficiencies are a host of difficulties encountered during the product design process.

Vertical Management Styles and Organization Structures. Organizational structures in which marketing, design engineering, manufacturing planning, manufacturing, tooling, maintenance, and other departments are isolated from each other are key contributors to the problem. Each of these organizational units has separate goals, personnel, procedures, and policies. Communication is more by chance than as part of routine operations.

Product Specification Changes. Marketing and sales do not understand the devastating impact of product specification changes on the design engineering process. "Sell, design, and specify" is too often the accepted sequence of events, as the sales staff tries to please the customer.

Short-Sighted Engineering Specialists. Design engineers have evolved into highly trained specialists, rewarded primarily for their creativity and technical genius, rather than for their attention to product cost and manufacturability.

Engineering Change Orders. Design engineers do not understand the extraordinary impact of ECOs on the manufacturing process. As described in Chapter 4, ECOs can devastate a manufacturing organization, resulting in overall disruption in the organization, longer lead times, poorer product quality, higher costs, and morale problems.

Ineffective and Inefficient Engineering Processes. The typical engineering design process is poorly managed. There is a blizzard of paper work. Data is entered several times in different formats into multiple databases. Design review and inspection positions exist only to catch errors *after* they are made. Communication is inadequate or nonexistent. Statistical tools and simulation techniques are not applied.

Limited Application of State-of-the-Art Technologies. For all the advances in CAD and CAE in the 1980s, design engineers still do not employ the technologies currently available to them. Leading CAD/CAE vendors offer systems to assist design engineers in managing their databases, simulating the impact of their designs, identifying and managing the specification constraints, and improving product introduction time while reducing product cost. Unfortunately, most design engineering organizations are still striving for full payback on their last CAD system and have failed to recognize either the changes in the demands being placed on them in the marketplace or the advances in technology available.

Fixing the Engineering Design Problems

Nothing short of a revolution in the design engineering function will fix these problems. The difficult and costly lessons that manufacturing has learned in the 1980s—the need for continuous improvement, elimination of waste, employee involvement, a focus on quality of work effort and product—must now be applied to the design engineering function. Five steps world-class companies are taking to correct these problems are described below.

1. **Develop a common understanding of the problems.** Both marketing and engineering organizations are improving their basic understanding of the negative effect of product specification changes (PSCs) on the design engineering process and engineering change orders on the manufacturing process. The effect of these "ineffectiveness" drivers is being analyzed in many organizations. Design engineering is more clearly defining the requirements for product specifications. Producible designs created through techniques like

design for manufacture (DFM), are becoming fundamental to engineering organizations.

2. **Replace vertical management styles with horizontal management practices.** Organizations are implementing management practices that integrate the various functions of the organization, tearing down the typical organizational barriers between the functions. An example of these *horizontal* management practices is the cross-functional work team that cuts through natural organizational barriers. Teams can be large or small, permanently constituted or ad hoc. The variables and attributes of the team depend on the nature of the product, the type of company, the industry, and the flexibility of the corporate culture. The cross-functional team approach leads to development of products that pull other functions and disciplines into the design process as required.
3. **Simplify administrative processes**. The administrative process associated with design engineering is being simplified. Organizations are focusing on necessary functions and eliminating all others, then improving the efficiency of the remaining operations. Companywide databases are being established, and paperwork is being streamlined or eliminated. Unnecessary reports are being discarded.
4. **Apply the latest CAD/CAE technology**. Leading manufacturers are applying the latest CAD/CAE technologies. The ability of the newest systems to simulate product and process designs provides companies with basic competitive advantages in both cost and time to market. The tools available in the 1970s resulted in a design process characterized by building the product or process first, then determining if it worked, and, finally, determining if it was the best approach. The tools available today allow design engineers to determine first if the product or process is the best approach and if it will work, *prior* to building it. Technology management is the subject of Chapter 10.
5. **Develop new rewards and incentives.** The reward and incentive systems used by leading manufacturers are being overhauled to foster teamwork and encourage excellent performance. These changes are affecting all employees, including design engineers. In Chapter 9, the new approaches to rewards and incentives are discussed in detail.

The Benefits of Restructuring the Design Engineering Process

Restructuring the design engineering process can result in:

- Improved product quality and functionality of design.
- Shorter flow time for prototyping and product release.
- Far fewer producibility errors which result in fewer engineering changes.
- Fewer shop floor rejects and less scrap and rework.
- Fewer customer returns and service or support problems.
- Quicker response to customer-requested design changes.

Clearly, the mounting competitive pressure on time to market, innovative designs, and low product cost necessitates improved design engineering processes. There is, however, no one right way to restructure. No single technique or best practice will suffice. What is required is thoughtful attention to the engineering process with an eye toward improving the effectiveness of the operation. U.S. manufacturers can ill afford to ignore the design engineering process in their quest to compete with their global rivals. Nor can they afford to waste resources attempting to define the single best practice as the ultimate solution to their engineering design problems. With this in mind, many manufacturers have turned their attention to simultaneous engineering as the rallying point in their efforts to restructure their design and manufacturing engineering organizations.

Simultaneous Engineering

Simultaneous engineering, also known as concurrent engineering, is a nonlinear product or project design approach during which all phases of manufacturing operate at the same time—simultaneously. W. David Lee, director of the Arthur D. Little Center for Product Development, defines the concept this way:

> Simultaneous engineering is the process in which key design engineering and manufacturing professionals provide input during the early design phase to reduce the downstream difficulties and to build in quality, cost reduction, and reliability at the outset. Successful simultaneous engineering requires a 'marriage' among team members and a multidisciplinary culture.

Henry W. Stoll, former manager of the Design for Manufacture program at the Industrial Technology Institute in Ann Arbor, Michigan, has developed what he calls the Four Cs of simultaneous engineering:

- **Concurrence**: Product and process design run in parallel and occur in the same time frame.
- **Constraints**: Process constraints are considered part of the product design. This ensures parts that are easy to fabricate, handle, and assemble and facilitates use of simple, cost-effective process, tooling, and material handling solutions.
- **Coordination**: Product and process are closely coordinated to achieve optimal matching of requirements for effective cost, quality, and delivery.
- **Consensus:** High-impact product and process decision making involve full team participation and consensus.

Figure 7-1 shows the interrelatedness of the elements of simultaneous engineering.

FIGURE 7-1 Simultaneous Engineering Model

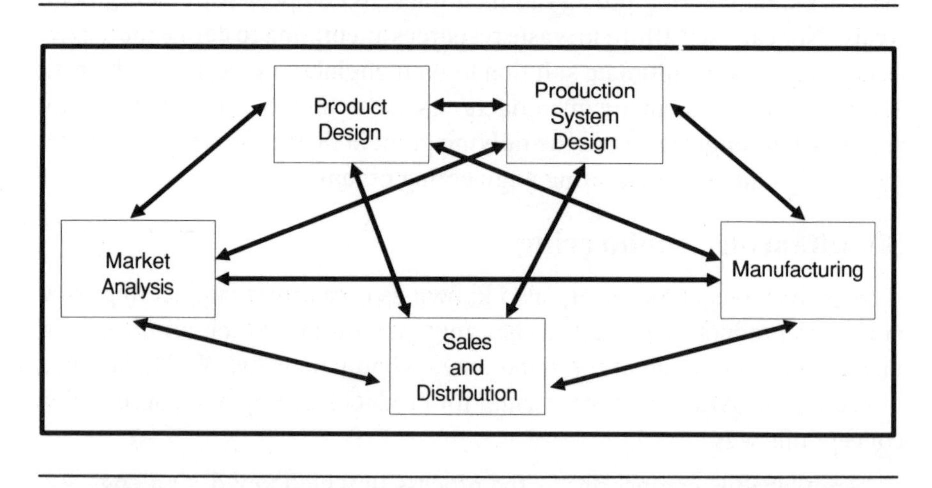

Source: Gary S. Vasilash, "Simultaneous Engineering," *Production*, July 1987, pp. 36-41.

The Traditional "Phased Project" Approach

The phased project approach, as shown in Figure 7-2, was formalized by NASA in the 1960s. It breaks projects down into a series of steps. Each step is accomplished sequentially, in a sort of relay-race fashion, by handing responsibility for a project from function to function. For example, product

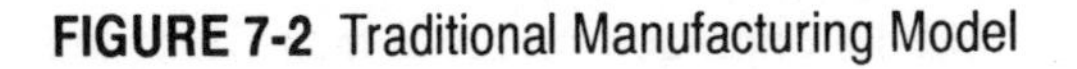

FIGURE 7-2 Traditional Manufacturing Model

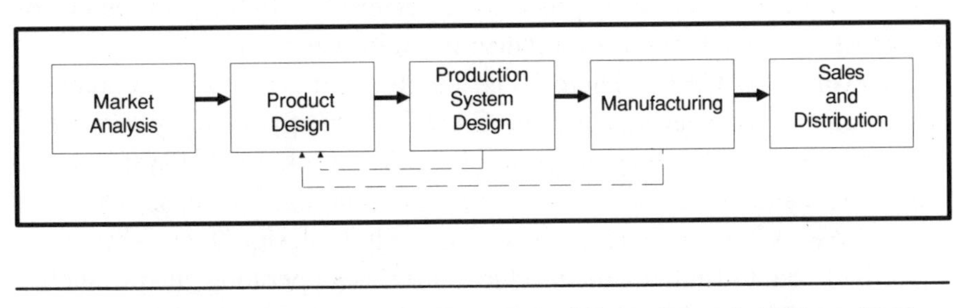

Source: Gary S. Vasilash, "Simultaneous Engineering," *Production*, July 1987, pp. 36-41.

planners hand a project to designers who hand it to manufacturers who finally hand it to salespeople and distributors.

The benefits of this phased approach are an orderly process and a tight functional job focus. The disadvantages include an extremely elongated conception-to-market lead time and products which are difficult to manufacture. Further, in dynamic markets, the products are often out of touch with the market by the time they arrive.

What Happens with Simultaneous Engineering

In contrast, a concurrent or simultaneous approach allows the company to respond more quickly to market changes and to reduce costs by reducing surprises and rework. In effect, the company does the work right the first time. Additional cost reductions are realized by standardizing product features and manufacturing processes and reducing the charges associated with tooling, fixturing, procurement, and the storage of parts. The process also creates a significant level of cooperation and enthusiasm within an organization.

Elements Crucial to Implementing Simultaneous Engineering

Many companies, however, are struggling with organizational barriers that traditionally define the day-to-day operating relationship of designers, manufacturers, and marketers and prevent the institution of simultaneous engineering.

A Clear Definition. Many companies lack a clear definition of simultaneous engineering. Many mistakenly look toward the technological solution first. They do not realize that the approach involves teamwork and revamping of the organization's culture as well as the way in which the phases of the manufacturing process are performed. In fact, computers are unnecessary for achieving simultaneous engineering.

Use of CAD/CAE. There is no single right way to implement the simultaneous engineering concept. CAD/CAE has given many companies the tools they need to encourage simultaneous work. CAD systems centralize data and make it available to multiple functional groups. Advanced CAD/CAE systems enable engineers not only to design the product more effectively but to simulate the product under both operating and production conditions. This enables the manufacturing process to be designed with less guesswork. Most companies in the U.S. that are serious about achieving simultaneous engineering capabilities are also committed to CAD/CAE systems.

Work Teams. The major hurdle to overcome in achieving concurrent capabilities is organizational. The key barrier is the traditional industry practice of segregating functions and sequentially conducting activities. This segregation mentality becomes ingrained in engineering school and continues throughout a person's career. Most experts agree that the key to successfully implementing concurrent capabilities is the interdisciplinary work team.

Simultaneous Engineering Success Stories

The Japanese have practiced simultaneous engineering for years, calling the process *parallel approach*. Hirotaka Takeuchi and Ikujiro Nonaka of Japan's Hitotsubashi University describe the process as more like a rugby match than a relay race. The two researchers have studied the methods of Honda, Canon, NEC, and Fuji-Xerox. They note that these companies pick a multidisciplinary team that stays with the project from start to finish, passing the ball back and forth as they all move downfield together toward the goal of product launch.[3]

The Japanese approach is based on work teams and breaking down traditional organizational barriers, not the use of computer technology. They have been successfully integrating functional departments.

The ranks of companies that have successfully implemented the simultaneous engineering concept are growing every day.

- Charles Lamb, director of Advanced Manufacturing and Engineering Technology at Emhart Corporation, reports that using simultaneous engineering techniques has reduced by one-third the time it takes a line of door hardware to reach the market.
- According to Robert L. Dorn, Cadillac Motor Car Division, "Simultaneous engineering is a process, not a project." With simultaneous engineering, the 1988 Cadillac Eldorado was restyled in 12 months instead of the 36 months typical in the industry.
- The impact of Ingersoll's implementation of simultaneous engineering was measured by comparing the design and manufacture of two similar transfer line projects, conducted 10 years apart, the latter using simultaneous engineering. The result was a reduction in engineering changes from 62 to 6, a total reduction in design and manufacturing time of 23 weeks, and a cost reduction of nearly $1 million.

The simultaneous engineering concept is not confined to internal company efforts. Suppliers of components and equipment are also being asked to sign up and work as partners. Many companies are attempting to capitalize on supplier product engineering expertise by implementing cooperative simultaneous engineering programs. The move to integrate suppliers into a company's operations is another step being taken by world-class companies. Just-in-Time manufacturing techniques are often the beginning point for supplier integration and as fundamental to developing world-class manufacturing capabilities as is restructuring the design engineering process.

Just-in-Time Manufacturing Practices

What is Just-in-Time (JIT)? JIT is a companywide philosophy oriented toward eliminating waste throughout all operational functions and improving materials throughput. To implement the philosophy in manufacturing, JIT techniques provide "the cost effective production and delivery of only the necessary quality parts in the right quantity, at the right time and place, while using a minimum of facilities, equipment, materials, and human resources."[4]

The JIT toolkit includes management rules of thumb and techniques for purchasing of raw materials and parts, delivery of materials and parts to the point of assembly, and the manufacturing process itself as well as manufacturing interfaces with nonmanufacturing areas. It can become a springboard for developing integrated policies for manufacturing products and purchasing materials and components.

The concept has received much publicity and has been evaluated or implemented by a large number of companies. As described in Chapter 4, throughput-time reduction is the single most effective method of increasing overall productivity. JIT is the prescription most often given to achieve throughput-time reduction. JIT and JIT-like concepts are among the best manufacturing practices. JIT is characterized by low levels of inventory and high levels of quality and customer service.

The Philosophy of JIT

While at the heart of the JIT movement is the basic philosophy of waste elimination throughout the organization, many U.S. managers see JIT as only a series of special techniques that are applied to the manufacturing environment to improve efficiencies. *JIT, however, must first be understood as an approach to business management and manufacturing from the perspective of better serving the customer.*

Taiichi Ohno, the father of the Toyota Production System, defines JIT in terms of eliminating seven wastes:

- **Waste of Overproduction:** Making more than is required is a waste. Efficient and effective operations are not those that run at high utilizations, but rather those that produce what is needed when it is needed.

- **Waste of Waiting:** Workers idled due to long setup times, delay of material movements, shortages of parts, and cumbersome processes and procedures are a waste.

- **Waste of Transportation:** Facility layouts and systems that require material to be transported significant distances are wasteful. Utilizing expensive advanced material handling systems is only a costly way of hiding the waste. The best solution is a manufacturing process designed to eliminate handling and material movement.

- **Waste of Processing:** Some processing of parts should not be done; sometimes even the part itself should not be made. Elimination of processes and components through the redesign of either product or process is one objective of JIT.

- **Waste of Stocks:** Stocking inventory at any stage of the process, from raw materials to work in process (WIP), is wasteful. Inventory stock is the by-product of an ineffective and inefficient system. Stock becomes the buffer a poor manufacturing operation uses to support itself.

- **Waste of Motion:** All motion should be considered wasteful unless it is absolutely necessary to add value to the product or service. Automating motion through the application of advanced process technology only masks the possibility that the motion is not be necessary at all.
- **Waste of Making Defective Products:** Engineering processes and manufacturing processes that permit defective products to be produced are wasteful. Whether the part or product must be reworked or scrapped, defective products as waste must be eliminated.

The context of JIT, then, is the elimination of anything that is unnecessary, i.e., all waste. When viewed from this perspective, JIT is a common-sense approach to dealing with business, in general, and improving competitive capabilities in all areas. Before attempting to implement the JIT techniques in this chapter, any manufacturer aspiring to become world-class must first adopt the basic philosophy of JIT—waste elimination and, effectively, quality improvement.

Benefits of Successful JIT Implementation

The application of JIT techniques has the potential for significant cost reductions.

- Direct labor can be reduced by as much as 20 percent, indirect and exempt labor by up to 45 percent, space by 40 percent, and as much as 50 percent additional production capacity can be made available.[5]
- When JIT techniques are applied to the full cycle of procurement, manufacturing, and delivery, quality defects can be reduced by as much as 60 percent, production time by as much as 90 percent, and capital expenditures by as much as 30 percent. Further, inventory reductions were perceived as the major benefit of JIT by 58 percent of the manufacturing executives surveyed.[6]

Harley-Davidson's reincarnation during the 1980s, using JIT concepts, resulted in an improvement of inventory turns from 5 to 20 times per year. Inventory levels were cut by 75 percent. Scrap and rework was reduced by 68 percent. Productivity improvements of 50 percent were achieved. Space requirements were reduced by 25 percent.[7] For Harley, inventory reduction was the primary benefit at the beginning of the implementation; significant cash that had been tied up in inventory was released and used to pay down the leveraged buyout debt and keep the company afloat.

The benefits of JIT can be significant and the philosophy is conceptually simple, but the course is not easy. While the techniques associated with JIT purchasing and manufacturing require little or no capital investment in machinery, implementation of JIT is dependent upon the stability of the user's scheduled requirements, supplier flexibility and incoming quality, and total employee involvement and teamwork. Implementation difficulties can be attributed to changes in supplier relationships and purchasing policies, total quality management, and the culture of the organization. These topics are discussed in the following sections.

JIT Purchasing

JIT purchasing means that materials are provided to the production facility just as they are required for use. The critical elements of JIT purchasing are:

- Reduced order quantities.
- Frequent and reliable delivery schedules.
- Reduced and highly predictable lead times.
- Consistently high quality levels for purchased materials.

The features of a successful JIT purchasing program are a major departure from those of traditional purchasing.

Features of JIT Purchasing

Limiting Suppliers. Sharp cutbacks in numbers of suppliers are underway everywhere that JIT is being implemented. This allows companies to concentrate on improving relationships with suppliers, improving supplier quality, and reducing the problems associated with managing dozens of suppliers.

- Xerox purchasing cut 4700 suppliers in one year. When the dust had settled, only 400 remained.
- Hewlett-Packard reduced suppliers from 35 to 5 for two component parts.
- AMP put 10 percent of its suppliers (several hundred) on inactive status in one year.

- Black and Decker adopted a *dominant supplier* strategy for all its major commodities. At the Tarboro, North Carolina, plant, the supplier base has been trimmed 50 percent over the last several years.
- General Motors evaluated more than 400 suppliers before paring the list down to 69 suppliers for a new engine.
- Nissan believes in single-sourcing parts, because the policy reinforces the idea that the supplier has total responsibility for the program.

Longer Term Relationships. Many companies, as described in Chapter 2, are now establishing longer term bonds with their suppliers, working with them as partners rather than as adversaries. Companies are initiating supplier certification programs to establish long-term relationships with suppliers willing to take total responsibility for the quality of their products. Allen Bradley's goal was to have all of their suppliers certified by the end of 1989. To date, they feel their certification program is a major reason that rejected lots from suppliers have dropped from 17 percent to 6 percent.[8] This trend is expected to become more prevalent in all areas if manufacturing over the next few years.

Cooperative Product Design. Price gaps between U.S. and Japanese products can often be attributed to product design differences. In a recent product review by Xerox, the Japanese enjoyed a substantial cost advantage over Xerox cost in the U.S. for plastic components. Forty-three percent of the difference was due to product over-design by U.S. engineers.[9]

Currently, many companies are looking for opportunities to utilize supplier expertise in the design of their product components. Simpson Industries is representative of suppliers conducting joint design efforts with their customers. Simpson has developed the balance shaft for General Motors' 3800 Series engine and the vibration isolator for Ford's four-cylinder Taurus/Sable engine. Since the supplier knows its equipment and processes best and will ultimately be responsible for producing the part, its assistance during the product design is invaluable.

Multidisciplinary Buying Teams. Some companies are increasing the amounts of *buying by team* they do. This is an attempt to bring the resources involved in the eventual use of the purchased component closer to the production source.

Frequent JIT Deliveries. JIT delivery is characterized by small orders, consistent order quantities, and firm order schedules. JIT delivery can result in suppliers arriving several times a day with raw material or component parts to feed the JIT production system. A survey of 23 auto industry suppliers found that the expected frequency of deliveries to their JIT customers would increase from an average of 2.8 times per month to nearly 12 times a month.[10]

Geographic Proximity. Trying to utilize suppliers located close to the original equipment manufacturer (OEM) plant site or attempting to convince suppliers to relocate closer to the OEM plant are part of JIT. Many JIT suppliers are establishing service centers near their customers to better support their JIT requirements. Some suppliers, like Simpson Industries, are placing engineers in the engineering departments of their largest customers.

Superior Product Quality. Improving quality continues to be a key issue in JIT. Poor supplier quality is the primary concern of companies attempting to implement JIT. In a JIT purchasing program, the ultimate goal is to eliminate incoming inspection altogether; the suppliers' production processes should be so good that there is no reason to inspect for defects.

Virtually all world-class manufacturers look to supplier training and assistance to improve quality. Ford now routinely teaches quality-oriented techniques, like those in the toolkit presented in Chapter 2, to help its suppliers meet the demands of a JIT environment. JIT manufacturers even expect suppliers to incorporate quality plans into their proposals on component design and continue to improve the levels of quality over time. These efforts have paid handsome dividends.

For example, Xerox had defect rates of about 10,000 parts per million (ppm) on purchased parts in 1983. Through implementation of JIT purchasing practices, Xerox reduced the defect rate of suppliers to 5000 ppm in 1984; to 1300 ppm in 1985; and to 900 ppm in 1986. The goal for 1987 was 325 ppm. Additionally, since 1984, Xerox has reduced its unscheduled maintenance by 40 percent and showed a 27 percent reduction in service response time.

Supplier Feedback. Effective supplier management also requires the company to provide timely feedback on quality, technology, responsiveness, dependability, and costs. Hewlett-Packard has quarterly, semi-annual, and annual reviews that give suppliers feedback on their performance relative to their competitors. In the interim periods, any detected defect is confronted and eliminated before further purchases from the supplier are made.

Impact of JIT on Purchasing

The purchasing functions of many companies have undergone substantial changes in recent years, as management has attempted to cope with increasing international competition. In addition to working to bolster operations internally, manufacturers now consider their suppliers to be visible extensions of Original Equipment Manufacturer (OEM) production processes. The supplier's last activity becomes the OEM's first. OEMs have found that they cannot achieve manufacturing excellence unless their suppliers are striving for excellence.

This recognition by U.S. manufacturers that their success is dependent on their suppliers has resulted in both cooperative efforts with suppliers and demands that suppliers comply with quality and delivery requirements.

General Motors' *Targets for Excellence* program was initially interpreted as a "do it or else" program by most of its suppliers. After participating in the evaluation process, however, most companies indicated that they benefitted significantly from the experience and now run their businesses with a focus on General Motors' targets.

Both Ford Motor Co. and Chrysler Corporation adopted similar programs that were also initially viewed as having hard-line messages. For example, Chrysler has mandated that all of its suppliers utilize electronic data interchange (EDI) to improve communications in support of their JIT practices. Most of their suppliers now recognize the benefits of the direction and assistance provided and are active participants in the process. As of 1988, 14.8 percent of the companies surveyed used EDI computer networks linking plant to subcontractors, suppliers, and customers. Within the next five years, the percent of establishments using EDI technology is expected to rise to more than 35 percent, with the transportation equipment industries and their suppliers taking the lead.[11]

Most of the adoption issues surrounding JIT involve educating supplier personnel in:

- The importance of JIT to both the OEM and the supplier.
- The benefit to the supplier of implementing JIT.
- The seriousness of the OEM in moving to a JIT operation.

Unfortunately, companies that force JIT on suppliers do not always lay solid groundwork. A recent Automotive Industry Action Group (AIAG) study indicates that there has been little real implementation of JIT in the internal operations of auto suppliers.[12] The early response of U.S. suppliers in Detroit to the forcing of JIT by the automakers was to revitalize the Detroit warehouse business – the very kind of thing JIT is supposed to eliminate.

When suppliers adopt JIT, whether through coercion or more cooperative approaches, the improvements are clearly measurable. In fact, JIT suppliers have consistently demonstrated better performance than non-JIT suppliers on a number of key performance measures, as described below.

CASE STUDY

HP Helps Deliver Just in Time

"Less than two years ago, only 21 percent of deliveries to the 50 manufacturing divisions of Hewlett-Packard were on time. We wasted many hours firefighting, trying to determine which parts would be late and devising schemes to keep production lines going anyway. Early deliveries, meanwhile, were costing us a fortune in inventory storage and control. Now 51 percent of deliveries arrive on time. Production is stopped less often, and reductions in early deliveries have reduced inventory expenses by $9 million.

"How are we achieving these improvements? One way out would have been to measure our suppliers' shipping performance and blame them. But even if suppliers performed perfectly, transit and receiving or other factors could slow delivery. Besides, orders for parts and materials begin with a communication from us to them. By carefully reviewing how we were working *with* our suppliers, we came to an unexpected and effective solution.

"I should point out that at HP we set an aggressive definition of on-time delivery: to be on time, a shipment has to be three days early to zero days late. After talking to the people on the production line, we determined that this window should under no circumstances be widened. How to proceed?

"As with most total quality control approaches, it was all deceptively simple. We broke the delivery process into three parts: first, our communication to the supplier and the supplier's communication to us regarding transit time; second, the supplier's manufacturing and shipping of parts; and, finally, the carrier's transporting of parts to us. Then we brainstormed about the possible causes of failure for each and collected data to test our theories.

"Regarding the first, communication, we measured the suppliers' expected ship dates and the expected transit times against the time that HP expected to receive the material. These should have been the same. But we found that buyers and suppliers were clear about what they had agreed on in only 40% of deliveries. In other words, 60% of the time a communications problem was contributing to late or early shipments.

"To measure suppliers' manufacturing and shipping, we compared the suppliers' expected ship dates with the actual ship dates. In our sample, fully 69% of the deliveries were shipped within the on-time range. Finally, we looked at failures in transit time and receiving. We measured the expected ship times against the difference between HP's receipt dates and the suppliers' actual ship dates. In 90% of the cases, the actual transit times were on time.

"Our study clearly revealed that *communication* was the chief culprit in on-time delivery failures—hardly a popular conclusion since it made us a

Case Study, concluded

primary cause of the trouble. We set out to take corrective action. It became apparent very soon that the supplier did not always understand whether the date on the purchase order was a shipment date or a delivery date. We changed the purchase order, labeling each date clearly.

"Expected transit time was another misunderstanding, and HP's hodgepodge routing instructions were at fault. We have tried to solve this with uniform routing guides. Finally, suppliers were manually subtracting the transit time from the delivery date to calculate the ship date. We have tried to preempt such errors—and other data entry errors—by installing electronic purchase orders that flow directly from HP's computers to the suppliers' open-order management systems.

"These simple changes have already made valuable contributions to our bottom line. But our job isn't finished. Perfection is the goal, after all, and we are continuing to use the total quality approach to understand our customers' needs and respond to them."

Source: Dan Marshall in D. Burt, "Managing Suppliers Up to Speed," *Harvard Business Review*, July-August 1989, p. 133.

JIT vs Non-JIT Supplier Characteristics

What are the characteristics of successful JIT suppliers versus suppliers that do not practice JIT techniques? A study of the operations of 21 U.S. suppliers to Japanese-managed companies in the United States in the electronics and automotive industries indicates the following:

- **Reduced Lead Times.** Seventy percent of the JIT suppliers have reduced lead times to better compete. Only 30 percent of the non-JIT suppliers have reduced lead times. Reducing lead times was found to be the primary factor affecting a supplier's ability to react quickly to changing customer requirements.
- **Lower Defect Rate.** JIT suppliers report, on average, a 50 percent lower defect rate to customers than non-JIT suppliers. Further, non-JIT suppliers are often forced to add extra inspection steps to ensure that the quality of the product they are supplying to the JIT customer is acceptable.
- **Less Raw Inventory.** Over 50 percent of the JIT suppliers currently have less raw materials inventory than in the past, versus only 9 percent of their non-JIT counterparts.

- **Less WIP Inventory.** Forty-three percent of the JIT suppliers currently have less Work in Progress inventory than in the past, compared to 9 percent of the non-JIT suppliers.
- **Less Finished Goods Inventory.** Twenty-nine percent of the JIT suppliers currently have less finished goods inventory than in the past, compared to only 18 percent of the non-JIT suppliers.
- **Flexible Workforce.** Sixty-seven percent of workers in JIT environments knew two or more jobs, versus only 54 percent of workers in non-JIT environments. Forty-five percent of JIT workers know three or more jobs, versus 34 percent of non-JIT workers. The benefits of cross-training workers have been found to include less worker idle time, reduced overhead, fewer layoffs due to downturns in specific product lines, and increased flexibility.[13]

JIT Manufacturing Techniques

Robert Hall, in his book *Attaining Manufacturing Excellence*, describes several techniques common to JIT manufacturing systems. However, not all the techniques described can be used by all companies, nor should they be. The objective is to achieve the benefits of JIT through a tailor-made system of common-sense, waste-elimination, and process-simplification approaches.

Features of JIT Manufacturing

Improved Work Place Organization. Hall advocates starting the implementation of a JIT manufacturing system with a reorganization of the physical work place through a five-step process:

1. Clearing and simplifying
2. Locating
3. Cleaning
4. Discipline
5. Participation

The result is a clean work place containing only absolutely necessary equipment and parts, located in the correct place, and *owned* and policed by the workforce without management intervention.

Visibility of Operations and Outcomes. Hall's second JIT technique deals with visibility. He argues that visibility "creates effective and immediate feedback." Common approaches to attaining visibility include posted schedules, signal lights, charts, visible logs, posted goals, U-shaped layouts, and limited visible inventory.

Kanban Production Control. The kanban card, signaling the approval to make more parts, is fundamental to the visible *pull* system approach to production control. Whether utilizing a one-card or a two-card system, the basic premise is to make more parts only when required—as signaled by the kanban—and never to make more parts than allowed by the standard kanban. The clearer the line of sight, the easier it is to implement the pull system. Hall cautions that in nonrepetitive and unique custom-engineered environments, the pull system and the use of kanban are not appropriate.

Steady-rate Timing. The best JIT environments are able to operate at a steady rate of production, with a steady consumption of materials and delivery of product to customers. Steady production rates allow the synchronization of production work centers using simple, visual techniques and algorithms. All direct and indirect activities can be timed to recur based upon the steady rate. Obviously, the nonrepetitive and unique custom-engineered environments make steady rate timing virtually impossible and, therefore, require the use of more sophisticated techniques often controlled by a computer. In many operations this translates into sequential production.

Flexibility. Providing for flexibility in the production system is critical to a responsive JIT environment. Flexibility is best achieved through lead-time reductions. Reducing setup and process times provides increased flexibility in a number of areas. Hall asserts that product mix flexibility, volume flexibility, people flexibility, engineering change flexibility, and new product introduction flexibility are all critical components of a JIT system. A manufacturer can achieve flexibility in these areas in a multitude of ways, but among the most common is the development of simple manufacturing processes that can be easily set up and changed and people added or subtracted to alter volume. These systems also eliminate or reduce work rules and promote generalist over specialist workers. Effective communication at all levels of the plant is also important.

Adoption of JIT in the U.S.

While JIT is a concept that has received much publicity and utilizes simple techniques, recent studies indicate that U.S. industry has not universally embraced and implemented JIT in day-to-day operations. For example:

- A survey of manufacturing managers conducted by *Electronic Business* magazine indicates that 51 percent saw JIT as only moderately useful, and 8 percent were negative about the concept.[14]
- Another survey of over 700 North American manufacturers indicates that 20 percent of respondents have no plans to implement JIT and 50 percent are still in the planning or setup stages.[15]
- A survey of 230 manufacturing vice-presidents and directors indicates that only 37 percent of their companies are currently involved in JIT activities.[16]

The rationale for this less than universal adoption rate focuses on three key areas:

- **Little understanding of benefits.** Most of the key executives who would be involved in initiating JIT efforts in their companies misunderstand the real benefits of JIT. They view JIT as an inventory reduction and control technique, rather than as a far-reaching operational philosophy of which inventory control is one benefit, and usually not the major one. In large part, this may be because much of the current literature is directed at inventory management rather than the total effects of JIT on the overall business.
- **Nonthreatening business environment.** Companies often must be pushed by some business-threatening force to change from current methods to new techniques. Many companies still are not feeling any pressure to alter their operations drastically. If there is no threat to a company's existence on the basis of cost or quality, then management teams have little reason to implement JIT or any new concept. This is especially true since JIT requires a formidable shift in manufacturing management and an upheaval in the practice of manufacturing. As one author puts it, "companies that are moving to embrace the entire JIT philosophy have management teams that are 'emotionally' ready to take action. They realize they must do something or go out of business."[17]
- **Not appropriate for all industries.** The broad concept of JIT, as a philosophy of waste elimination and throughput, is applicable to all industries. The widespread applicability of JIT techniques for materials management may have been oversold or misunderstood. JIT has been shown to work in some process-oriented industries, such as pharmaceuticals, chemicals, plastics, and petroleum, or even in

low-volume job shops, but these are not the norm. Companies in different industries value JIT differently, as shown in Figure 7-3.

In general, JIT fits well in high-volume batch production operations, including metal products, automotives, and electronics manufacturing. It does not fare as well in process-oriented industries. Process-oriented industries are more capital and machinery intensive; therefore, they have less flexibility in modifying current operations. They also have less work in process, as there are fewer opportunities for interoperative queues.

FIGURE 7-3 The Strategic Importance of JIT Significantly Varies by Industry

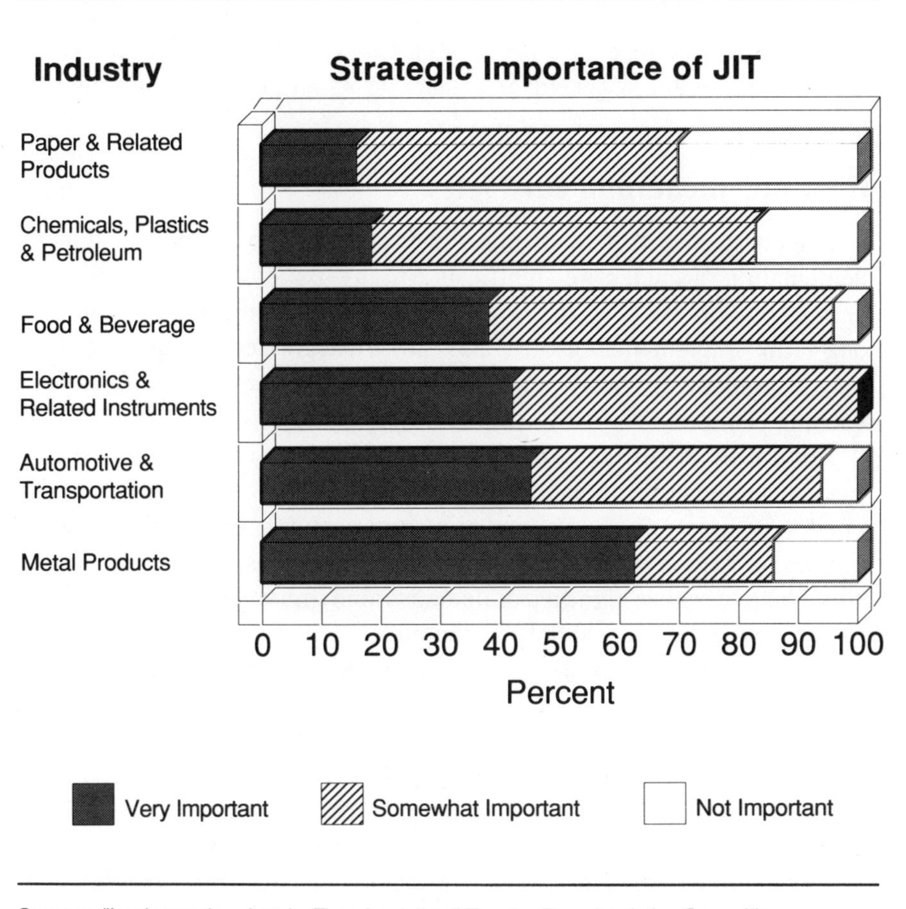

Source: "Implementing Just-In-Time Logistics," Touche Ross Logistics Consulting Services, 1988 National Survey on Progress, Obstacles and Results.

Additionally, process industries still use cyclical production runs to build inventory for peak periods. Note that later in this chapter, we show how JIT can be used with a more standard production planning and control technique: Material Requirements/Resource Planning (MRP).

Some companies have become discouraged by a concept that "looked too good to be true." Other companies have found that the structure and operation of their companies do not allow full JIT implementation. For example, they may not be able to change their culture on a large scale or gain a commitment to the philosophy from their entire organization.

While many companies indicate they are not using JIT inventory techniques, they are nevertheless pursuing the JIT concepts, including cross-training, reducing lead times, and automating communications with customers that make sense in their business. This indicates a general agreement with the necessity to improve operations and at least a partial agreement with the concept of JIT, even though a general reluctance to adopt JIT as a philosophy of operation continues.

JIT Implementation Obstacles

For companies that overcome their reluctance to adopt JIT, the implementation obstacles are significant. Many are attributable to management problems and management buy-in and commitment over the long-term. Implementation obstacles can be categorized as requiring a long time to implement and demonstrate results or requiring significant far-reaching changes.

Long time required to implement and demonstrate results. A professor at a university in Japan reported recently on Japanese companies that failed to complete JIT implementations successfully. One of the principal reasons for failure was that top management underestimated how long it would take to realize significant, demonstrable improvements to the financials. Top management simply lost interest.

The "quick fix" mentality of U.S. managers in particular makes JIT a difficult technique to apply. Results are unlikely to be seen on next quarter's financial statements. A survey of companies implementing JIT indicates that only 11 percent have implemented JIT into all areas of the company. Of those, 59 percent are six months to three years away from full implementation, while 26 percent are more than three years away from complete implementation.[18] While results should start to be visible within the first year, it must be recognized that it took Toyota Motor Company 30 years to get to where it is today.

Significant far-reaching changes required. Long time frames required for complete implementation of JIT are daunting, but the *magnitude* of change required to implement JIT in most organizations can be overwhelming. JIT affects all levels of the organization, including supplier relationships. In fact, the implementation of JIT usually causes factories to become focused, as described in Chapter 8.

JIT is not a program that can be layered onto an existing organization. It must be applied from the ground up, changing the way the processes of purchasing and manufacturing are managed. Common changes are:

- Most businesses must undergo a major cultural change as the fundamental philosophies and business ground rules used for decades are challenged by JIT principles.
- JIT requires a high level of cooperation and interdependence, both between departments and between companies.
- Systems support, often in the form of real-time information systems supporting inventory management, order entry, purchasing, and order tracking, must be obtained.
- JIT requires a certain level of environmental stability. It also requires manufacturing process stability; e.g., processes are capable of performing to specifications and are in a state of statistical control, and the product and process designs are robust. Fundamental to JIT's successful implementation is guaranteed conformance quality, every time. This implies that JIT manufacturers and their suppliers are adhering rigorously to the statistical quality control standards.
- Manufacturing must often struggle with manufacturing capabilities and production flow. Many facilities were designed around economies of scale to achieve high volume relying on large WIP inventory, rather than around economies of scope, which achieve superior quality and volume and product flexibility along with a small WIP inventory.
- Purchased goods quality often must be dramatically improved. In many instances, each part is bought from a single supplier, in exactly the right quantities at the time. JIT assembly operations require perfect incoming quality, or the line stops. OEMs often find they must train their vendors in JIT and specific quality control programs. These are often a major time and resource constraint for JIT implementation.

- Quality must be engineered into the production process. This requires increased engineering time as well as a large amount of time to train workers and identify and troubleshoot defects as they arise.
- Reducing setup time and cost to permit the running of smaller batch sizes requires an investment in engineering and retraining time for production personnel who must be organized around day-to-day activities.

A wide array of obstacles bars the successful implementation of JIT. *The major goal in JIT implementation, however, is to gain and maintain top management support for the long period it takes to make the transition to JIT.* If the obstacles of support, level of impact, and time can be overcome, then JIT is both an achievable and beneficial manufacturing practice.

Often a company's first encounter with JIT is in the purchasing area. Companies are either trying to reduce their raw material inventory levels and improve vendor lead-time performance, or their customers are demanding that they perform as JIT suppliers. This may not be the best approach.

Even with measurable benefits, JIT inventory techniques are not for everyone. Some environments are not conducive to the JIT techniques described, and attempting across-the-board JIT implementations in such environments leads to poorer performance. The answer for these complex environments may be MRP or some form of an MRP/JIT hybrid system.

Material Requirements/Resources Planning

Material Requirements Planning (MRP) and Manufacturing Resource Planning (MRP II) are computerized manufacturing control systems that were originally developed by Joseph Orlicky in the early 1960s.

Promoted and spread through the nation by the likes of Oliver Wight and George Plossl, MRP has become the single most widely-implemented production planning and control technique in use in the U.S. today. The MRP process starts when items on the bills of material are "exploded," or multiplied, by a master schedule to determine material requirements on a time-phased basis. Next, stock on hand and material on order are subtracted from the exploded requirements to determine what else should be ordered. Although MRP logic is straightforward, a computer is required to manipulate the data. *MRP is a push system, meaning items are produced at the time required by a predetermined schedule.*

MRP II, like JIT, is much more than a system. It is a way to introduce an integrated style of management into a company. In fact, some view the integration of materials management and production control as a prerequisite to successful implementation. MRP II focuses all areas of operation on the same set of numbers to obtain synergy in operations. This synergy leads to the identification of corporate goals and an opportunity to translate those goals into individual targets.

While MRP refers only to the systems aspects of production planning and control, an MRP II system can bring a company to a stage where it has control over all plantwide activities. Most literature today refers to MRP and MRP II concepts interchangeably. In the discussion in this chapter, MRP means MRP II techniques.

MRP's Fall from Grace

MRP started its fall from grace in the eyes of U.S. managers when it became clear that many MRP implementations ended in failure. It hit the ground when the Japanese invaded the U.S. marketplace in the 1980s. U.S. manufacturers were unable to combat the responsiveness and flexibility of the Japanese JIT systems with MRP explosions and the huge investments in the mainframe computers required to crank out MRP answers.

Over the last 25 years, MRP implementations have suffered from a number of problems. The three most common are described below.

- **Length of time and magnitude of effort to implement**. Like JIT, the length of time required to implement a complete MRP system is significant. Experts agree that 18 months is short; 24 to 36 months is not unusual. In addition, the level of effort required to implement the system is extraordinary. Clean inventory records, clean routings, and clean bills of materials are mandatory. These efforts alone can require many man-years of effort and significant elapsed times. This is only the beginning. Training and education in the fundamentals of MRP are essential for all employees. Project teams become armies that include everyone from computer programmers to shop personnel. It is not uncommon to expend 20-, 30-, or 40-man-years of effort on the implementation process. This is why many companies start with a vision of MRP for the whole organization and end up using it only on the shop floor.
- **Cost to implement and technical requirements**. Today's MRP systems can be extremely expensive. The cost of the large computers and the database and application software required to implement

MRP can easily exceed $2 million in large installations. This is without considering the cost of the manpower required to implement the system. Software for performing distributed processing is not even available. While today's microcomputer-based MRP systems are feasible and far more attractive from a cost perspective, when the dollars are added up, the cost is still substantial. This cost aspect, coupled with the length of time and level of effort required to implement MRP, causes many companies to avoid it and others to abort their implementations mid-stream.

- **Maintenance discipline required.** Those manufacturers that get the MRP system up and running find MRP a demanding and unforgiving mate. Exception messages must be responded to. Record accuracy must be maintained. Master schedules must be reflective of actual plans and adhered to. Computer schedules must be met. Failure to follow the rules results in computer-generated confetti. For many manufacturers, the rigor required to use the system properly has proved too much to handle.

Despite these implementation obstacles, MRP continues to be an important method for planning, controlling, and executing production in complex environments. The reality is that manufacturers have no choice about which system they should use; their environment makes the determination for them.

The Manufacturing Environment Dictates the Use of MRP and JIT

Should MRP and JIT be used together in the manufacturing environment? The strength of MRP is planning; the strength of JIT is execution and quality improvement. Many companies believe that they must use either MRP or JIT, exclusively, without realizing that, together, the two approaches are more powerful. The manufacturing environment itself will dictate the degree to which these techniques can and should be integrated.

Figure 7-4 shows the environments most appropriate for MRP and JIT. The critical characteristics that determine which system should be used include:

- The variability of the lead times
- The stability of the production rates
- The production volume
- The number of unique parts being produced

- The engineered content of the products
- The complexity of the bills of materials

Complicated Custom-engineered Environments. MRP, with its sophisticated logic and ability to apply significant computer resources, provides superior performance in these job shops. The complexity of the bills of materials makes MRP logic essential to time phase the exploded part requirements, while the variability of lead times makes shop floor execution difficult without constant feedback regarding order status. MRP is well-suited to this environment. JIT's simple view of the world will be overwhelmed and result in material shortages and large buildups of WIP, poor

FIGURE 7-4 MRP vs. JIT: Environment Dictates Solution

Environment Characteristics	Job Shop Custom-Engineered	Batch Production	Repetitive Continuous Flow
Lead Times	Highly Variable	Moderately Variable	Constant
Production Rates	Variable	Stable	Constant
Production Volumes	Low	Moderate to High	High
Number of Products	High	Moderate to Low	Low
Engineered Content	High	Moderate	Low
Bills of Material	Deep/Complex	Simple to Complex	Simple/Flat
PLANNING SYSTEM	MRP	MRP	JIT
CONTROL SYSTEM	MRP	MRP or JIT	JIT
EXECUTION SYSTEM	MRP	MRP or JIT	JIT

Source: Adapted from Karmarkar Uday, "Getting Control of Just-In-Time," *Harvard Business Review,* September-October 1989.

schedule performance, and an overall inability to manage the production environment.

Batch Production Environments. MRP and JIT can be combined in some batch production environments. MRP handles the planning and order control and release functions. JIT handles the shop floor execution. The key to using either MRP or JIT is knowing what the manufacturing environment requires from the planning, control, and execution perspectives.

Repetitive, Continuous Flow Environments. In those environments with simple, flat bills of materials and stable lead times and production rates, JIT can be used for the planning, control, and shop floor execution. Application of MRP concepts in this simple and steady environment should be restricted to longer range material requirements, if used at all. MRP in this environment will perform poorly, always lagging several steps behind the actual production situation. In this type of environment, MRP will be unable to keep up with the volume of production transactions and the overall pace of the environment.

Adapting MRP and JIT to Work Together

Some firms are using both MRP and JIT. For example, the Foxboro Company is using MRP for planning and understanding material requirements and JIT for controlling the execution of the shop floor. Kanbans are used to pull material through the shop. For the two approaches to operate concurrently, some changes in MRP operation must be made to accommodate JIT.

- JIT tends to flatten bills of material because it eliminates levels of subassembly. On some MRP systems, a *pseudo,* or phantom, item capability may be required to service engineering and production costing needs adequately.
- JIT can increase the need for *planning* bills of materials. JIT encourages companies to design their own product structures so that a small number of materials and parts can be purchased, fabricated, and assembled into components.
- The MRP time-phased planning of a JIT product should use a zero lead time. Since the JIT product moves through the shop at such a rapid rate, zero lead times are essential for MRP to have any chance of determining when materials will be needed.

- The order release function for JIT-controlled items must be disabled. Other activities associated with releasing orders may also need to be disabled.
- Relief of inventory in JIT is usually accomplished by back flushing or post deducting inventory withdrawals based on the quantity of product made, rather than by making transaction-based withdrawals for each component used.
- Since JIT emphasizes line stocking at point of use, MRP must have the ability to handle multiple stocking locations for an item to keep track of perpetual inventories.
- In JIT operations machines from several different work centers are often grouped to form a manufacturing cell (the cell should be loaded so that its capacity is equal to the demand placed on it). This requires a change in capacity requirements planning logic.
- The need for manufacturing routings in JIT is greatly reduced or eliminated. However, most companies feel it necessary to maintain routings in some form as a reference.
- In a JIT order control and execution environment, work orders are eliminated. Kanbans are used to move orders and material through the shop. Labor and material are accounted for in a simplified fashion.

In the future, the battle between MRP and JIT will subside, and world-class companies will utilize the best method available for their manufacturing environment. The best method available for most manufacturers will be some MRP/JIT hybrid for the planning, control, and execution of the shop. However, at the root of the successful application of either technique are simplified process and improved quality.

Product and Process Simplification with Value Analysis/Value Engineering

Value Analysis (VA) originated in 1947 under the leadership of L.D. Miles of the General Electric Company. In 1954, the U.S. Navy's Bureau of Ships instituted the first Value Engineering (VE) program in the military forces. It was adopted in 1955 by the U.S. Army. Value Analysis and Value

Engineering are used together (VA/VE) to reduce the complexity associated with product design, which is a critical element of improving quality.

Currently, many companies in the U.S. are developing or renewing their interest in VA/VE in response to competitive pressures. However, overall adoption rates in the U.S. continue to be low; no more than 25 percent of the companies responding to a recent survey indicated they were using VA/VE. The direct benefits of applying VA/VE can be very impressive:

- Phillips Industries in Dayton, Ohio, a $1 billion a year producer of components supplied to the recreational vehicle, manufactured housing, and transportation equipment industries, has achieved more than $62 million in savings using VA/VE since 1981. More than 1500 employees have participated in VA programs, and every dollar spent on VA activities returns $10.30 in savings.
- Textron Lycoming, producer of the Army's M1 AGT 1500 gas turbine engine, has saved more than $37 million in just two years by working with its suppliers to apply VA to its purchased parts.
- Delco Remy held its first VA workshop in 1980. Since then, the Value Analysis staff has saved more than $11 million.
- Hitachi, the electronics giant in Japan, saved 5 percent, or $500 million, of the company's $10 billion operations costs in just one year through the application of VA/VE.

Although Value Analysis and Value Engineering have been in limited use for over 30 years in the U.S., they have been used quite extensively in Japan, where they have found wide acceptance. The Japanese think so highly of VA/VE that, in 1983, the Society of Japanese Value Engineers developed the Miles Award to be given annually to companies that show the most benefit from the use of VA methodology. Further, 17 of the top 20 Japanese industrial firms have VA/VE vice presidents.

Value Analysis and Value Engineering Defined

VA/VE can be defined as "an organized, systematic study of the function of a material component, product, or service with the objective of yielding value improvement through the ability to accomplish the desired function at the lowest cost without degradation of quality."[19] It is an organized, systematic method of cost reduction designed to counter the profit squeeze brought on by increasing labor and materials costs and intensifying market competition.

The objective of VA/VE is to achieve equivalent or better performance at a lower cost while maintaining all functional and quality requirements defined by customers through market research or quality function deployment. This is done largely by identifying and eliminating unnecessary costs. VA/VE is concerned with the total value and costs involved—from conception to retirement of a product.

How does this work? VA analyzes existing products and the specifications and requirements shown in production documents and purchase requests. It is used as a cost-reduction technique. VE, while similar in concept to VA, applies to the research and development, design, and proposal stages of the product. VE occurs *before* the production stage and, therefore, is considered a cost-avoidance or cost-prevention technique.

While, technically, VE is applied in the design stage and VA in the production or procurement stage, the line between them is often blurred. If VE were properly applied in the concept and design stages of each product, VA could be eliminated. In practice, however, this does not often happen. New developments in materials, processes, product applications, or marketing volume require the application of VA techniques to products previously treated by VE in the design stage. Therefore, both VE and VA applications are carried on concurrently within most organizations. Both techniques can be described as disciplines for the removal of unnecessary costs without impairing desired function and quality of the product.

Implementing Value Analysis and Value Engineering

The implementation of a VA/VE process can have any number of steps. Westinghouse implemented VA/VE using a five-phase problem-solving model:

1. **Information Phase:** Study performance parameters and information about manufacturing methods, quality, objectives for the product, and related areas.

2. **Analysis Phase:** Allocate costs to functions, conduct a Pareto analysis to separate the vital few from the trivial many considerations, define functions, analyze costs, and identify problem/opportunity areas.

3. **Creativity Phase:** Write down ideas for or brainstorm every opportunity and problem area. Seek a large number and wide variety of ideas; do not judge the value of the ideas.

4. **Evaluation Phase:** Select ideas to be developed, considering the impact on cost, performance, and quality. Determine the approaches

which should be used to maximize results given the available resources. Compare the approaches to the objectives for the process. If necessary, repeat the creativity phase.

5. **Implementation Phase:** Present proposals to management, develop an implementation plan, and obtain the necessary commitment of resources.[20]

The VA/VE process brings many techniques together. VA is not just a cost-reduction tool or a quality-improvement process; it is an organized process intended to accomplish what is desired for a minimum expenditure of resources. Few issues surround the use of VA/VE, other than to make sure the techniques are used for their intended purpose, rather than just as a cost-reduction exercise.

Many American product design engineers view VA/VE very narrowly, as nothing more than an effort to cheapen the product. Therefore, it is necessary to educate employees and communicate regularly to prevent this attitude from developing. VA/VE should be used to increase value to customers.

VA/VE offers so much promise at so little risk that it is hard to imagine that it will not become a critical component of the best manufacturing practices of world-class companies. Appropriate use and realistic expectations, combined with honest assessments of the value returned, are the only considerations for any company implementing VA/VE.

Maintaining the Production Machine

Maintenance started as a simple function charged with repairing broken equipment. It evolved over time to become the responsibility of specialized professionals segregated from the production department. The purpose of maintenance management is to make equipment and facilities available when production needs them. However, a balance must be achieved between machine availability and overall maintenance costs. Most organizations have great difficulty in achieving this balance and, as a result, are dissatisfied with their return on investment.

Maintenance management may well be the biggest challenge facing companies today, especially those attempting to implement TQM and, especially, computer-integrated manufacturing (CIM) and JIT. In a CIM or JIT environment, unplanned downtime is not only more visible, but significantly more costly as well. As companies have aggressively pursued both

JIT and CIM over the past decade, the problems associated with the maintenance function have become far more obvious.

The maintenance function in most organizations has not kept pace with changes in manufacturing theory, philosophy, and technology. Few companies have seriously worked to establish good maintenance programs; maintenance has always tended to fall to the bottom of the management concern list. Even in companies that have put formal maintenance management programs in place, serious management attention has rarely been devoted to this area. Fifty percent of companies recently surveyed had some sort of maintenance management program in place, but only 5 percent felt their programs were working.[21] There is confusion over the goals and objectives of maintenance management as well as how to achieve them.

Maintenance Management Development Stages

The management of the plant maintenance function can be broken down into five stages of development:

- **Stage 1**: Informal system (when it breaks, fix it).
- **Stage 2**: Semi-formal system (some manual systems).
- **Stage 3**: Formal system (computerized maintenance control system).
- **Stage 4**: Formal system plus predictive maintenance programs.
- **Stage 5**: Formal system plus predictive expert systems programs.[22]

With the emergence of new manufacturing techniques and technologies and the emphasis on plant automation, maintenance management is moving from a *reactionary* role (Stage 1) to a *preventive* or *predictive* role (Stage 5).

Preventive Maintenance (PM), as opposed to breakdown repair, "consists of periodic inspection or checking of existing facilities to uncover conditions leading to production breakdowns or harmful depreciation, and the correction of these conditions while they are still in the minor stage."[23] The intent is to move maintenance operations from being unplanned, reactive work to include a high percentage of anticipatory work.

Artificial intelligence systems, especially expert systems, are providing solutions when conventional maintenance management software cannot diagnose equipment or direct repair procedures. An expert system is different from a step-by-step procedure. It follows a logical set of general procedures for finding solutions to maintenance problems, usually assisting a trained operator.

Total Productive Maintenance

Total Productive Maintenance (TPM) is viewed as a sort of Total Quality Control methodology for maintenance. TPM was created by GE in the 1950s and exported to Japan where it flourished as part of the overall JIT movement. Under TPM, production employees are responsible for their own equipment and perform regular lubrication and minor repairs. They also detect potential problems before they result in major equipment failures.

TPM is meant to attack the six major losses in manufacturing:

- Equipment failure
- Setup and adjustment time
- Idling and minor stoppages
- Reduced speed
- Defects in process
- Reduced yield

In 1971, the Japan Institute for Plant Maintenance (formerly the Japan Institute of Plant Engineers) established five goals for TPM:

- Maximize equipment effectiveness and improve overall efficiency.
- Develop a system of productive maintenance for the life of the equipment.
- Involve all departments that plan, design, use, or maintain equipment in the TPM program.
- Involve all employees, from shop workers to top management, in the program.
- Actively promote TPM through small and autonomous group activities.

Successful TPM programs can yield tremendous benefits:

- At the Nissan Tochigi auto plant in Japan, equipment breakdowns have dropped by 80 percent, and the product reject rate has fallen from 0.6 percent to 0.1 percent. The plant also claims a 20 percent man-hour reduction, which includes reducing the number of maintenance personnel from 800 to 700.
- In another Japanese plant affiliated with Toyota, machine failures have been reduced from 5000 per month to 50 per month. Rework on products has been reduced from 0.45 percent of products to 0.15 percent of products.

- The Japanese Dai Nippon plant, the world's largest printing plant, spent $2.1 million implementing TPM and saved $5.5 million within three years. They broke even on the investment in six months.[24]

A key component of successful TPM programs is continual training. For operators to maintain their own equipment, they must know how. Education is a long process, taking years for some operators and some of the more advanced equipment. Existing plant maintenance departments should play a role in the training process. The expertise regarding the equipment has always rested there.

Active, on-going training programs should be set up for the operators. Management must work hard to dispel the notion that TPM is simply another way to eliminate skilled craftspeople. Over time, a number of skilled craftspeople may be reduced through attrition and as other plant openings become available, but TPM should never be introduced as a way of "getting rid of the trades people" or "forcing the operators to do more." This is the best way to ensure failure of the program.

Conclusion

The fundamentals of successful manufacturing in the future will consist of the best manufacturing practices outlined in this chapter. The practices will look different from company to company, and industry to industry, as competitive variables change. Adherence to any singular practice will not result in success over the long term.

It is hard to imagine a manufacturer that applies JIT without innovative human resource practices, superior equipment maintenance, or outstanding quality. Likewise, superior product quality alone will not yield long-term success if material and production are poorly planned, resulting in long lead times and missed schedule dates.

These key action programs are part of a holistic manufacturing strategy. Taken together, they form a set of mutually reinforcing manufacturing choices that are being driven by the changing competitive environments. Time and flexibility are becoming critical dimensions of competition, with quality at the foundation.

The current alphabet soup of best manufacturing practices, beginning with the TQM and the quality revolution, and, followed by JIT, VA/VE, and TPM, are perhaps the manifestation of the changing basis for competing. Compressing time and space while simultaneously increasing quality and lowering costs are the real benefits of the combined application of best manufacturing practices.

Case-by-case examples show that the firms that successfully execute these practices as an integrated whole are likely to derive multiple benefits: customer satisfaction, asset productivity, and speed. When coupled with innovative products and keen insight into customer requirements, they form the basis for long-term overall business success. In the next chapter, we describe the design of the new manufacturing organization, the cornerstone of manufacturing.

CHAPTER 8

RESTRUCTURING THE AMERICAN WORKPLACE

Restructuring is usually a corporation's response to some sort of shock or a reflection of its aging process. The corporation views it as a route to better performance. During the 1980s, restructuring had a profound effect on corporate America. Increasing numbers of companies changed the shape and operation of their organizations in response to increased competition, mergers and acquisitions, and the injection of advanced technology. Even the corporate elite were significantly affected. No less than 56 percent of the Fortune 500 industrial companies embarked on a slimming down process.[1] American manufacturing is spinning toward decentralization of authority. Consider the experiences of these companies:

Ford Motor Company. Ford's major restructuring in the 1980s had as much to do with its recent significant financial success as did the introduction of the Sable and Taurus models. Ford cut back its workforce by over 30 percent, saving over $5 billion dollars a year. The 10 percent of total cost represented by this cutback lowered Ford's breakeven point by 40 percent in terms of the number of automobiles it had to sell. Fifteen manufacturing facilities were closed. Those that remained were further restructured internally to utilize JIT manufacturing techniques and total quality management approaches. The result: Ford's stock skyrocketed and, for the first time in this generation, Ford became more profitable than General Motors.

Eastman Kodak. Kodak's restructuring, after several failed attempts in the early 1980s, included a cutback in both personnel and products. The staff cutbacks were targeted at 10 percent, while more than 10,000 products were eliminated from product lines. Kodak also organized into 24 strategic business units, focused around product lines, departing from a long history of being functionally organized.

Ford and Kodak restructured voluntarily. Other companies have been forced into it, restructuring through declaring Chapter 11 or just by getting close to bankruptcy.

Navistar International (International Harvester). From 1980 through 1983, International Harvester lost over $2.9 billion. This catastrophe followed the company's best year in history. The International Harvester that emerged from restructuring, renamed Navistar International, cut its workforce by more than 80,000 people and trimmed its manufacturing plants from 48 worldwide to six located only in the U.S. and Canada.

Still other companies restructure to avert disasters not yet upon them.

Baxter. For Baxter Healthcare, restructuring has been almost continuous from 1985 through 1990. Baxter's restructuring problems stem from its hostile takeover of American Hospital Supply Corporation. Within a year, Baxter Chairman Vernon R. Loucks, Jr., announced plans to combine the two companies' operating units, cutting 6000 jobs. The latest reshuffling, in January 1990, resulted in the abandonment of low-growth and low-margin businesses. Loucks argues that Baxter's restructurings are decisive responses to problems before they become severe.

Restructuring of a company is not a one-time event; it is a never-ending process. Cutting costs, selling marginal businesses, and buying back shares of stock can make a company look good for a while. But companies must reinvest money in labor-saving machinery, in maintenance of a strong R&D effort, and in looking for opportunities to grow through new products and markets to ensure long-term health.

Companies do not restructure just for the sake of reorganizing. The current drive for efficiency appears to be a "do it or else" proposition. Employees not psychologically attuned to restructuring efforts do not stay around, no matter how loyal they have been. Companies that have resisted restructuring have either disappeared or been taken over by corporate raiders. It is likely that restructuring efforts will continue into the 1990s as more companies are forced to become more efficient from both a financial and an operating perspective.

Ford, Kodak, Navistar, and Baxter are examples of companies that met the competitive challenges they faced with significant restructuring. Many other American companies are following the same approach, reinventing themselves rather than becoming extinct. Still other companies are targeting their restructuring efforts on specific pockets of employees, such as the corporate staff.

The Shrinking Role of Headquarters Staff

Corporate staff, those individuals at headquarters whose role it is to support the line operations are disappearing in many organizations. Downsizing the corporate staff is an increasingly popular route to greater efficiency and is changing traditional organizational designs. Many companies are eliminating their collections of headquarters planners, economists, marketers, central purchasing agents, real estate managers, human resource specialists, futurists, and deep thinkers.

When Lawrence G. Rawl assumed the positions of chairman and chief executive officer of Exxon in the mid-1980s, he proceeded to downsize the company significantly. Rawl cleared out many layers of management, consolidated Exxon's worldwide operations, and trimmed his workforce by 30 percent in a two-year period. He studied organization charts of IBM, Royal Dutch/Shell, and British Petroleum and concluded that Exxon had far too many headquarters staff for what was being accomplished. In addition, he found that the system was producing reams of paper and volumes of information—far more than any individual or group of executives could effectively use. Talented people were being kept busy generating the paper required to feed the system. Rawl changed that by eliminating the people and processes that required the reams of paper and volumes of information.

Is Corporate Downsizing a Good Idea?

The value of corporate staff downsizing, other than cost reduction, is that decisions are made faster and managers are closer to their markets. The disadvantage is that the corporate staff is often the glue that holds the corporation together, and arbitrary cuts can do more harm than executives can assess. Further, it can take some time before the negative impacts of the corporate staff cuts are visible. Corporate staff downsizing should be viewed as a way of making the organization more effective, not as a way of saving money.

Replace the concept of downsizing with the idea of rightsizing. While achieving the *right size* may be the appropriate goal, the concept of *rightsize* often becomes overwhelmed by the downsizing effort itself. Many companies that aggressively pursued downsizing efforts in the 1980s found this to be true.

For example, Exxon's downsizing efforts resulted in a dramatic change in culture and contributed to widespread morale problems initially. However, Exxon also became far more profitable through its efforts, achieving profitability per employee levels unmatched by any other major U.S. company.

While Exxon experienced significant negative publicity over the staff reductions and reorganization, its downsizing efforts allowed it to become more focused on its markets and more effective in its operations; managers spend less time in meetings, decisions are made quicker, and executives spend more time in the field.

Depending on Outside Service Organizations

In the past, large corporations created value by recruiting and training talent and providing capital to invest in operations. Today, however, sophisticated headhunters, investment bankers, and consultants can do just about anything a corporation needs—without the expense of maintaining a full-time staff.

Some corporations believe that value is created only in their operating units and that the role of a corporation is to act as a demanding and omniscient parent, but not one that tries to do everything the operating units need. American Standard Corporation operates within this philosophy. With a very small corporate staff, American Standard views its operating units as being self-sufficient. The corporate executive staff sets the tone, establishes the overall objectives, and then counsels the operating units.

Some corporations are allowing operating managers to decide what corporate services (out of their own budgets) they are willing to pay for. Another trend involves thinking of corporate staff as a small merchant bank or holding company which invests capital among various enterprises, monitors profitability, replaces underachieving top managers, and looks for new investment opportunities.[2]

The Many Shapes of Restructuring

The fascination of U.S. manufacturers with restructuring, whether brought on by the merger and acquisition craze of the last decade or the simple lack of competitiveness, will surely continue. Too many manufacturers have experienced too much success with restructuring for it not to be a common theme into the twenty-first century. Restructuring takes many shapes.

Strategic Business Adjustments

Much of the restructuring activity today is aimed at developing a more focused strategic business unit (SBU) concept. In a strategic business unit, management and labor efforts are directed toward a narrower purpose that is in line with the dynamic market conditions affecting specific products. Many companies facing significant competitive pressure in the 1980s reorganized into SBUs to develop business strategies, as described in Chapter 3.

For example, AT&T, long organized by function, found itself losing market share in the long distance market in 1988. It also was going nowhere with its computer business, as its joint venture with Olivetti began to fall apart. After a careful introspective examination, AT&T executives and new chairman and CEO Bob Allen concluded that AT&T needed to become more aggressive, more focused, and move more quickly in reacting to the dynamic market conditions characteristic of the global markets AT&T had chosen to compete in. AT&T reorganized into 19 strategic business units organized around product lines. Each business unit has responsibility for its own profit and loss, pricing, marketing, and product development. AT&T chose the SBU approach to bring its businesses closer to its customers and to create a more responsive customer-driven organization.

Motorola has created a separate business unit for the Japanese market, which oversees sales and gathers market intelligence. To compete effectively with Japan, Motorola has consistently paid special attention to selling in the Japanese market and provides each of Motorola's business units with key data.

Often, the business unit accomplishes its mission by focusing its manufacturing organization in a way that is structurally consistent with that of the business.

Focused Manufacturing

Focused manufacturing organizations bring the business closer to its customers and harness the power of the organization by concentrating resources on the manufacturing task required for market success. At the plant level, *focused factories* have consistently outperformed unfocused, conventionally laid-out manufacturing facilities.[3] This happens in part because equipment, supporting systems, and procedures can be targeted on a single set of tasks for a particular market niche or on a particular production process technology, instead of attempting to satisfy a broader set of goals. The strategic implications of focused factories were discussed in Chapter 3.

At a broader level, the manufacturing organization, itself, must be structured relative to the competitive priorities of the business unit.[4] Which should be the focus, the product or the process? The process-focused organization is one which is usually built around a specialized production technology. The emphasis is on applying technical expertise to processes or materials. Process focus is commonly applied where scale of economies are required and/or in continuous flow manufacturing such as chemicals or paper. Some process-focused manufacturing organizations have more than one plant, each one with a dominant process that adds value sequentially to the product. In environments where flexibility and innovation are more important than tight control and extremely high plant utilization, product focus has the edge. Product-focused facilities allow close coordination between plant activities and customer and market requirements.

Because the movement toward customer-driven strategy and decentralization has coincided with developments in flexible process and advanced communications technology, there has been a shift toward focused manufacturing. More specifically, a focus around product lines to improve quality and organizational effectiveness is being observed. Some examples of the shift toward product focus are:

- **Mead Corporation** organizes its massive facilities around product lines, rather than similar manufacturing processes, to better serve the customer. Plants are focused around different paper product characteristics such as commodity grade paper products, premium-coated paper products, carbonless copy paper, and so on. Product focus is a key element of being close to customers.
- **Eaton's Cutler-Hammer** operations have a long history of stripping product lines out into focused factories to improve quality and operating efficiencies. When specific lines of motor control products become commercially viable on a large scale, the products are removed from their place of origin, usually the Milwaukee, Wisconsin, headquarters, to new plants dedicated to their manufacture.
- **The Foxboro Company** developed a completely new production facility using JIT manufacturing techniques to introduce its *Intelligent Automation* series of process control equipment. The competitive advantage offered by the new product line was enhanced by developing a new plant dedicated to its manufacture. The new plant uses JIT production control techniques and new methods of workforce management. It allows Foxboro to cut lead times, improve quality, and focus on customer requirements. On a broader scale, the Foxboro Company is reevaluating existing product lines to define "factories within factories" and focus existing resources.

- **Pratt & Whitney's** operations in East Hartford, Connecticut, exemplify the way in which larger facilities apply the concept of focused factories around product lines and component families using a factory-within-a-factory approach. While Pratt & Whitney's engine facility in East Hartford houses more than 20,000 people under one roof, 200-person manufacturing units are being implemented. The actual size depends on technical complexity. Often these focused units have physical walls put up around them to control the environment. Pratt & Whitney attempts to make product decisions as low in the organization as possible. The goal: to leverage the technical expertise of the employees.

Regardless of the approach, product or process, focusing the manufacturing organization is the key. Reducing complexity and driving out confusion is paramount in an effective manufacturing organization. *Many organizations have found that focus dramatically enhances their ability to improve the manufacturing and engineering processes concurrently.* When all employees in the focused organization are concentrating on the same product lines, the process improves naturally.

The focused organization concept is enhanced by close integration with suppliers. Consequently, component and equipment suppliers are being asked to sign up and work as partners. The move to integrate suppliers into a company's operations is yet another step being taken by world-class companies.

Vertical Control through Supplier Integration

The type of relationship a manufacturer has with its suppliers determines whether *vertical control* or *vertical integration* is the best approach to the management of incoming parts and material. With vertical control, there is no formal ownership of the supplier, but the relationship is more than a loose coupling bounded by a single contract. It implies a "hold" on the supplier through long-term quality management programs and other collaborative arrangements like those adopted for Just-in-Time manufacturing. Vertical integration is defined in the traditional sense by corporate ownership of the supplier.

Vertical control is an increasingly important component of a manufacturing strategy due to a number of factors:

- Vertical control allows the company to focus on its own core production processes and products.
- Vertical integration, an approach whereby a company extends in-house production capabilities to lower and lower product levels through internal development or acquisition, is becoming obsolete.

- Vertical control allows a company to extend in-house production capabilities to lower product component levels through the establishment of very strong relationships with outside suppliers.

All major companies practicing JIT have formed strong supplier relationships or implemented supplier integration and are exercising ever-increasing control over their suppliers. This increasing dependence on outside suppliers reverses a trend that has been in evidence for almost a century. Reasons for this reversal are:

- Companies are finding it cheaper to outsource than to own all stages of the production process. Outsourcing can provide immediate cost reductions without the significant capital investment required to completely reinvent its own production processes and products.
- The increasing level of international competition makes traditional vertical integration difficult to achieve on a global scale.
- Many companies have found that the application of vertical integration can lead to a rigid cost structure – not a lowest cost alternative. They become locked into the overhead structure required to support the vertically integrated operations and have difficulty responding quickly when rapid changes in cost structure are required.[5]

General Motors' *Targets for Excellence* program, which qualifies suppliers through detailed on-site reviews by General Motors-sponsored review teams, is a prime example of the techniques major companies are using to increase their vertical control. In fact, all of the Big Three have similar programs: Ford's *Q1* program and Chrysler's *Quality Excellence* and *Pentastar* awards. Other companies, such as Eaton and Honda, also have highly visible programs.

Traditional vertical integration is still being used by companies who are fine-tuning their operating strategies. For example, when IBM introduced its personal computer, it outsourced many components. Now that the product has become widely accepted and extensively redesigned, IBM is reverting to making its own components.

While vertical control of suppliers is becoming more viable as a way to ensure the quality and delivery required for world-class manufacturing, many other companies have turned to partnerships to exploit market and technologies. Joint ventures/strategic alliances and R&D consortia are surfacing to improve competitiveness.

Joint Ventures

Joint ventures and strategic alliances have become common ways for companies, that recognize the need and opportunity, to leverage their existing assets into new markets and technology development. The number of joint ventures between companies is increasing sharply. Joint ventures are viewed as ways of teaming up to share risk and extend capital. In the past, pride kept many companies from initiating joint ventures; they went forward alone. In the new, riskier business climate, however, U.S. companies have adopted joint ventures and alliances as the best new strategies for competing abroad and, in some cases, at home.

An alliance of West Germany's Daimler-Benz AG and Japan's Mitsubishi Corporation will capitalize on new technology and markets for the two businesses currently involved in aerospace, electronics, automotive, and services. The Daimler-Benz state-of-the-art assembly plant in Bremen will now be available to Mitsubishi. The companies will fit their technologies by tailoring their governance structures to meet the needs of the particular situation.

In the U.S., seemingly unrelated Ford Motor Company and International Business Machines Corporation have developed alliances for technology sharing. General Motors and Toyota have teamed up to build small cars in the U.S. General Motors has also teamed with Daewoo in Korea and Isuzu and Suzuki in Japan. Ford and Mazda have teamed up, while Chrysler and Mitsubishi continue their venture. Numerous firms, including Johnson & Johnson, Gillette, Heinz, Procter & Gamble, GE, Xerox, and Allied-Signal, have entered the South Korean and Chinese markets with joint ventures.

Guidelines for Successful Joint Ventures

Before determining that a joint venture is the right solution, managers must examine many critical elements. Guidelines for establishing successful joint ventures include:

- **The joint venture must help a company expand its competitive capabilities.** This expansion of capabilities must be critical to competition. Joint ventures should not be the primary vehicle for competing in foreign countries unless the venture is a necessity, because, for example, a company cannot exploit its competitive capabilities on its own. For AMC, the joint venture with Beijing

CASE STUDY

AMC's Disaster in China

Joint ventures are not without problems. The Chrysler/American Motors (AMC) and Beijing Automotive Works joint venture is a case in point. While different from most joint ventures, due to the close involvement of the Chinese government, it, nevertheless, is illustrative of the problems companies can have with joint ventures.

AMC held a 31 percent stake in Beijing Jeep Company, resulting from a $16 million investment—50 percent in cash and 50 percent in technology. The Chinese company, Beijing Automotive Works, agreed to invest $31 million in cash to control the other 69 percent of the company. The joint venture was to produce a new generation of Jeep vehicles for the Chinese and, eventually, for export. After disagreements over the management staffing, the management pay scales, and the design of the product (AMC wanted to produce its new Cherokee line while the Chinese wanted a new military Jeep vehicle), the Jeeps (AMC's new Cherokee) finally began to roll off the line in 1985.

The venture ran into problems when AMC found it could not get paid for the Jeep kits that were being assembled. Very quickly the venture went broke. The Chinese wanted to exercise total control over the venture, to the point of determining if and when they paid for part kits and vehicles sold. After several startups and shutdowns due to cash and control problems, the venture began to receive widespread coverage in the international media, forcing both sides into compromises. The Chinese realized they were in a fish bowl; the whole world was watching to see what Chinese business ventures were going to be like for Western companies considering investment. While the compromises resulted in clearer definition of the terms of payment and the number of vehicles that would be produced, the effort was scaled back considerably from what AMC had planned initially and was far less profitable. All U.S. personnel departed China when the Tienanmen Square incident occurred and the hard-line Communist Chinese took control of the country.

While the problems AMC experienced were unusual and the involvement of the Chinese government was significant, problems with joint ventures occur when one party expects too much control. Joint ventures allow for everything but control. They must grow out of mutual need, and both parties must benefit. Unfortunately for AMC, they did not have the basics of who benefits and at what price (control) worked out before they made their investments.

Adapted from Jim Mann's *Fortune* article, (November 6, 1989) entitled "One Company's China Debacle" and Mann's book *Beijing Jeep* (New York: Simon & Schuster, 1989).

Automotive was a necessity to compete in China. AMC had no other effective method of entry into China.

- **All companies involved must benefit.** Over-control leads to disaster. All parties must have equal say in the venture's working relationship. Mistrust and broken promises by any party lead to a slow and painful disintegration of the joint venture, as occurred with AMC-Beijing Automotive Works.
- **Details associated with each party's commitment must be clearly stated.** They should include: the exact structure of the operations and staffing, the level of profit expected and the method for transferring funds, and the level of product quality and customer service required. *They must agree upfront what each company is bringing to the table and how*. In this regard AMC failed miserably. The details associated with the venture in China were an afterthought. Once into the day-to-day operations, AMC found that its expectations were neither common nor realistic.
- **The host government's restrictions and expectations must be clearly understood.** Ownership restrictions must be recognized, and the companies must evaluate whether they can live within such restrictions. What the host government expects to get out of the joint venture and what it is willing to give up are critical elements that often determine the success or failure of the venture. Further, the long-term stability of the host government and its effect on a company's investment must be considered. AMC failed to understand the motivating force for the Chinese government; it wanted a new military vehicle. Nor did AMC properly assess the stability of the Chinese political climate and its potential impact on its venture.
- **Joint ventures should not be considered permanent.** The joint venture provides a company with an approach to solving an intermediate-term tactical problem. Managers must plan and be prepared for the day when the venture is no longer useful. They must work actively to close the capability gap that required the joint venture in the first place. Unfortunately for AMC/Chrysler, the venture did not last long enough to make worrying about planning for the future necessary.

Joint ventures are not a viable solution for all companies. A joint venture best suits a particular point in a company's evolution. Knowing when to utilize a joint venture, how long to maintain the relationship, and what realistic business objectives are for the venture are the keys to success.

In addition to the joint venture, many organizations have turned to R&D consortia to leverage their existing assets and to effectively extend the capabilities of their current organizations.

R&D Consortia

R&D consortia differ from joint ventures in one critical respect: R&D consortia include direct competitors, joint ventures do not. R&D consortia are more loosely coupled than are joint ventures.[6] In fact, the R&D consortium harks back to the nineteenth century, when U.S. companies within a given industry joined together to pursue their collective interests until the Sherman Antitrust Act of 1890 made such collaborations illegal.

The National Cooperative Research Act (NCRA) was passed by Congress in 1984, in direct response to increased competitive pressures from Japanese and European companies. This act allows companies to organize into consortia in which competing members of the same industry can perform joint research and development. The R&D consortium represents one of the newest, most promising, and, perhaps, most complex organizational forms companies can utilize in improving their competitive positions in the 1990s.

Since the passage of the NCRA, 137 consortia have filed with the Department of Justice. These consortia allow member companies, which are actually competing organizations, to invest capital, technology, facilities, and management time and talent. The member companies work together to define common agendas for directing their R&D efforts. These efforts can be focused on specific technologies and manufacturing processes or be more broadly based. For example, the National Center for Manufacturing Sciences (NCMS) is a consortium, and this book is part of its research and development effort.

The NCRA encourages cooperative R&D efforts at the precompetitive stage of production by limiting the antitrust exposure of consortia that file with the Department of Justice. Precompetitive research encompasses:

- Experimentation and study of phenomena and observable facts.
- Development and testing of engineering techniques.
- Development of prototypes and models.
- Collection and exchange of research information.

The U.S. government retains the right to adjudicate the legality of R&D consortia. Although consortia members can be found guilty of antitrust violations, by filing they protect themselves from paying treble damages. They can only be held liable for actual costs. In the five years since passage of the act, no consortium has been the subject of an antitrust lawsuit.

Advantages of Consortia

R&D consortia present tremendous opportunities to their participants. Among its benefits, an R&D consortium can:

- Help members make better informed decisions concerning research needs and priorities, thereby allowing them to leverage their financial and business resources to obtain a more strategic focus.
- Provide both the forum and the organizational structure within which to identify and allocate research projects and initiate shared research among members.
- Provide a cooperative arena for discussion of mutual concerns of members where antitrust laws previously impeded cooperation.
- Maintain a pipeline of information among members to alert them to government developments, critical industry news, and new information of concern to them, including state-of-the-art developments.
- Address issues that affect industry, such as anti-trust laws, tax laws, and technological piracy.
- Facilitate transfer of technology at all levels. This may include locating and making available to members research that resides on shelves in university, industry, and government laboratories across the country and from foreign countries.
- Organize workshops and demonstrations of new and existing technologies.
- Establish teaching factories or other laboratory facilities to provide demonstrations and training in new and existing technologies.
- Provide a library and research services which are accessible to all members.

Examples of R&D Consortia

Sizes and agendas of R&D consortia vary. Ninety of the existing consortia have between 3 and 25 members; 7 have more than 25 members.[7]

- **The National Center for Manufacturing Sciences (NCMS)** – the largest manufacturing consortium – is recognized as a catalyst in developing a partnership between government and the manufacturing community. NCMS is working at the national level to assist the Bush Administration, Congress, and other organizations in establishing a manufacturing agenda for the nation, including the development of policies and programs that will enhance the U.S. manufacturing base.

  The NCMS assists its more than 100 U.S. and Canadian member companies in their efforts to become world-class manufacturers. It includes very large companies such as General Motors, Ford, AT&T, Texas Instruments, United Technologies, and Digital Equipment as well as such smaller manufacturers as Kingsbury Corporation, Sheffield Machine Tool Company, and Manuflex.
- **The Plastics Recycling Foundation (PRF)** is developing techniques to recycle plastic bottles. It has 41 members including DuPont, Exxon, Coca-Cola, Pepsi Cola, Procter & Gamble, and Kraft.
- **The Microelectronic and Computer Technology Corporation (MCC)**, formed prior to the passage of the NCRA, includes leading electronics manufacturers conducting long-term research into areas such as artificial intelligence, CAD/CAM software, chip packaging, and superconductivity. The MCC has 19 members.
- **The Center of Advanced Television Studies (CATS)** is another consortium. CATS has 12 members attempting to keep abreast of developments, such as high definition television (HDTV), as opposed to directly conducting joint product R&D.

The majority of the consortia that have been formed are open only to U.S. firms, but several also welcome foreign companies. The NRCA does not prohibit foreign firms from participating.

Factors for Success

While R&D consortia present tremendous opportunities for U.S. firms, they are not without their challenges. For example:

- **Active participation is required.** To establish meaningful research agendas and effectively transfer technology, consortia managers have found that the active participation of members must be maintained. The more active the participation, the more momentum is gained, and the more all members benefit.

- **Demonstrated results are essential.** Annual member fees for R&D consortia can be substantial, but so can the results and payback for involvement. Consortium managers have found that establishing R&D agendas with high probabilities of achieving demonstrated results is paramount to satisfying the pragmatic concerns of their members.
- **Legal concerns are always nearby.** When working with new technologies, attaining approval from regulatory agencies is a common concern for many consortia. While no consortium member has yet been charged with antitrust violations, consortium managers must watch for potential programs and applications that might result in such accusations.
- **Developing consensus.** Reaching a consensus among the large number of members that may belong to a consortium often proves difficult. A board of directors sets overall policies and direction. Specific projects and decisions regarding those projects, however, are left to an array of committees and subcommittees, which vary from one consortium to another. Skilled consortium managers, with expertise in conflict resolution and facilitating joint decision making, are very important for R&D consortia.[8]

The R&D consortium as an organizational form will clearly survive and prosper into the twenty-first century. The tremendous potential for improving competitiveness through joint R&D and technology transfer will make the R&D consortium a critical organizational form for competing in the global marketplace.

The Process of Organizational Change

Whether through staff reductions, focused manufacturing, vertical control, joint ventures and strategic alliances, or R&D consortia, most companies today are undergoing some degree of change in their organizational structure. More than 3.2 million jobs were eliminated by Fortune 500 industrial companies in the 1980s. The actual number of reorganizations and corporate restructuring that must have surrounded these losses is beyond comprehension.

Changes also occur because of technology injection, reaction to competitive pressures, and as part of the merging of two organizations during an acquisition. Change in any organization causes problems; it places an organization under stress, resulting in at least a temporary drop in productivity and, in some cases, a failure of the intended change due to mishandling of the change process.

Organizational change has tended to follow one of two paths over the last decade:

- **The Slash and Burn Path:** This path is characterized by reliance on cutting labor costs by laying off workers, moving manufacturing facilities to lower wage areas, and farming out work to low-cost suppliers.
- **The Yellow Brick Road:** This path is characterized by increasing labor value, continuously retraining employees, accepting flexible job classifications and work rules, linking wage rates to profits and productivity improvements, and creating more permanent relationships with all of those who have a stake in the firm.[9]

Organizational Change Affects Employees

Research has shown that the stress potential of an event is determined as much by the amount of change it implies as by whether the people affected view the change as positive or negative. In most organizations, the stress resulting from the trend toward mergers, acquisitions, spinoffs, buyouts, downsizing, and restructuring has been staggering.

Unfortunately, most of those companies that followed the slash and burn path did not get the results they were seeking.

- *Fortune* magazine states that: "More than half the 1468 restructured companies surveyed by the Society for Human Resource Management reported that employee productivity either stayed the same or deteriorated after the layoffs." The same article reported that "senior managers at recently downsized companies said that their workers had low morale, feared future cutbacks, and distrusted management." [10]
- J.F. O'Reilly, CEO of Heinz, recognized for his expertise at downsizing, found that while his downsizing efforts during the 1980s achieved their financial objectives, they alienated workers and negatively affected the quality of Heinz products.

While organizations must change if they want to meet the demands of the fast changing international marketplace, the success of these changes is dependent on people. For some time, Tom Peters has argued that organizations are becoming more people-intensive and less management-intensive. In moving organizations over to a new style or philosophy of operation, it is important to consider and plan for the effect of change on the people who make up the organization.

Helping Employees through Change

Changes in technology and policy can occur quickly. People, however, rarely are able to change from old to new ways without a period in which they mentally let go of the past and accept the new ways of doing things. People are also often resistant to change. For example:

- Workers may resist transfers even though the transfers represent continued employment.
- Supervisors feel threatened when training programs increase workforce competence in a way that makes the old style of supervision unnecessary and counterproductive.
- Rumors can be debilitating. For example, a rumor that all employees over the age of 50 will be forced to retire can cause older workers to resist training the reassigned workers they erroneously view as their replacements.

A long period of time and a high level of energy are required to implement changes in organizations. Research has shown "that to achieve even half of the change a company attempts, it typically must spend an amount equivalent to between 5 to 10 percent of its annual budget for the personnel whose behaviors are supposed to change."[11]

What Employees Expect

As difficult and expensive to accomplish as organizational change may be, what employees expect is relatively simple, according to organizational behavior experts:

- Employees expect management to tell them what it will take for the company to succeed and how they fit into the puzzle.
- Employees expect the organization to provide the financial, physical, and human resources they need to do their jobs.
- Employees expect honest feedback about their own performance, the performance of their work unit, and the performance of the company.

- Employees want to be recognized and rewarded for their contributions.[12]

To effectively implement change in an organization, people must understand the change envisioned, the transition process, and the effects the transition will have on each of them. One given in any organization's future is that changes will occur. Helping employees through the transition is part of the process of remaining competitive. Other ways to motivate employees are described in Chapter 9.

Organizations and New Technology

The concept of injecting technology into the manufacturing environment is not new. What is new, however, is advanced manufacturing equipment that can operate without human intervention, drastically increase productivity, and change the fundamental way in which day-to-day business is performed.

Extraordinary changes in everyday life are attributable to advances in technology. The swift spread of microprocessors, the convergence of computers and telecommunications, and the creation of artificial intelligence and other technical advances are accompanied by important social, demographic, and political changes. All of these changes affect traditional management, values, cultures, organizational procedures, and organizational forms. *Organizations that are unreceptive to new technologies and fail to absorb them will become outdated and no longer be competitive.* Aggressive strategic management of technology will be a major tool of world-class manufacturers in the 1990s.

Becoming More Receptive to New Technologies

Manufacturing must become both more receptive to the adoption of advanced technology and ensure that the performance of these technologies is not hindered by the culture and systems into which they are injected. Because the organizational culture can facilitate or inhibit the adoption of new technologies, the strategic management of corporate culture is crucial.

Four key factors must be considered when implementing new technology:

- Effective implementation of technology is influenced by the attributes of the new technology, the perception of top management regarding their personal link to the technology, and the specific implementation steps taken to prepare the organizational culture for the introduction of the technology. Staff development is crucial.

- The attributes of the new technology, including the degree of complexity, specialization, and technological uncertainty, in large part determine the magnitude of the impact on the organization.
- The objectives and functions of new technologies determine the magnitude and rate of change in the dynamics of the organization.
- Implementing new technologies affects all elements of corporate culture; therefore, it is essential, in adopting any new technology, to strategically manage organizational culture.

These factors are discussed in detail in Chapter 10. This last consideration, the impact on and role of corporate culture, can be among the most important and most difficult for organizations to deal with. Those organizations with "old and proud" cultures typically have the most difficulty changing their cultures to adapt to the implementation of new technology. Further, *changing the culture only achieves a strategic advantage when it cannot be imitated by a competitor and when it can be linked to performance.*

Changing Culture to Support the New Technology

Culture change in organizations can be positively influenced by consistent, thoughtful managerial action. Changing culture is best accomplished by gradually reducing perceived differences between current norms and the new norm. *Planning for change is crucial.* Managers cannot be expected to change their tasks, relationships, and management styles until they clearly understand the change and their roles in it.

Guidelines for Managing Cultural Change

The research conducted in the 1980s on the effects of corporate culture on performance points to the same conclusion: business success is directly related to the strength of the company's culture. Terrence E. Deal of Harvard University and Allan A. Kennedy of McKinsey & Co. identified the following guidelines for managing cultural change successfully:

- **Actively involve leaders.** They bring to the team a sense of unswerving dedication and commitment that becomes infectious and inspires belief in the change process.
- **Use the temporary culture of task forces and team approaches** to provide a means of letting go of the old and attaching to the new.
- **Train all employees in the basic values and expected behavior patterns of the new culture.** Provide adequate structure and time

to learn about the new work activities, management processes, and interaction patterns that accompany the new culture.

- **Utilize outside consultants** to bridge the gap between old and new and assist in the change management process. Management consultants are most effective in facilitating change and acting as rallying points for the organization.
- **Develop tangible structural symbols** to signify the organization's new direction and management's commitment to it. For example, the installation of new and expensive advanced manufacturing equipment in refurbished surroundings is a tangible structural symbol of management's commitment to changes in the culture.
- **Guarantee the job security of the workers involved.** Keeping people employed is often the most important ingredient in successfully implementing change.[13]

The effective implementation of new technology is linked very closely with change in the organization's culture. It is through the culture that technology affects the organization's bottom line. *Increasing decentralization, a cultural advantage over competitors, and decreased communication costs are hallmarks of the restructured American workplace.* It becomes clearer now how important human resources are to world-class manufacturing performance. As the rate of technology injections into organizations continues to increase, effective organizations will be those that have a cultural capacity to adapt and change. Organizations that are not receptive to and accepting of new technologies will become obsolete.

Conclusion

The changes manufacturers have undergone in the 1980s and those they face in the 1990s require alterations in the structure of their organizations. The new manufacturing organizations are developing layers of teams. Companies, such as Ford in its documented manufacturing strategy, have indicated that the challenge of becoming world class requires teaming up North American and European organizations to globalize resources and minimize the not-invented-here syndrome. The first level, businesses joining with other companies and even competitors to take advantage of new markets and technologies, is rapidly emerging. R&D consortia provide new opportunities for collaborative efforts between small and large companies—both users and suppliers—on research and the transfer of technology. Through these cooperative efforts, members gather the strength to restore America's competitive posture in world markets.

At the second level, companies are decentralizing and refocusing operations so that the requirements for internal teamwork are accelerating. These changes necessitate a new type of organization, horizontally characterized by small cross-functional work teams and task forces with both financial and managerial responsibility. These teams are connected to their corporations through technology and the strength of their corporate values and are empowered to satisfy customer requirements without regard to the boundaries that have historically limited them.

Executives will lead their teams in the application of the tools discussed in this chapter, i.e., focused manufacturing, joint ventures, R&D consortia, and supplier integration. *Executive leadership and vision are mandatory for world-class manufacturing as traditional management styles become obsolete.*

CHAPTER 9

DEVELOPING HUMAN ASSETS

Although people are the basis of world-class manufacturing performance, human assets have not been a major focus of most manufacturers in the past. Many U.S. manufacturers are still merely giving lip service to the notion that people make the difference. Yet, all manufacturing improvement is ultimately dependent on the people who execute and maintain those improvements. The firm that successfully transforms its values, attitudes, and management practices to world-class performance levels has achieved a sustainable competitive advantage that is difficult for competitors to emulate. One company that has accomplished this feat is Mead Corporation.

In the 1970s, Mead Corporation set out to build a paper mill like no other in the world. In a small town in Alabama, Mead invested in some of the most technologically sophisticated mill equipment money could buy. The goal was to develop the most efficient and productive facility of its kind. The result is a facility that produces virtually no scrap and has, perhaps, the highest utilization of any mill in the world. It also operates with a very small number of employees, compared with other facilities of its size, providing Mead with a cost advantage virtually impossible for competitors to match.

How did Mead accomplish this? With people. At the Stevenson Mill, people make the difference. The management process is participative; employees are *team members* not just workers. There is no union at this facility; the workers cannot imagine why they would need one. They establish the policies and procedures and produce the results. Rules and regulations are minimal, by their choice. Members interview and hire their own co-members, they cover for each other when someone is sick, and they share in the profits. While the Mead management team admits that the technology is important, they argue that it is the approach to human resource management that has made the difference in the world-class competitiveness of Stevenson Mill.

As globalization diminishes national advantages in technology, natural resources, and strategic location, the development of superior human resources becomes ever more important. This chapter addresses several of the most critical human resources issues:

- Current demographic trends
- The education crisis
- Recruitment, selection, and training and development of workers
- Approaches to employee involvement
- The issues surrounding engineering resources
- The declining role of the supervisor in American factories
- Reward and incentive system

The Changing Workforce

The demographic profile of the U.S. workforce is undergoing its greatest change since the 1930s. The supply of workers and the composition of the workforce are changing dramatically, and the failure of the U.S. education system to keep pace with the rapidly changing needs of society is resulting in workers ill-prepared to perform the jobs of the 1990s. These factors add up to a tremendous challenge for manufacturing companies.

Implications for Manufacturers

The changes in the demographics of the U.S. workforce and the failure of the U.S. education system to keep pace have significant implications for manufacturers. Among these are:

- **More creativity required in recruiting.** Most of the people who will make up the workforce of the next century are already in the workforce. Companies that have relied on a constant stream of qualified young employees will have to adapt to the new sources and skill levels of labor force entrants. Higher requirements for technically literate employees will place additional emphasis on recruiting, training, and retention.
- **Difficulty implementing new technologies.** Because the average employee will be older and more experienced, he or she may find it more difficult to adapt to changing technologies and competitive environments. This will have an impact on manufacturers, which must implement advanced process technologies to remain competitive.

- **Cost structure affected by rising wages.** The higher wages that older workers command will affect the cost structure of all manufacturers, forcing them to make better use of their human resources.
- **Growing pool of obsolete workers.** As job requirements become more specialized, many of the skills developed by employees will be outdated. Older workers and those with obsolete skills will have increased difficulty finding and holding jobs within the companies where they have worked for many years.
- **Reduction in the management ranks.** The reduction of the number of people in management and staff positions throughout America's industrial sectors, particularly in the middle management levels, will require major shifts in job responsibilities. As organizations restructure and traditional career paths based on hierarchies dissolve, competition for the remaining positions will be even more intense.
- **Significant training and education requirements.** The training and education that will be required to provide this ever-changing workforce with the necessary knowledge and skills to be productive will be staggering. To be competitive in the twenty-first century, manufacturers will have to step up to the challenge of this "knowledge building" requirement, which will have to be an integral part of the strategic plan and the operating plan, with support coming directly from the budget as a major cost of doing business.
- **More tolerance of language and cultural differences.** As the ethnic diversity of the workforce grows, companies will need to develop environments more tolerant of language differences and cultural variations. A large portion of those entering the workforce will be women and minorities. Companies that cling to the rigid work environments of the past will not be able to compete effectively for the eclectic worker of the future.

These changes will have a revolutionary effect on the way manufacturers hire, train, and maintain their workforces. A clear understanding of these demographic trends is essential for all manufacturers as they compete for human assets in the twenty-first century.

The Baby Boom Goes Bust

The growth rates of the U.S. population and the labor force for twentieth century are illustrated in Figure 9-1. The population growth of the post World

FIGURE 9-1 Population and Labor Force Growth Will Drop by Year 2000 (Average Annual Gain)

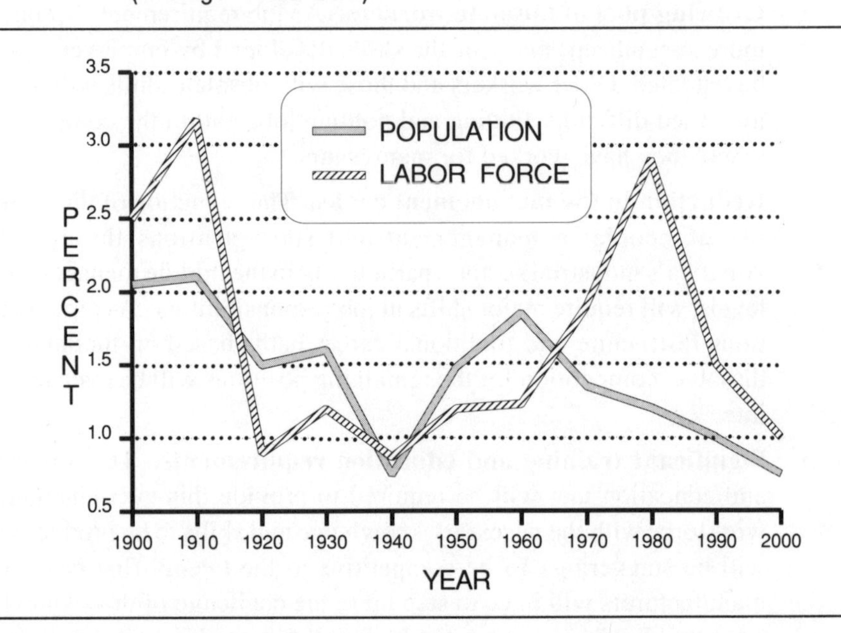

Source: William B. Johnston and Arnold H. Parker, *Workforce 2000: Work and Workers for the 21st Century* (Indianapolis, Indiana: Hudson Institute, 1987).

War II period peaked in 1960. These Baby Boomers entered the workforce during the 1960s and 70s, causing an increase in the annual growth rate of the labor force from 1.3 percent to a 2.8 percent peak by 1980.

Following the Baby Boom generation, the birthrate was expected to decrease until the boomers reached childbearing age. The expected increase in birthrates in the 1970s and 1980s, however, did not occur. The post-war generation postponed marriage and childbearing and limited its family size. Following the peak in 1960, the birthrate has declined steadily. Since the growth rate of the labor force lags behind the birthrate by 20 to 24 years, growth of the labor force is expected to decrease to 1 percent annually by the year 2000 as it tracks the declining rate of population growth.

The rate at which entry-level workers are joining the workforce is shown in Figure 9-2. The percentage of the workforce represented by entry-level workers in the total U.S. population is declining and will decrease further throughout the 1990s.

FIGURE 9-2 U.S. Entry-level Workers (Percentage of Total Population)

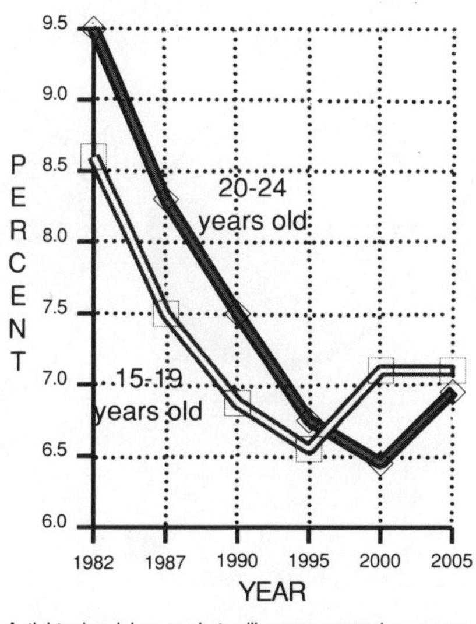

A tightening labor market will mean upward pressure on wages and more automation wherever possible.

Source: "Plugging the Gaps in the U.S. Labor Market," reprinted from *Electronic Business*, 8/15/88, © 1988 Reed Publishing USA.

A Graying Workforce

The declining numbers of young people entering the workforce has another significant effect: the average age of American workers is increasing, as shown in Figure 9-3. Commissioned by the U.S. Government, the Hudson Institute's 1987 Workforce 2000 predicts that by the year 2000, the median age of the population will reach 36—six years older than at any other time in history. Further, over 50 percent of the workforce will be over 55 years old.

Ethnic Potpourri

Along with the aging of the U.S. workforce comes a significant increase in the diversity of ethnic backgrounds in the workforce—a diversity that has not been seen since the early 1900s. The white male, once the almost exclusive worker group of the U.S. workforce, will represent no more than

FIGURE 9-3 The Middle Aging of the Workforce

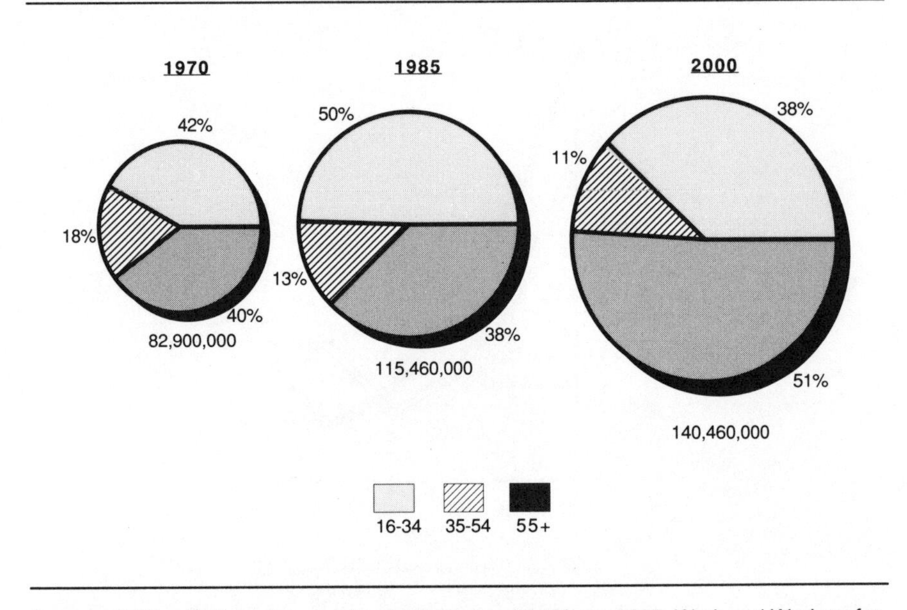

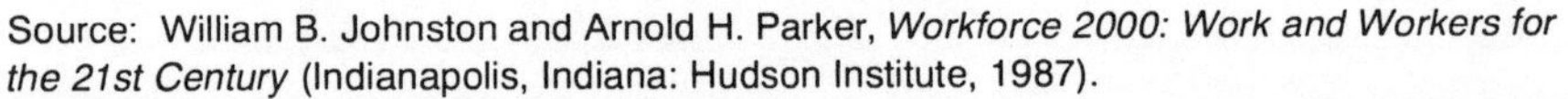

Source: William B. Johnston and Arnold H. Parker, *Workforce 2000: Work and Workers for the 21st Century* (Indianapolis, Indiana: Hudson Institute, 1987).

45 percent of the workforce by the year 2000. Further, studies suggest that the white male will represent only 32 percent of the entries into the workforce during the 1990s.

Minorities and immigrants will make up a hefty 26 percent of the workforce by the year 2000, while the white female will hold a 39 percent share. During the 1990s, the greatest gains in the workforce will be made by women and minorities. Hispanics, as a percent of the workforce, will increase nearly 75 percent during the 1990s, showing the largest gain of any single ethnic group. Minorities, immigrants, and women will account for 80 percent of the future entrants into the labor force.

The Education Crisis

Notwithstanding the shrinking pool of young entry-level workers and the diversity of ethnic backgrounds, manufacturing firms will find it increasingly difficult to prepare new employees for the jobs available. A combination of low educational achievement and rising job requirements will reduce the pool of employable workers. Of the jobs created over the next 10 years, more than 50 percent will require education beyond high school. Over 30 percent will require college degrees. The U.S. education system is not meeting these demands.

As the number of skilled manufacturing jobs increases, the quality of American public education continues to plod along with antiquated systems based on a model of society as it was over 100 years ago. A recent National Assessment of Educational Progress, undertaken by the U.S. Department of Education, found that among 21- to-25-year-olds:

- Only 60 percent of whites, 40 percent of Hispanics, and 25 percent of blacks could locate information in a news article or an almanac.
- Only 25 percent of whites, 7 percent of Hispanics, and 3 percent of blacks could decipher a bus schedule.
- Only 44 percent of whites, 20 percent of Hispanics, and 8 percent of blacks could correctly determine the change due from the purchase of a two-item restaurant meal.[1]

In a recent international study of achievement in science by school children, the U.S. ranked no better than 8th out of 15. The ranking declined steadily as the students increased in age. In standardized tests between 1983 and 1986, high school seniors in the U.S. came in last in biology, 11th in chemistry, and 9th in physics among 13 countries. Similar results have been found in comparisons of mathematics skills.

Compounding these poor results is the fact that school children in the U.S. spend less time in school than do children in other industrialized nations. The agrarian calendar is still used almost exclusively by schools in the U.S., although the days of tending the fields in the summer have long since passed for American school children. Recent reports indicate that Japan, the U.S.S.R., Hong Kong, Britain, and Canada all have longer school years than the U.S. does. The Japanese lead the world with a 243-day school year calendar compared to the average U.S. calendar of only 180 days. In the Soviet Union, Japan, and parts of West Germany, children also go to school on Saturdays, a requirement unheard of in the U.S.

While U.S. colleges and universities outrank the entire world in quality of education, the number of students that actually make it to institutions of higher education is staggeringly low: fully 25 percent of students drop out of high school. In 1988, of the over 3.8 million 18-year-olds in the U.S., over 700,000 had dropped out of school and another 700,000 could not read their high school diplomas. Experts estimate that some 27 million adults are functionally illiterate and that the number is growing at a rate of 2.3

million each year. Other estimates indicate that the combination of functionally illiterate individuals and those individuals that are barely competent in reading, math, and basic reasoning represents fully 30 percent of the potential workforce. Successful companies of the 1990s will need to cope with these staggering statistics as the illiterate and undereducated enter their plants and factories.

Manufacturers will face many human resource challenges throughout the 1990s. Among the most difficult will be recruitment and selection of new workers and the training and education of these workers.

Manufacturers' Responses to the Changing Workforce

Human resource professionals are moving away from traditional personnel management toward a more comprehensive and active Human Resource Management (HRM) role. This new HRM role facilitates the new methods required for recruiting and selecting new employees and helps companies define new methods of educating, training, and managing employees.

New Selection Processes

Companies can no longer hire employees off the street, take chest X-rays, make a few reference checks, and send them out to work in their plants. Many companies now scrutinize potential employees much more closely to ensure that they have not only the right aptitudes, skills, and education, but also the right attitudes.

Most of the Japanese-owned or operated plants in the U.S. put all job applicants, from entry level production worker through general manager, through a lengthy selection process.

- Applicants at Toyota's Georgetown, Kentucky assembly plant must pass 14 hours of screening. More than 45,000 people applied for 3000 job openings when the plant opened.
- Diamond Star Motors Corp., a joint venture between Chrysler and Mitsubishi Motors Corp., puts applicants through a 16- to 24-hour selection process.
- Nippondenso, one of the world's largest automotive parts suppliers, screened 2000 candidates for positions at their new Jackson, Michigan, automotive compressor facility and were able to qualify only

seven for training. A restructured selection process was then initiated to provide remedial education for the next highest qualifiers prior to submitting them to specific training programs, which included being trained in Japan facilities.

U.S. plants have also increased the scrutiny given to the hiring and selection process:

- Applicants at Colgate Palmolive's liquid detergent plant in Cambridge, Ohio, go through a 14-hour screening process.
- At Rohm and Haas Bayport, a subsidiary of Rohm and Haas, 155 applicants were interviewed before a mere 16 were hired when the plant opened.

In most cases, the employee selection process involves a battery of tests, several highly structured interviews, and costly and time-consuming assessment center exercises and job simulations.[2] The cost of this level of scrutiny can be significant.

- When the Toyota Georgetown plant was opening, applicants went through a four-phased screening process including general aptitude testing, structured interviews, job simulations, and group exercises. Not all applicants went through all phases, as some were weeded out during the process. Toyota clearly invests heavily to find the right workers for its jobs.
- At Nissan's Smyrna, Tennessee, facility, which opened in 1983, more than 200,000 people applied for 3200 jobs. Applicants who made it to a certain point in the screening process were required to participate in a 60-hour pre-employment training program conducted by the Tennessee Industrial Training Service. Participation did not guarantee employment. The only guarantee Nissan made was that it would hire only from those that successfully completed the program.

Do the results merit the effort? Florida Power and Light, after adopting a similarly rigorous screening and selection program for its groundmen, has demonstrated that the effort is worth it. After carefully tracking the results of their groundman selection program for four years, Florida Power found that:

- Turnover dropped from nearly 50 percent to under 5 percent, saving the company over $1.2 million in total employee replacement costs.
- The more qualified workers allowed the company to expand its customer base over 4 percent per year with 16 percent fewer workers. Further, the

more qualified groundmen become journeymen in an average of only four years compared to more than seven years previously.[3]

At the Rohm and Haas plant described earlier, similar results have been seen. While only 7 of the 16 technicians originally hired remained after two years, after further refining of the hiring and selection process, the turnover rate dropped from 50 percent to less than 5 percent.

Certainly, world-class companies are finding that attracting and hiring the right people is critical to success. The results can be counted in the dollars and cents associated with lower turnover, more productive workforces, higher product quality, and improved safety records.

Nontraditional Recruiting

The present and anticipated shortages of young labor force entrants is forcing manufacturing companies to alter their recruitment, selection, and employment practices. As older workers, disabled workers, women, and minorities become the growing demographic groups of the 1990s, companies will have to devise programs to include them in their operations. To be effective, these recruitment efforts must be publicized to the target employment prospects.

- During its recruiting efforts, Xerox used reprints of a *Black Enterprise* magazine article rating the company as a good place for African Americans to work.
- Hewlett-Packard tells prospective women candidates about its high rating in *Working Woman*, a widely distributed business publication.
- Procter & Gamble has been able to recruit minority and women engineers at a higher rate than those groups are represented in universities by increasing on-campus recruiting sessions and using their own minorities and women to do a significant amount of the on-campus visitation.
- Other companies have reported success in attracting and retaining high-quality female employees by establishing day-care facilities for their children.
- Still other progressive companies are using flexible schedules and calendars to attract mothers of young children, enabling them to be home for school holidays and to match daily school schedules.
- A Johnson & Johnson plant targets senior citizens to supplement its full-time workforce.

- High Quality Manufacturing, a supplier of emissions control harnesses for the Big Three auto companies, has recruited its workforce from inner city mothers who have never held a job before. The program, part of Focus:HOPE, a civil rights umbrella organization in Detroit, Michigan, trains them and provides full day-care facilities for their children. Additionally, Focus:HOPE recruits inner city high school graduates into its Fast Track program. This program provides remedial education prior to the graduate's acceptance into its Machinist Training Institute or into another program for further training.
- Many companies have established cooperative work experience programs with colleges and universities, where the work is part of the students' curricula. These programs provide an excellent opportunity for the companies to observe high-potential students in action.

Recruitment and selection problems are compounded by the poor educational background of the new entries into the workforce. In Detroit, for example, the average high school graduate has math and reading skills somewhere between the seventh and eighth grade level. Supplementing basic education with remedial training will become a consideration for all companies.

Meeting the Training and Retraining Challenge

To turn inadequately educated labor force entrants into productive employees, manufacturers will have to shoulder some of the burden of providing them with basic educational skills. While a number of significant efforts are underway to improve the quality of American public education, it is unlikely that these efforts will have a significant effect within the next decade.

In fact, more and more workers across the nation are finding themselves displaced as technology drastically changes the way manufacturing work is performed. *As many as two million jobs vanish each year in the wake of automation and foreign competition in the U.S.* The new jobs being created in the manufacturing sector require much higher levels of basic education and technical skill and knowledge than traditional manufacturing jobs. The displaced workers from traditional low-skill jobs do not have the qualifications for these new positions. Some 400,000 workers need extensive retraining to find new jobs each year.

The increasing need for technological skills and the less than required education level of new hires will force most manufacturing companies to institute extensive basic education and technical training programs for their employees. *Leading manufacturers will increasingly become involved in two primary activities: providing remedial basic education for those employees lacking fundamental skills, and cooperating with and assisting local school boards to promote higher quality education.*

Provide Employees with Basic Educational Skills

Providing basic education for current employees can be addressed in several ways. Some companies have set up in-house courses to improve mathematics, science, and reading skills, and many have been successful in encouraging employees to earn their high school diplomas by taking the high school equivalency test. In 1989, companies with 100 or more employees planned to spend over $44 billion on direct training programs. The American Society for Training and Development estimates that by the year 2000, over 75 percent of all workers will need to upgrade their basic career skills.

According to conservative estimates, employer investment in employee education and training will increase 25 to 30 percent by 1995. The actual amount of the increase needs to be much greater. This increase will be fueled by technological change, by pressure to compete in world markets, by greatly increased learning needs of the workforce, and by public and private recognition that training is a good investment.[4]

Companies are finding that without basic employee skills, they are unable to effectively upgrade their manufacturing technologies.

- Ingersoll-Rand found that it had to provide basic remedial education and training for workers at its Athens, Pennsylvania plant after spending some $24 million to implement advanced process technology. Workers were unable to handle the new job requirements, and more than half lacked the basic skills needed to learn their new jobs. Ingersoll developed a long-term program to upgrade hourly workers' basic skills in math, reading, and interpersonal communication. Ingersoll found that the program not only resulted in a more competent and productive workforce, but also improved the morale of the workers.

- At Motorola, the need for better educated workers has lead to the establishment of minimum skill levels for workers and remedial education programs. Motorola has a minimum requirement of a fifth-grade level in math and a seventh-grade level in reading. Between 700 and 1000 employees are actively involved in the basic skills building and remedial education program at any given time.[5]

- General Motors' investment in people came after the company invested billions of dollars in advanced manufacturing process technology. General Motors found, as did Ingersoll-Rand, that employees needed to have their skills upgraded before the process technology investments could be utilized. General Motors is spending $1 billion each year on workplace training that covers everything from basic literacy and high school completion courses to courses on advanced process technologies, such as robots and lasers. At General Motors' Windmere school, a former elementary school in Lansing, Michigan, more than 12,000 workers attend more than 550 classes. Many of the courses are mandatory, and General Motors encourages workers to go to school on company time for certain courses. Windmere is not unlike dozens of other facilities General Motors has in its more than 150 domestic plants.

- Kodak is experimenting with accelerated learning programs to reduce the total time required to train workers in statistical process control, MRP, and other production methods.

The U.S. education system continues to turn out prospective employees unable to cope with the demands of a competitive and sophisticated global economy. This situation demands that companies place heavy emphasis on retraining in the short run and on bringing about change in the education system over the long run.

Cooperate with Educational Institutions

Two-thirds of the respondents in a recent study of corporate community involvement activities pointed to participation in primary and secondary education. This figure is up from about 40 percent two years ago. The respondents also felt that education would be the most pressing community issue over the next two to five years.[6]

Particularly effective programs involve company sponsorships and scholarships for students who complete high school. Other approaches include contracts between businesses and public education programs.

- In Lowndes County, Alabama, GE has committed $1 million over the next several years to triple the number of college-bound high school students.

- In 1982, more than 600 Boston companies formed the Boston Compact, agreeing to reserve a quota of jobs for high school graduates if schools improved attendance levels and reading skills. While Boston's schools are still experiencing high dropout rates, one-third of the graduating students find jobs through the program.

Other similar programs have been initiated in seven other cities—Albuquerque, Cincinnati, Indianapolis, Louisville, Memphis, San Diego, and Seattle.

- Chicago's Corporate/Community School was started in 1988 by 16 leading organizations in the Chicago area including Sears, Baxter Healthcare, and United Airlines. These companies invested over $2 million in a model school to demonstrate that public education could succeed. Two hundred minority children from pre-school to fifth grade, drawn by lottery from Chicago's most underprivileged neighborhoods, attend. The approach is all business, with a strong focus on critical goals, such as improved reading and math skills in the student population. The approach, which involves one-on-one attention and high levels of participation in team learning settings, appears to be an overwhelming success.
- The Fannie Mae (Federal National Mortgage Program) program provides another example of how business leaders can get involved and make a difference. In 1988, Fannie Mae allocated over $1 million to develop a 10-year mentor program called the Futures 500 Club at Woodson High School in Washington, D.C. Students who earn all As or Bs on their report cards, are admitted into the program. They are assigned a Fannie Mae mentor and helped to find jobs with Fannie Mae and other businesses. Each student is given $500, which is put into an account earmarked for college tuition, for each semester he or she is in the program. During the first 18 months of the program, the number of students receiving all As or Bs at Woodson High School increased from 33 to 130.[7]

Once companies find and train employees, they must address the labor management policies that might allow these valuable assets to be temporarily or permanently laid off. New methods for managing the workforce are required to protect the financial investment made in human resources through education and training.

New Methods of Workforce Management

In the past, American manufacturers viewed their workforce as a variable expense that could be laid off and rehired at will. Limited knowledge requirements, along with generous compensation in many sectors, made this a viable policy. Companies that still use these traditional practices are finding that fewer employees are willing to return from layoff, many having found other employment. Employees with technically valuable and marketable skills are the least likely to return.

When Cleveland-based KT-Swasey's business picked up enough to add a second shift, it recalled many of the employees it had laid off during downsizing of its operations. Most of the workers had found other jobs, often with competitors. KT-Swasey had to hire new, unskilled workers and train them at significant cost.

Companies that invest heavily to hire unskilled workers and train them must modify their workforce management practices to avoid layoffs and avoid letting their valuable human resources escape into the competitive labor market. World-class companies constantly challenge themselves to find innovative ways to keep their best workers employed. One of the most effective methods is to provide cross training, which allows workers to move both vertically and horizontally to alternate jobs.

There is mounting empirical evidence, such as S.J. Havlovic's research, that worker participation has a significant impact on labor productivity. Increasingly, leading manufacturers have adopted more progressive human resource policies to attract and retain the best people. Almost without exception, these policies revolve around programs aimed at enhancing the level and quality of employee involvement.

Employee Involvement

Over most of the twentieth century, American industrial management has viewed employees as factors of production motivated primarily by economic considerations. A workplace culture developed in which managers defined unskilled and narrow jobs requiring little more than physical effort from the employees. As a result, workers confined their demands to economic issues and working conditions. In an international economy in which U.S. manufacturing had enormous advantages in technology, a mass home market, and abundant natural resources, these wasteful human resources policies were tolerated. In the 1990s, American manufacturers can no longer afford them.

Having observed the success of Japanese approaches to industrial management, U.S. managers have accepted the need for making wide-ranging change in their employee involvement programs. U.S. companies' mixed record of success with the Quality Circles concept has motivated them to experiment with other methods of worker motivation and involvement. In fact, two-thirds of the Fortune 500 have used quality circles or other participatory methods since the early 1970s.[8] More details concerning employee involvement and quality improvement are presented in Chapter 2.

Making Involvement Programs Succeed Is Not Easy

Current research suggests that North American firms believe work teams, participatory management, and suggestion programs are easily implemented solutions that yield immediate and clear benefits. According to Ford's Donald Peterson, however, the process at Ford started off with a significant amount of cynicism and doubt and picked up speed only after a significant period of time.

Japanese firms take a longer term view and use a process of *fusion*. The company identifies both the factors of worker motivation and the desired employer goals. Once these are known, the firm develops a management program that can satisfy both employer and employees. Identifying and aligning and realigning goals is a continual process, and the periodic resurveying of employee motivators is necessary to ensure long-term success. A major key to success is identifying the employee motivators.

Identifying Employee Motivators Is Important

Employees of Japanese corporations surveyed in 1974, 1977, and 1980 revealed these primary motivations:

- To improve themselves through their present jobs.
- To work optimistically in their shops, unconcerned with company objectives and aims.
- To overcome various frustrations in their present jobs caused by excessive direction from a supervisor and needless company regulations.

The desire to "improve themselves through their present jobs" was the strongest desire expressed in each of the survey years.

Most American companies would expect the desire to "overcome various frustrations in their present jobs caused by excessive direction . . . . " to be a top motivator of their employees. They would be less inclined to believe that self improvement and a desire to work optimistically would enter the conversation. However, those American firms that have taken a long-term perspective on human resources and have been successful in their implementation of enhanced employee involvement programs have found that similar motivators are indeed a reality. Production and management systems must reflect these employee desires. In the remainder of this section, American experiences with work teams, participatory management, and suggestion programs are reviewed in further detail.

Work Teams and Participatory Management

Work teams are groups of two or more employees organized to solve problems or to accomplish work-related tasks requiring coordination. The composition of teams can vary from single to multifunctional groups. Chapter 2 detailed the practices of companies using a variety of different team structures to tackle continuous quality improvement. Teams also vary considerably in degree of employee participation, including:

- The degree of employee involvement in decision making, specification of employee contracts, and participation on management boards.
- The degree of self-management or autonomy in directing, controlling, and evaluating daily activities. For example, the superteams described in Chapter 2 have the greatest degree of autonomy.

Companies attempting to implement work teams often struggle to find the best structure for those teams. These examples illustrate the various structures and levels within the organization with which teams operate:

- **The A.O. Smith Corporation** has a three-level organizational structure. A Policy Committee of senior members of management and unions sets up the overall process, plans the joint problem-solving procedure, and handles companywide matters. An Advisory Committee of operations middle management, union officials and stewards, informal work group leaders, and supervisors develops plantwide improvement programs, works on problem solving, and handles companywide issues. At a third level, 90 work teams of volunteers address specific plant issues.

  A.O. Smith has also started operator control programs to improve quality. Formerly, operators were held responsible for quality, but were not given adequate training or equipment to monitor quality. The company is now investing in quality training. Employees have been given both the responsibility and the authority to make on-line quality decisions. Operators and inspectors work together in teams.[9]
- **General Foods** has organized interfunctional work teams. Each team has an appointed leader, and each holds total responsibility for operating a segment of the business.
- **Rohm and Haas' Bayport Company plant** in LaPorte, Texas, was redesigned for participatory management and team approaches.

Hierarchical levels were reduced, and the remaining managers were given wider spans of control. The plant manager has four direct reporting relationships: financial, personnel, and two from manufacturing, each responsible for a mini-plant operation.[10]

- **The Northwest Forest Operations (NWFO) division of ITT Rayonier** has a contract relationship with employee work groups. Under this system, the independent work groups receive monthly payments based on revenues generated and costs incurred. Employee groups have the right of first refusal on contracts. Union agreements are maintained, and safety standards are upheld. The contract process is managed by an operations committee of management and employee group representatives.[11]

High-performance teams are those which consistently demonstrate high productivity, problem-solving capabilities, and leadership. *While the composition of teams can vary from single to multifunction groups, five interrelated preconditions characterize most high-performance teams*:

- Vision of the goal.
- Belief in and commitment to the goal by the leader.
- Teammates' ability to achieve the goal.
- Coordination of team resources.
- Ability to adjust and adapt to evolving situations.[12]

The key ingredient is a strengthened sense of ownership, involvement, and responsibility for business results on the part of all team members. *High-performance teams demonstrate these six additional operating characteristics*:

- Goals are clear.
- Roles sort themselves out, based on members' skills and the problem to be addressed.
- Leadership is by a champion of the team.
- Team relations are solid; rules and norms of behavior are openly discussed and agreed upon.
- For self-regulated work teams to be successful, team members must be skilled at a variety of tasks.
- Rewards and recognition are handed out to members who contribute to success.

Companies that have instituted work teams and participatory management are spending considerable time training team members. Team-building sessions are held to create a sense of teamwork and unity. Unions and companies have been retraining both managers and workers in Japanese-style production methods with the goals of increasing productivity and reducing defect levels. In fact, one element of the 1982 Ford contract was the agreement to institute training, retraining, and education for hourly employees.

Performance Implications of Using Work Teams

The performance implications of using work teams and participatory management are both quantitative and qualitative.

- One team at a leading manufacturer reduced changeover time between models from hours to less than 20 minutes, saving over $100,000 per year.
- Contracting with employee groups has resulted in a 20 percent cost reduction for ITT Rayonier , Northwest Forest Operations, as well as greatly increased productivity. Contracting also resulted in:
  - Improved management-labor relations.
  - Better-kept equipment, holding down maintenance costs.
  - Elimination of half of the management team.
  - Emergence of a team attitude.
- General Foods found that work teams are the most effective way to stimulate member participation and involvement. Additional benefits include:
  - More sharing and integration of individual skills and resources.
  - Tapping of previously unrecognized skills.
  - More stimulation, energy, and endurance by team members.
  - Emotional support between members.
  - More ideas.
  - More commitment and ownership by team members.

Participatory Management Issues

Participatory management techniques involve a systematic change in the way everyone in the enterprise relates and operates. Donald F. Peterson says, "I think it is important today to develop in all employees a good understanding of just how interdependent we are as a world of nations and how truly international the competition is."[13]

The crafting of a future vision is a cooperative effort of management, employees, and unions. Without the support of any one group, efforts are doomed to failure. The vision must be of a better, achievable future, rooted in workplace realities.

Traditional organizational structures do not lend themselves to participatory management and work groups. Embedding work teams in these hierarchical and political environments results in worker rejection. To ease the implementation of new organizational structures, detailed planning and preparation must be undertaken well in advance. The first task is to address desired attitudes and expectations. Management, labor, and unions must understand and accept the need for change. Ideally, participative team management eliminates political behavior and focuses on common production goals.

Teams take time. Like the propagation of delicate plants, actual implementation of work teams and participative management is most successful when a temporary greenhouse can hold bad weather at bay; initial team development must be well-planned and structures must be in place to ensure success. This provides a model on which to build future teams. Work groups will not be firmly established in organizations experiencing high levels of uncertainty due to rapid growth or decline, merger rumors, issues of job security, or major changes in top management. The nurturing of effective teams takes time as the organization learns and as stability is established. Teams need to be together long enough to begin operating at their optimum performance levels, so continued monitoring and guidance is necessary as the learning curve progresses.

Teams change career paths. Team environments change the personal and professional growth patterns of the team members. With the reduction in organizational hierarchy and the widening of spans of control, promotions from team member to team leader are slow in coming. Without new directions for growth, the apparent lack of promotion opportunity could become a problem. Most employees, however, will be pleased with the organization, enjoy the sense of responsibility, and savor the opportunity to learn and master a variety of tasks, if career paths are identified and opportunities for advancement are known.

Not everyone is a good team member. Not all employees can function easily in a team setting. When Rohm and Haas Bayport opened a new plant using self-regulated work teams, turnover jumped to 50 percent in the first two years. Since then, the firm has learned how to identify the personality traits of successful team members in prospective employees. These traits include:

- Responsibility

- Motivational match
- Versatility
- Learning ability
- Honesty
- Initiative
- Cooperation
- Communicativeness
- Openness
- Tact/Sensitivity [14]

Suggestion Programs

Where they are given a meaningful role, suggestion programs are an important means of increasing productivity and competitiveness. They generate ideas for improvements and play a role in the development of a corporate culture supportive of world-class performance. Effective suggestion programs support innovation, motivation, and a sense of self-worth for employees. The fact that employee suggestions are important enough to the management of the enterprise to warrant interest and discussion sends a clear message to all.

Japanese firms have relied heavily on suggestion programs to support continuous improvement goals. Japanese preeminence in process technology implementation is ample evidence of the potential of suggestion programs to aid the learning process. By contrast, U.S. industry has ignored suggestion programs and relied heavily on laboratory-based research and development to identify places where improvement may be realized. While the U.S. approach has been beneficial in terms of scientific development, the U.S. has been less than successful in adapting that learning to production realities.

The success of suggestion programs depends entirely on management commitment and helping employees feel comfortable with making suggestions, which may reflect negatively on management's performance. *The Japanese even have books that describe how to write suggestions.* In a comparison of 100 American and Japanese companies, the better Japanese performance was attributed to higher numbers of suggestions per worker. While quality circles are used much more extensively in Japan, and those plants with quality circles have high suggestion rates, participation in the circles is not strongly linked to suggestion rates. The prime motivators are cash awards and lifetime earnings.[15]

Despite some remarkable successes, low participation rates by U.S. workers in suggestion programs may be an indicator of attitudes and cultures. The traditional industrial culture has not bound workers to the firm for a long-term relationship of mutual consideration and concern. *Alienated employees have little interest in suggestion programs.* They are most successful where management combines them with a holistic participatory management style, which ensures at least some consideration of the suggestion and has a feedback mechanism.

How Employees View These Problems

While enhanced employee involvement programs are viewed positively, they have not been totally successful in involving employees outside their primary job functions. *A large number of organizations reported active employee involvement in team suggestion programs, quality circles, quality of work life programs, and cross-functional employee task forces, but the median percentage for participation by employees was only 15 percent.* The majority of employees themselves doubt the value of such programs and view them as fads that will not be supported by management over the long term.[16]

When an enhanced employee involvement philosophy is successfully implemented, it has a profound impact on the supervisors in the organization. The next section discusses the typical supervisory environment, the factors that are making traditional supervisory practices obsolete, and what American manufacturers can do to address these issues.

The Extinction of the "American Made" Supervisor

> American employees have lost confidence in their supervisors and managers. There is a pressing need for a serious evaluation of both leadership and supervision in corporations today. Today's supervisors and middle managers are likely to be at the crux of our most important problems or solutions as we hurdle through the decade. They can have a dramatic effect on how most workers perceive the goals and purpose of the organization.[17]

Historically, the first-line supervisor has played a pivotal role in manufacturing. He or she has been charged with the interpretation and execution of management's plans. With such a critical mission, it should not be surprising that supervisors have a strong impact on worker motivation, performance, and satisfaction, not to mention overall company business performance.

Traditionally, supervisors have been rewarded primarily for achieving high labor and machine productivity. Human relations considerations have taken a distant second priority. The immediacy and pressure of quantitative production measures shaped a culture in which too many first-line supervisors exercised command and control through fear and intimidation. All too often, poor product quality, low employee morale, delayed innovation, and polarized labor-management relations have been the results of this type of management system.

Changing Expectations, Changing Requirements

Changing social attitudes are affecting the social structures and practices of American manufacturing firms. The ingrained management-labor cultures in many enterprises were formed by the Great Depression and the Second World War. Most of today's workers grew to adulthood in an unprecedented era of economic growth and expanding civil rights. Increased emphasis on justice, equality, and self-fulfillment has created very different work expectations for these post-war workers. Today's employees look to their jobs to provide a sense of personal pride, accomplishment, and self-respect. Their higher levels of formal education have also contributed to these expectations.

Increasing complexity of technology is also affecting supervisory roles. As firms become less labor intensive and employee skills increase relative to those of supervisors, authoritarian management styles are less effective.

The National Opinion Research Center survey concluded that "the challenge for supervision in the 1980s is to release the substantial potential for employee commitment that there is in today's workforce, and to provide the circumstances that will allow this commitment to be directed toward both individual and organizational goals."[18]

The Supervisor of the 1990s

Will we need supervisors in the twenty-first century? With the rise of teams and superteams, do supervisors represent unnecessary overhead? The answer is an emphatic NO! Supervisors will function as intermediary resource facilitators, coordinators, and coaches. These roles are necessary regardless of organizational design. *In fact, research has shown that "top-notch performers (supervisors) do their jobs in much the same way, regardless of the work system (participative versus traditional hierarchy) or their formal title."*[19] However, these new roles will likely require some additional skills, even for top supervisors.

The first-line supervisor of the 1990s fills a multidimensional position. As a leader, he or she is expected to build a team and motivate employees to carry out the company's mission. The supervisor also plays a crucial role in the development of individual employees. Classroom education and training must be reinforced by rigorous and detailed on-the-job training. The role the supervisor plays in the development of appropriate work attitudes and behaviors is extremely important in building an environment of cooperation and trust.

This expanded understanding of the supervisor's importance to world-class manufacturing will lead to changes in how supervisors are selected and trained. Management must carefully select individuals with both leadership talent and detailed technical knowledge and experience, and they should not overlook their existing workforce in their search for high performers. Highly effective team facilitators come directly from the production floor. Leadership talent can be developed through training and practice. Technical knowledge is the result of both training and experience. The supervisor must develop both skill sets to win the respect of his or her people and to be an effective leader. *Training in group dynamics, problem solving, and group presentation skills is a must.*

Production workers and the issues surrounding their personal growth, development, and supervision are basic issues for managing human resources into the twenty-first century. Equally as pressing for American industry is the growing shortage of qualified engineers.

The Quandary Over Engineering Resources

The adequacy of technical and engineering resources is of growing concern in many companies. The increasing sophistication of products and processes, the emergence of a global manufacturing environment, and a multitude of social and economic changes have contributed to this concern. Two factors are at work, one quantitative, and one qualitative. Not enough young people are attracted to technical careers, and there is a high incidence of engineers leaving technical careers to pursue other career options.

Engineering by the Numbers

The declining interest and competence of high school students in mathematics and science has manifested itself in a reduction in the number of students entering engineering and technical schools. Coupled with declining numbers of young entry-level workers in general, this further squeezes companies that want to develop advanced technical staffs.

Manufacturing companies have not made engineering careers attractive. In most companies, there is no clear career path for engineers. It is not unusual for an engineer to reach the top of the technical ladder after only five to seven years. The perceived lack of future opportunity in engineering and the relative appeal of finance, marketing, and general management career ladders, curtail many promising technical careers. And many of the best technical people are lost by promoting them into management positions. In contrast, America's foreign competitors employ many more engineers as a proportion of total employees. Most notably, Japanese schools produce two to three times the number of engineers the U.S. produces. Numbers offset the loss of engineers to management positions.

To make the *Factory of the Future* a reality, manufacturing and industrial engineers will be required to demonstrate a high degree of technical knowledge and broad abilities to integrate technical knowledge across functions. It is estimated that manufacturing engineers will be required to triple the number of technologies they use by the year 2000 (see Figure 9-4).

A study conducted for the Society of Manufacturing Engineers on the outlook for manufacturing engineers in the twenty-first century identified four important roles for engineering resources:

- Manufacturing engineers will serve as "operations integrators" developing and coordinating the entire manufacturing process, from product design to service and support.
- Manufacturing engineers will serve as quarterbacks who will call signals for the entire manufacturing team.
- Manufacturing engineers will advise management and develop strategic manufacturing plans to meet company business goals.
- Companies will increase their use of outside services or contingency employees. No individual or company will be able to master all aspects of the changing technologies or staff engineering departments with every type of expert.[20]

FIGURE 9-4 Forecast of Technologies Used by Manufacturing Engineers

Technologies	Percent Currently Required to Use	Percent Required to Use in 2000
CAD,CAE,CAPP, or CAM	56%	69%
Expert systems, artificial intelligence and networking	11%	47%
Composite materials	16%	36%
Automated material handling	23%	58%
Simulation	17%	40%
Sensor technology, such as machine vision, adaptive control, and voice recognition	16%	51%
Flexible manufacturing systems	32%	56%
Laser applications, including welding/soldering, heat treating and inspection	17%	51%
Advanced inspection technologies, including on-machine inspection & clean-room technology	32%	57%
Integrated manufacturing systems	27%	57%

Source: "Countdown to the Future: The Manufacturing Engineer in the 21st Century," Profile 21 Executive Summary, © 1988 Society of Manufacturing Engineers.

Engineering Obsolescence

In the new manufacturing environment, the ability to understand and effectively employ advanced technologies is a prerequisite to effective competition. *Too many of America's engineers are working with obsolete bodies of technical knowledge.* For some time, a large portion of America's engineers have not had the level of skills required to do their jobs. As it takes several years of neglect to product an obsolete factory, an erosion has taken place in the knowledge base of the engineering community.

In firms where the phenomenon of success through *technology avoidance* is practiced, it is not surprising that engineering knowledge has been allowed to become obsolete.

- In the automobile industry, more than half of product and manufacturing engineers are operating with a knowledge base older than 10 years.[21]
- An aerospace company recently retired most of its engineering staff because it was too far out-of-date. Even more disturbing is that it took upper management a long time to notice.
- In the same Society of Manufacturing Engineers study cited above, less than half of the CEOs surveyed gave manufacturing engineers a "good" to "excellent" rating in supporting their companies' business objective.
- Some executives have discovered that orders they gave as many as five years ago to automate their production systems have not been carried out. In some cases, plant engineers did not know how to do it. In others, unwillingness to accept technological risk as a condition of technological advancement prevented modernization efforts. The last generation of manufacturing engineers insisted on risk-free solutions; Japanese engineers, meanwhile, were willing to adopt less than perfect solutions.[22]

Just as they have been reluctant to invest in factory hardware, companies have failed to invest in developing adequate engineering resources. All too often, companies have responded to training needs by sending a few engineers to a "short course" and then adding the task to their job descriptions.

The result is an engineering staff with obsolete skills, stripped of its most promising performers by more attractive opportunities in other fields. Companies are now facing an ever-widening gap between current capabilities and future needs.

The Solution to the Engineering Dilemma

The solution for America's engineering dilemma is not sending engineers to short courses. World-class manufacturers must actively grow and develop their engineers. Three challenges must be surmounted.

- **The U.S. educational system requires an overhaul, including a more significant focus on math and science skill development.** Education at the elementary, secondary, and high school levels must turn out a larger pool of qualified raw talent for American universities.

Further, American universities must encourage the attainment of better developed skills through advanced degree programs, while broadening the basic business and engineering education of undergraduates.

- **Employers must develop a commitment to continuing education and pick up where colleges and universities leave off.** Education is not inexpensive, nor is it prudent to ignore. Advances in technology require that engineers view their active education as a never-ending process. Manufacturers must redefine their budgets and invest in their technical resources with the same commitment with which they invest in stock plans and new plants.
- **Career paths for engineers must be redefined, providing more incentive for them to stay practicing engineers.** Hewlett-Packard has actively worked to return to the days of the "long thin engineer" rather than the corner office executive as the primary role model. Companies must make it clear that it is highly desirable to be a practicing engineer. They must create paths of advancement for engineers, not paths of isolation. Innovative career pathing combined with philosophical views of the engineer as company cult hero and compensation systems to match are vital to attracting and retaining the best engineers.

As manufacturers struggle with the engineering resource, one thing will remain constant. Engineers, supervisors, and production workers will continue to share a keen interest in the reward system utilized by their employers. No subject can inspire a more emotional discussion than reward and incentive systems for employees.

The final section of this chapter addresses employee reward systems. Because of the complex nature of this topic, the discussion is limited to reward systems for those involved directly with production.

Rewards and Recognition

At its most basic level, a company's reward system influences behavior. The reward system should be designed to synergistically align personal needs (physiological, safety and security, social, esteem, and self-actualization) with the company's profitability, productivity, quality, and innovation goals. An effective reward system must be embedded in every aspect of company management and policy.

This discussion focuses on innovative uses of compensation as a form of reward. Without questioning the power of financial rewards, managers must also recognize the equally important need to reward employees in nonfinancial ways. Recognition, praise, coaching, and showing concern for subordinates are all vital forms of reward that must never be neglected. Both company and employee benefit when the reward system induces entrepreneurial *esprit de corps*, a sense of pride and craftsmanship, and the ability to use one's talents to the fullest.

As part of a total reward system, compensation has a major role to play. Surveys conducted over the past 25 years have consistently found that:

- Most people in all types of jobs and industries believe in "pay for performance."
- Monetary awards are highly valued by virtually all employees.
- Most people feel they could perform better if they really wanted to, if they felt that such behavior would "pay off" in some way.
- Most employees see little connection between their pay and their performance.[23]

As U.S. companies struggled to improve their competitive capabilities in the 1980s, they increasingly turned their attention to nontraditional forms of compensation and reward. This section explores some of the most popular and innovative compensation systems put in place by leading manufacturers and the issues surrounding their use. The primary types of programs being implemented, their focus, and their advantages and disadvantages are summarized in Figure 9-5.

Merit Pay Programs

In use for decades, these programs award base pay increases on the basis of individual employee performance. The effectiveness and fairness of merit pay programs are highly dependent on the ability of management to accurately and objectively observe and evaluate performance. An accurate performance appraisal system is essential to an equitable and effective merit pay program.

When company goals are clearly defined, it is easier to identify behavior that contributes to company success. However, differentiating between clusters of good and outstanding individual performance can be very difficult. Many companies struggle with the implementation of such programs. The less specific the criteria to warrant merit pay, the greater the potential for problems. Often companies utilize merit pay for very specific

FIGURE 9-5 Comparison of Common Reward/Incentive Schemes

Type of Program	Application	Advantages	Disadvantages
Merit Pay	Individual	• Allows management to target specific behavior and to easily evolve criteria over time.	• Criteria tends toward arbitrary and unbiased when incorrectly administered. • Often not clearly tied to business goals.
Stock Purchase Plans	Individual (Applied across groups)	• Build employees stake in business; develops sense of ownership. • Supports retirement programs.	• Does little to motivate individual behavior or group behavior or performance.
Profit Sharing	Group	• Ties business performance to employee reward.	• Often individual or group behavior is not correlated to business performance.
Gains-Sharing	Group	• Specific group performance directly tied to employee reward.	• Often focuses excessively on cost control. • More applicable for tactical improvements than strategic changes.
Lump-Sum Bonuses and Individual Bonuses	Either	• Allows management to vary criteria and magnitude of reward; able to target specific actions and behavior.	• Often used for and seen as deferred compensation. • Not always a tie to business goals or performance.
Pay-for-Knowledge	Individual	• Allows management to target specific types of skills and personal growth.	• May not impact business performance unless management targets correct skills and applies new skills effectively.

and easily measured *event achievement*—the best individual quality rating or the highest number of parts produced, for example.

Merit pay will be used increasingly as a reward system into the twenty-first century. Many companies will use improvement as a criterion for merit recognition. This approach has the potential for becoming a driver to move the company toward the goal of increased competitiveness.

Stock Purchase Plans

Stock purchase plans are intended to provide the employee with both a real and psychological investment in the company's future. Typical plans entitle employees to purchase company common stock up to a value of 6 to 10 percent of their annual salary. Most broad-based plans provide for some degree of corporate matching of employee contributions. While these plans may contribute to a sense of ownership of the business, they provide no explicit link between company performance and individual effort. Poor management and a dysfunctional culture will still induce poor performance from subordinates, even with a significant stock purchase plan.

Profit-sharing Plans

Profit-sharing plans pay bonuses based on the firm's profitability. These payments may be in the form of lump-sum cash payments, time-phased payouts, stock bonuses, or investment in a profit-sharing fund. The underlying concept of all plans is the same. If profits are generated, there will be payouts; if not, there will be no payouts. The plans are designed to motivate employees to achieve short-term goals, to improve productivity, and to provide competitive compensation.

Profit-sharing programs are similar to stock purchase plans in that employees feel as though they have a stake in the company's financial success. For the company, a key benefit of profit sharing is that the fixed-cost component of pay programs is minimized.

As with all group incentive plans, profit-sharing plans minimize individual performance as a compensation factor. The singular dependence on company profitability may also impair the firm's long-term competitive ability. Firms with high amounts of profit sharing may be reluctant to risk innovation if the cost of failure is perceived as too high in individual terms.

Gains-sharing Plans

Similar to profit sharing, gains-sharing plans pay awards based on group or organizational results. The key difference is that gains-sharing programs are tied to labor cost savings. Gains-sharing programs are popular because they reduce the fixed-cost component of compensation. They motivate efforts to improve labor productivity and seem to have a greater effect on day-to-day improvement than does profit sharing.

The major difficulty with gains-sharing concerns the complexity of labor cost calculations and the equity of gains-sharing awards. Productivity measures are very difficult to develop in some areas; the potential for disagreement is great. In some systems, the potential gains-sharing bonus is highly dependent on the degree of inefficiency *before* the system was implemented. For example, in an inefficient operation, the potential payoff to workers and supervisors is greater than in an area that has already been made efficient by the efforts of its workers and supervisors. Since most plans are based on work group performance, there is no differentiation in awards for the best and the worst employees. Considerable difficulty in deciding to authorize awards is experienced when labor productivity improves but profitability and other performance measures decline.[24] In general, installing gains-sharing and profit-sharing plans has the potential of undermining motivation and altruistic behavior.

Lump-sum Bonus Programs

Lump-sum bonus programs use multiple measures to assess the level of organization and work group performance. The sizes of the awards are different, based on individual and organizational performance. The relationship between individual performance and individual contribution to company success is best understood by employees when the lump-sum award is preestablished and widely communicated.

A problem with this type of plan, however, is that when the business experiences a downturn for any extended period, employees begin to pressure to change the system. For example, in the years 1983 to 1987, Ford paid a substantial lump-sum reward in each year to Ford employees, while General Motors paid lump-sum amounts of only 30 to 40 percent of the Ford awards in the first three years and did not give any awards in 1986 and 1987.

Individual Incentives or Bonuses

Individual incentives or bonuses have been given primarily to managers and professional employees. Since the bonuses are based on individual contributions and efforts, they are frequently paid regardless of the firm's profitability. Bonuses are often awarded throughout periods when profit- or gains-sharing program awards are not made to the rest of the employees. The unfortunate result of this practice is a sense of resentment and inequity among those who worked hard but received nothing. The question frequently posed is, "If the company is profitable enough to pay executive bonuses, why isn't it profitable enough to pay profit shares?" Management rarely has an answer that satisfies everyone.

Individual incentives and bonuses are typically easier for managers to implement in that they can be based on very specific performance criteria. However, when the criteria established become unclear and individual bonuses become expected, they begin to resemble deferred compensation and have little real impact on behavior. Like merit pay programs, individual incentives and bonuses will continue to be a common and popular reward mechanism for management ranks in the future. However, this form of compensation will also be directed toward production workers as they assume more management and decision-making functions. An organizational structure built on team performance will allow more individual and small group evaluation. This will help develop more specific criteria for the recognition process.

Pay-for-Knowledge Programs

Pay-for-knowledge programs compensate employees based on the number of jobs they can do, rather than the job actually performed on any given day. These plans enhance production flexibility through the development of broader worker skills. Most often used in facilities operating under a team concept, these plans usually have three levels:

1. Starting at the entry level, employees receive pay increases as they acquire new skills.
2. When employees master the skills required within their own teams, they can start to learn skills outside their immediate areas of responsibility.
3. The third level is the all-plant level at which the employees acquire the skills necessary to move throughout the plant.[25]

Other types of systems include earned-time-off programs, all-salaried programs, and recognition programs. The recognition programs include Employee of the Month programs and the distribution of merchandise, monetary awards, or symbolic awards.

Pay-for-Performance and Knowledge – The Incentives of the 90s

With the exception of the merit pay and individual bonus programs for managers, most performance-based pay systems have been adopted quite recently. For example, more companies installed gains sharing and small group incentive plans since 1980 than in the previous 20 years.

A 1987 survey conducted by the American Productivity and Quality Center (APQC) of 1600 Canadian and American companies showed that 75 percent are using nontraditional reward systems somewhere in their organizations. According to the survey, 30 percent of respondents used lump-sum bonuses, followed by small group incentives (14 percent), gains sharing (13 percent), and pay-for-knowledge (5 percent). Gains-sharing programs grew at an unprecedented rate in the 1980s, with over 73 percent of the currently existing programs being implemented since 1980. Many Fortune 500 firms such as General Motors, Bank of America, and AT&T are moving from merit pay systems to systems that more directly tie pay to performance.

Most of the companies that have tried pay-for-performance programs have had positive experiences. Survey results show that 89 percent of those with pay-for-knowledge systems, 81 percent of those with gains sharing systems, 75 percent of those with small group incentives, and 66 percent of those with lump-sum bonus systems are "positive" or "very positive" about their system. Obviously, this requires that performance management systems in the twenty-first century be tightly coupled with the nonfinancial drivers of manufacturing performance defined in Chapter 6.

Productivity gains of 20 to 30 percent are not uncommon under such programs. Pay-for-performance and pay-for-knowledge programs can result in flexibility in staffing, skills, and operating systems to respond to shifting customer demands. The five top benefits of multiple skill or pay-for-knowledge programs are:

1. Increased flexibility and efficiency.
2. Lower staffing levels.
3. Higher quality of output at lower cost.
4. Lower absenteeism and turnover.
5. Greater long-term productivity.[26]

More than 90 percent of Motorola's U.S. employees are eligible to earn bonuses of up to 42 percent of base pay through some type of pay-for-performance or pay-for-knowledge program. In a typical Motorola location, scrap has been cut by two-thirds and WIP and employee turnover by half.[27]

Planning Issues

Before compensation and reward planning can take place, management must have a clear vision of where the firm wants to be in terms of human resources. Once this vision is clarified, compensation and reward programs can be custom-designed to propel the company toward its destination.

Reward systems are an integral part of how the entire enterprise is managed. Compensation planning must be approached as a major strategic investment, since the company's human resources are its longest-term assets. When viewed over a working lifetime, they represent the single largest investment the firm makes. Management cannot afford to view compensation planning as a personnel department issue or something that only requires a periodic quick fix.

Individual vs. Group Incentive?

Once the environment is deemed suitable for performance incentives, the advantages and disadvantages of individual and group incentives must be considered. Individual incentives focus on quantity versus quality. Without diminishing the strong link between effort and reward for individual incentives, most modern industrial jobs are interrelated. The more complex technologies become, the more difficult it becomes to measure individual performance.

As a company moves from traditional to cellular or JIT manufacturing, many previous incentive systems are rendered obsolete. Individual piecework incentives are counterproductive to the cooperation and teamwork needed to make these new production systems successful. On balance, small group incentives may be the most effective, given the increasing complexity of most manufacturing environments and the movement toward a team organizational structure.

Implementation Is Difficult

Implementation of new incentive plans is very difficult. It is a time-consuming process involving extensive consultation and agreement between the workforce and management before the plan is decided upon and implemented. One of the most difficult transitions to make is from an individual to a group incentive plan. Sharing information is critical to successful operation of a nontraditional system. A poorly designed and implemented plan has the potential to destroy many years of careful planning and progress.

The company must also evaluate whether the current job structures are compatible with performance incentives.

- Jobs must have the potential for meaningful performance variation.
- While self-paced workers can vary their output and quality, machine-paced employees have less influence. The employees must have, and believe that they have, the ability to perform at higher levels.
- The relationship between reward and effort must be perceived as equitable, and the system must be seen to be fairly administered.

Criteria for Success

The key components of any program that hopes to motivate employee behavior toward improving corporate competitiveness are:

- Program goals must be clearly established, and these goals must be reviewed regularly to ensure congruence with critical business goals and objectives, e.g., quality, flexibility, and cost.
- The specific criteria established for achieving the rewards and incentives must be clear, fair, achievable, and consistent with the overall program goals. Good or desired behavior should be easily identified by an examination of the criteria required to achieve the rewards.
- Whenever possible, team-based programs should have a significant element of "contingent upon" structure, in which the reward is contingent upon the achievement of specific criteria.
- The employee should perceive the reward as resulting from his or her individual and team contributions and thus must be clearly linked to the achievement of the firm's goals.[28]

During the 1990s, many more new and creative methods of compensation and reward will be developed. These new programs will continue to promote increased competitiveness and the achievement of goals and objectives. Over time, team-based programs will become the most common and popular. Manufacturing environments that lack any form of nontraditional rewards and incentives will be in the minority.

Conclusion

Throughout the 1980s, American companies struggled with improving their fortunes in the face of significant international competition. Many companies tried and failed at improving their lot by implementing advanced process technology. *For those companies that failed with technology, the typical common denominator was a lack of attention to the human element.* Successful companies have found that attention to the human element can provide a competitive advantage so significant that competitors are often unable to respond. Without question, the human edge spells the difference between competitors with a similar strategic focus and comparable manufacturing capabilities and technologies.

As manufacturing becomes more global and more technologically sophisticated, the human element becomes ever more important. *The twenty-first century will be marked by the human resource renaissance.* Successful companies will actively cultivate their human resources by carefully selecting the best and the brightest employees, implementing innovative team-based employee involvement programs, developing genuinely participative management approaches, and promoting desired behavior with creative reward mechanisms. They will continually train and retrain their employees.

Those manufacturers that first develop their human assets will have an edge in becoming world-class companies. Managers, as the catalysts of change in the twenty-first century, will be required to:

- Build unity of purpose by communicating and strengthening commitment to the company's vision.
- Lead and execute a customer-driven manufacturing strategy.
- Become creative in developing innovative human resource policies and rewards that are consistent with the business thrust.
- Develop cooperation, good employee relationships, and teamwork within and across functional units.
- Match the human resource strategy with the technological strategy of the firm.
- Lower the total costs of investing in human resources while simultaneously increasing value through better selection, technology, employee relations, and the application of accelerated learning techniques.

CHAPTER 10

TECHNOLOGY INVESTMENT AND REALIZATION

Now is the time to take longer strides . . . time for a great new American enterprise . . . time for this nation to take a clearly leading role in space achievement, which in many ways may hold the key to our future on earth.

John Fitzgerald Kennedy
May 25, 1961

Alarmed by the continuing erosion of America's position in world markets, the U.S. government is throwing money at "big science," relying on ambitious and highly visible research projects to boost national competitiveness in global high-technology markets. These projects include:

- A $4.4 billion plan for a superconducting particle accelerator.
- Billions for a space station.
- A $30 million research program for high-definition television.

The goal is to bolster U.S. technological competitiveness with the same cold war strategy used to beat the Russians to the moon. Twenty years ago, the National Aeronautics and Space Administration (NASA) was spending $4 billion a year on big science to put a man on the moon. Unfortunately, the emphasis on big science may be the wrong strategy today, because it fails to attack the root cause of America's competitive dilemma, namely, the lack of ability, resources, and government support to rapidly commercialize research and development.

Increased spending on big science research and development will accelerate the siphoning off of American ingenuity and know-how by global competitors. R. B. Reich, in *Scientific American*, describes the macro influences of U.S. policies on technological competitiveness and the steps the government may take to lead the U.S. back to technological preeminence. The U.S. already leads the world not only in the quality and quantity of research but also in spending.

However, attesting to the demise of U.S. competitiveness in global high-technology markets are the following failures:

- The U.S. share of the world semiconductor market declined from 50 percent to 37 percent between 1984 to 1988.
- U.S. companies no longer sell dynamic random-access memory (DRAM) chips on the open market.
- The market share of U.S. machine tool manufacturers has declined from 25 percent to 10 percent; Japan now dominates the numerically controlled machine tool industry.
- The U.S. has virtually no presence in the global consumer electronics market. The U.S. has no share in the videocassette recorder, motor-driven 35 mm autofocus camera, compact disc player, or fax machine markets.
- Over the last decade, the Japanese presence in the automobile market has expanded dramatically; Japan now claims 26 percent of the U.S. market.[1]

Most of these products were discovered or invented in the United States. Foreign competitors then reverse-engineered the products and refined them. *Technology transfer out of this country, in comparison with the amount of technology being imported or the speed at which new technology can be generated, is occurring at an alarming rate.*

The National Challenge

Our national challenge is to build a strong competitive base. The current national infrastructure, however, supports the continued erosion of the U.S. technology base.

- The U.S. laws and regulations governing financial markets reward short-sighted business planning. Demands for short-term profits by investors curtail investments that have long-term impacts on research and development, capital, and human resource investments.
- There are large gaps in U.S. industry's ability to access and apply state-of-the-art research findings on a global basis. In Japan, for example, foreign journals and research publications are translated into Japanese by the government within days of publication. Even access to U.S. government big science research is limited for commercial enterprises.

- The educational crisis in the U.S. is one of the most significant technological challenges of the twenty-first century. Even if employers invest in educating their workforce, American workers lack the mathematical and scientific sophistication to effectively absorb technology and develop new insights. This problem exacerbates the structural misalignment between available technology and the workforce.

Support for U.S. competitiveness at the national level requires a new orientation about the cause of the problem and a redeployment of resources. Figure 10-1 outlines six steps the United States can take to return to technological preeminence. They require investment in worker skills at all levels as well as innovations tied to specific commercial goals.

In the environment described in Figure 10-1, technology plays a key role in national competitiveness. But what role does technology play in manufacturing?

Manufacturing Technology

In the traditional sense of industrial engineering, technology is the comprehensive way in which people, machines, and materials interact to accomplish something. This is a limited view.

Manufacturing technology embodies five *interdependent* dimensions:

- **Physical Production Processes:** Design and layout, type and mix of equipment, movement and flow of people and materials, degree of automation, computer hardware, inspection, and simulation.
- **Product/Process Design:** Planning software to facilitate the design of products, including materials, parts, components, and features as well as design of processes and their interconnection with products.
- **Information Systems:** Software for communication, integration and coordination, intelligence, and production control.
- **Management Technology:** "Orgware" that supports the transformation process, including administration, communication, integration, coordination, knowledge capture, learning, process control, and rewards systems.
- **Product Materials Technology:** Core materials, attributes, part interconnection, and functions.

FIGURE 10-1 Six Steps Back to National Technological Preeminence with U.S. Government Support

1. **Scan the globe for new insights.**
   - Foster international conferences and trade shows.
   - Systematically gather and translate information from scientific journals, trade publications, and newspapers from around the world and make it readily accessible to industry.
   - Support U.S. students studying abroad or visiting foreign companies and university research laboratories.
2. **Integrate government-funded research with commercial production.**
   - Use industry-led R&D consortia to glean insights from publicly funded research and transfer that technology to American industry. Defense research has been an inefficient source of commercial spinoffs; basic research is often far removed from commercial applications.
   - Establish networks via consortia to provide technical assistance and joint research support to small- and medium-sized firms.
   - Provide incentives for individual firms to link R&D with commercialization and support incremental improvements in designs.
3. **Integrate corporate R&D with commercial production.**
   - Existing products and process should undergo continuous review as researchers, engineers, and technicians gain insights for improvement.
4. **Coordinate early adoption of industrywide standards so that emerging technologies will be compatible with each other and speed commercial acceptance.**
   - Allow competitors to vie with each other to establish a standard, but coordinate efforts to assure the efficient adoption of new technologies.
5. **Invest in technological learning.**
   - Provide incentives to individuals to invest in their own technological education.
   - Provide incentives to businesses to invest in the technological development of engineers and workers.
   - Provide support for development and dissemination of innovative training and development material, such as accelerated learning programs.
   - Offer technological assistance in the application of manufacturing technologies to small businesses.

Figure, concluded

6. Improve the basic education of all citizens.

- Overhaul the curriculum content and delivery format of primary and secondary education.
- Tighten science and math requirements for high school graduation.
- Support on-the-job training in new technologies.
- Provide incentives and support to teachers for learning about technology and revamping classroom curricula to stimulate critical thinking.
- Develop and disseminate materials to parents and communities concerning their contribution and role in the national educational mission.

Source: This material is drawn from the work of R.B. Reich, "The Quiet Path to Technological Preeminence," *Scientific American*, 261 (4), October 1989, pp. 41-47.

Manufacturing technology, in the broadest sense, describes the way in which these five dimensions interact to transform resources into products and services that accomplish a purpose. The purpose of manufacturing is to satisfy customers. Technology on any of the five dimensions may or may not be innovative — something new, state-of-the-art, or advanced.

Firms vary considerably as to the specific types of technology deployed within each category and the degree of innovation inherent in each. For example, many firms use state-of-the-art computerized numerical control (CNC) machines to create so-called *islands of automation.* Not many use integrating software to link machines together. There is wide diversity in the application of state-of-the-art technologies. NeXT Inc. applies multiple state-of-the-art technologies — JIT, CAD, CAM, TQC, intelligent software — to create a state-of-the-art factory and product.

Manufacturing Technology Policy

The importance of this holistic perspective on product and process technology is that the technology strategy determines the company's distinctive competency and, ultimately, its potential competitive advantage. A firm's *manufacturing technology policy* has been defined as the "long-range plan for modernization of the productive core of an organization and its key interface and support functions like engineering and materials handling or logistics."[2]

In a modernization effort, a successful program matches management and process innovation. The relative success or failure of advanced manufacturing process technology may be misaligned on other technology dimensions—especially management technology. Two examples of advanced manufacturing technologies, the benefits of which have not been fully realized, are flexible manufacturing systems (FMS) and computer-integrated manufacturing (CIM).

Flexible manufacturing systems are computer-controlled groups of semi-independent workstations linked by automated material handling systems. A landmark study comparing technically similar FMSs in the U.S. and Japan showed that the Japanese were significantly more successful in their application of the process technology. The study attributes the difference to the competence of management and the technical literacy of the workforce.[3] Management technology in the U.S. remained stagnant. Management treated the FMS as vintage technology for high volume, standardized production. Policies for integration of workers with proper skills to match the new technology were seriously lacking.

CIM was intended to be the union or fusing of information and manufacturing technology; it was to result in complete automation of the manufacturing process and integration with the finance, marketing engineering, distribution, and strategic planning processes within an organization. This union has occurred rarely—so rarely, in fact, that the typical measure of progress toward CIM has degenerated into assessments of the use of the individual components of advanced manufacturing technologies. Even by this piecemeal definition, the adoption of advanced manufacturing technology, as evidenced by changes to the production process that are based on a new science, has been very limited.

A 1988 capital spending survey of North American manufacturers by *Production Magazine* found a continuing trend away from investments in more sophisticated technologies like automated manufacturing cells, FMS, robots, and lasers. Findings included:

- The percent indicating plans to purchase a manufacturing cell (14.4 percent) was at its lowest point since 1984 (13.7 percent).
- The same held true for plans to purchase FMS technology; for example, 11.2 percent in 1988 vs. the next previous low of 11.5 percent in 1984.
- There has been a steady decline in plans to purchase lasers; from 16 percent in 1985 to 11 percent in 1988.

- The purchase plan decline for robots has been the most significant; 37 percent had plans to purchase in 1985 vs. 23 percent in 1988.

It is likely that significant mismatches exist among the various dimensions of technology. If this hypothesis is true, then discernible differences should be observed in the relative percentage of manufacturers who have moderate to state-of-the-art experience with product and process technologies. Furthermore, advances in product, process, and design technologies far outpace management's ability to understand and implement them. In the next section, current progress on various technological applications shows that the relative percentages of companies that are state-of-the-art are small.

Current Progress

In an age known for technological miracles, the progress of technology implementation in U.S. factories has been startlingly and disappointingly modest. The 1989 Report of the President's Commission on Industrial Competitiveness concludes: "Perhaps the most glaring deficiency in America's technological capabilities has been our failure to devote enough attention to manufacturing or process technology."

In 1988, the U.S. Department of Commerce reported on the adoption of 17 advanced manufacturing technologies in five industrial sectors: fabricated metal products, industrial machinery and equipment, electronic and other electric equipment, transportation equipment, and instruments and related products. About 68 percent of the manufacturers reported using at least one of the 17 advanced technologies; 23.1 percent indicated that they use five or more advanced technologies. The most commonly used technologies were NC/CNC, CAD/CAE, and programmable controllers.[4] These were unevenly applied.

Similar, but more detailed results were reported in the 1989 North American Manufacturing Technology Survey; however, because the study was biased toward leading manufacturers, the relative percentages of manufacturers who reported using these technologies in the 1989 study were higher than those reported by the Department of Commerce (see Figure 10-2).

Relatively old and unsophisticated technologies characterize the U.S. technological base as limited and immature. Technologies like flexible manufacturing systems and robotics have received little attention in most of the U.S. manufacturing community. U.S. manufacturers have very limited experience with system integration and the basic tools of manufac-

FIGURE 10-2 America's Use of Technology Is Limited

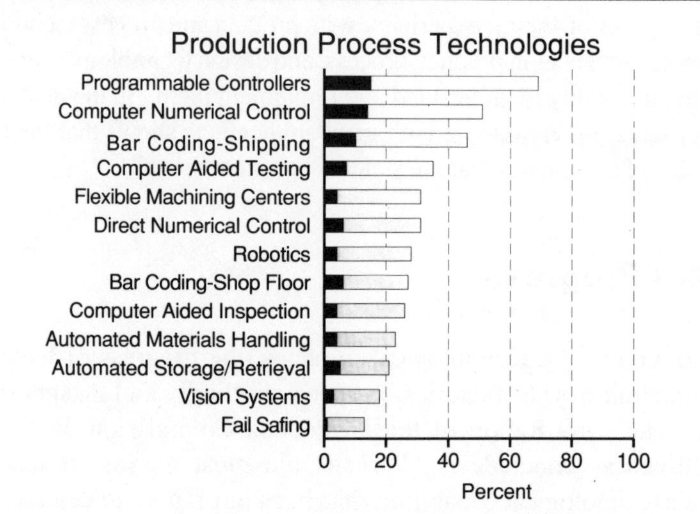

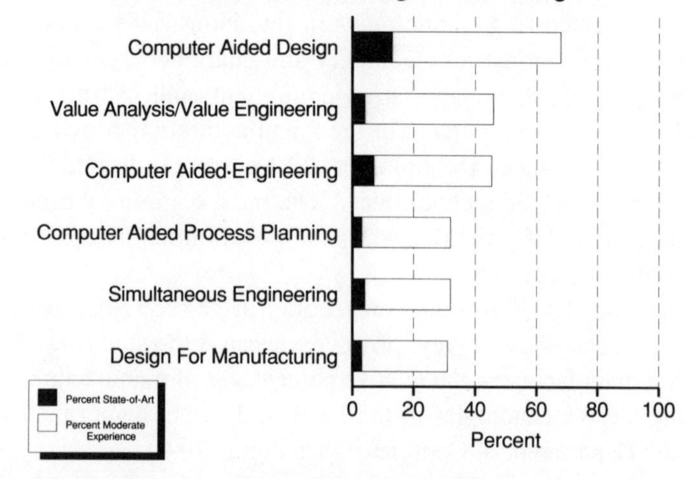

Source: C.A. Giffi and A.V. Roth, "Making the Grade in the 1990s," Deloitte & Touche Third Annual Survey of North American Manufacturing Technology, 1989.

turing integration. Local area networks (LANs), electronic data interchange, relational database management systems (RDBMS), manufacturing automation protocol (MAP), and Ethernet networks are unfamiliar to the typical U.S. manufacturer (see Figure 10-3).

FIGURE 10-3 America's Use of Information Technologies Is Limited

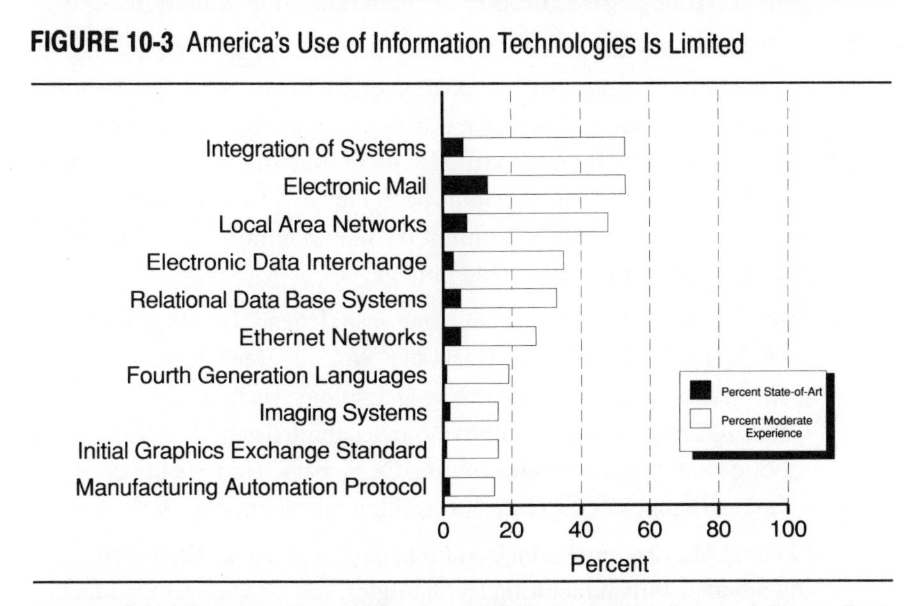

Source: C.A. Giffi and A.V. Roth, "Making the Grade in the 1990s," Deloitte & Touche Third Annual Survey of North American Manufacturing Technology, 1989.

There has been some technological progress over the last decade, especially in those industries and by those companies that are under competitive attack. The aircraft and aerospace industry is the most experienced with advanced techniques and technologies; Boeing, McDonnell-Douglas, Hughes, Lockheed, GE, and Pratt & Whitney have been among the most technologically advanced organizations. Pioneers in other industries have also forged ahead.

- Allen-Bradley's motor contactor facility in Milwaukee is a showcase for CIM. The World Contactor Center is a 45,000-square-foot factory within a factory that is capable of producing over 1000 variations of contactors at speeds reaching 600 completed contactors per hour. This facility requires just six people to oversee all operations.

- IBM developed a state-of-the-art facility in a 27-year-old plant in Kentucky. IBM's Lexington, Kentucky, typewriter plant was revitalized with an investment of more than $350 million. An automated materials handling system moves parts over seven miles, while automated parts retrieval systems select the correct parts and more than 200 robots are utilized in the manufacturing operations.
- NeXT Inc. has developed a factory that comes as close to "lights out" manufacturing as any in existence. At Steve Jobs's new facility, the ultra sophisticated NeXT computer is manufactured in a continuous process that requires virtually no human intervention. Robots using the latest in vision systems pack the NeXT motherboard more densely than any other commercial manufacturer can, providing NeXT with an unparalleled advantage.
- Deere & Co. has become an American legend for its success in revitalizing its John Deere Harvester Works. Deere's management team made steady investments in technology such as CAD, CAM, CNC, on-line gauging systems, automated guided vehicle systems, and robotic paint systems while bringing Deere back to competitiveness in an industry that saw extreme competition during the 1980s.
- General Motors, often criticized for being the first to fail in implementing advanced manufacturing technologies, has since developed facilities like its Saginaw Vanguard plant, which is among the most technologically advanced in the world. The Saginaw facility includes 23 manufacturing cells, more than 50 robots, and 12 assembly cells, and utilizes automated guided vehicle systems as well as an automated parts storage and retrieval system.

While the single largest category of capital investments being made by U.S. manufacturers is advanced manufacturing technology, the overall pattern for manufacturers' investments in technology indicates disdain for and distrust of technology by executives.

Europe is leading the world in creating and installing manufacturing cells. South Korea has the potential to become the next industrial miracle through its application of technology. And the Japanese are beginning to invest heavily in technology. The likely outcome: a widening of the competitive advantages the Japanese already hold in so many of our markets.

CASE STUDY

Gillette: Fighting Competition from Disposables with New Technology

Ten years in the making, Gillette Company's new Sensor razor cost more than $200 million to develop and start manufacturing. Add to that $110 million in advertising in its first year alone, and you have a product that had better be snapped up by millions.

If the new razor is a hit, it will cap the company's recent comeback and vindicate management's claims that its strategies would pay off for its shareholders over the long run. If the razor flops, charges that Gillette executives entrenched themselves at the expense of stockholders will be renewed.

Gillette, the repeated target of corporate raiders, would like nothing more than to score a victory for mainstream corporate America over Wall Street raiders. Takeover artists offered short-term profits; management promised a bigger long-term payoff—if only shareholders were willing to wait for promising new products. "If we don't get sizable returns" on Sensor, concedes Gillette vice chairman Alfred M. Zelen, "we ought to be criticized for having wasted lots of money."

Gillette is gambling on its investment in new technology to halt a 15-year trend toward inexpensive disposable razors. Gillette has made huge investments in new products and manufacturing efficiencies that curtailed earnings in the 1980s. But executives expect those investments to pay off in the 1990s.

Energized by its close shave with corporate raiders, Gillette slashed employment 8 percent and dumped mediocre divisions and subsidiaries. "But what we didn't do was change the fundamental way we do business," states Zelen.

For all its other businesses, Gillette means one thing to consumers worldwide: shaving. Gillette dominates the U.S. with a market share of 64 percent. Razors and blades account for one-third of sales and two-thirds of profits.

Disposables, once used by relatively few, now are the razor of choice for more than half of all American shavers. But profits on them are thin, because they cost more to make yet sell for less than the cartridges used in permanent razors.

Sensor, Gillette believes, can reverse the trend by using new technology to provide a better shave. Sensor has a novel suspension system that allows two blades, each with tiny springs soldered onto it, to move separately. Thus, the blades move up, down, and sideways with the contours of the face.

Gillette had hoped to introduce the new product in 1984 or 1985, but it could not figure out how to mass produce the cartridges. The biggest

Case Study, concluded

obstacle was attaching the springs to the blades on a high-speed assembly line that make 2.5 cartridges a second. No commercial laser could operate at the speed Gillette needed to weld the springs to the blade, so Gillette spent two years and $150 million to develop its own laser and manufacturing equipment for mass production.

Though competitors may be able to license the technology in the future, they may be hampered by the high cost and complexity of setting up manufacturing facilities. "To replicate Sensor would be a pretty tough undertaking," says one Gillette source.

Adapted by Francine Hyman from Lawrence Ingrassia, "A Recovering Gillette Hopes for Vindication in a High-Tech Razor," *Wall Street Journal*, 28 September 1989. pp. 1,3.

What Has Caused Our "Falling Out" with Advanced Manufacturing Technology?

The availability of equipment and the existence of essential technological ideas, concepts, and mechanisms has far outrun the ability and willingness of the modern factory system to adopt them.[5] Manufacturers simply do not believe they have been deriving benefits from their investments; the North American Manufacturing Technology Survey has shown a steady decline from 1987 to 1989 in the percent who believe they are receiving significant benefits from the application of advanced manufacturing technology. In fact, the percentage has dropped by half! (See Figure 10-4.) Further, as shown in Figure 10-5, all categories of benefits have tapered off over the last three years.

Why are U.S. manufacturers so frustrated by their attempts to derive benefits from their investments? In addition to the misalignment of technology dimensions, previously discussed, there are three major contributing factors:

- **Manufacturers continue to focus their investments on reducing the direct labor content of their products.** Overwhelmingly (2 to 1), the 1988 Capital Spending Survey found, direct labor cost reduction is the number one benefit expected from manufacturers' capital spending plans. This trend continues, despite the fact that the labor content of the typical U.S. product now contributes less than 15 percent of the total product cost.[6] It would take nearly a 50

FIGURE 10-4 Manufacturers Souring on Technology?

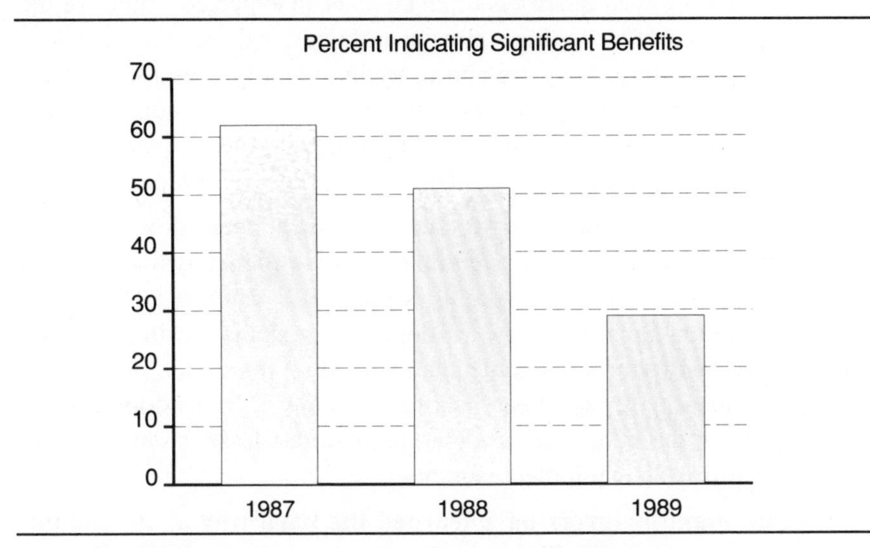

Source: C.A. Giffi and A.V. Roth, "Making the Grade in the 1990s," Deloitte & Touche Third Annual Survey of North American Manufacturing Technology, 1989.

FIGURE 10-5 Benefits of Technology Continue to Fade

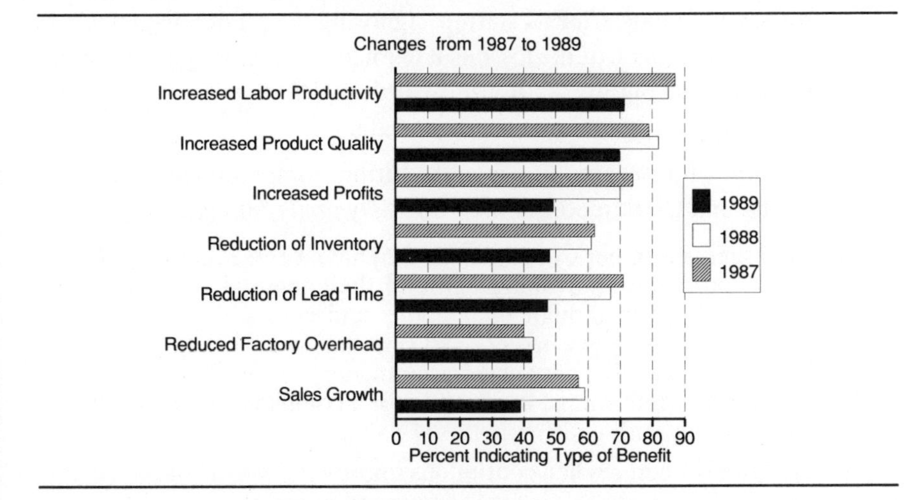

Source: C.A. Giffi and A.V. Roth, "Making the Grade in the 1990s," Deloitte & Touche Third Annual Survey of North American Manufacturing Technology, 1989.

percent reduction in total labor cost to yield a modest 6 percent improvement in gross margin for the typical manufacturer.

While improved quality is often cited as an expected benefit of the application of technology, most manufacturers do not measure their costs of quality. They are unprepared to identify even rudimentary benefits from the impact of technology on their product quality. Other benefits, including the strategic impacts of their investments, are typically either not considered or not analyzed properly.

- **Attempts to implement technology have been slowed down.** Progress has been slow due to the gaps in product availability and an inability to integrate products properly. Vendors and systems integrators have been slow to adopt open systems architectures and continue to squabble about standards. More important, basic product functionality has been lacking for many U.S. vendors, driving many of the manufacturers that do automate to seek out Japanese and European equipment vendors.
- **U.S. manufacturers have learned the hard way about the importance of a sound infrastructure prior to automation.** General Motors aggressively invested in advanced manufacturing technology throughout the 1980s but experienced several setbacks, including the much publicized problems at the Hamtramck Cadillac plant. Problems with robots, software, and integration with the workforce left Hamtramck one of the least productive auto plants in the country for several years after its startup. Ignoring the public display General Motors made of the lessons it learned in attempting to automate without first putting its house in order, U.S. manufacturers have repeatedly charged down the same path.

After years of research into manufacturing competitiveness, Hayes, Wheelwright, and Clark report in their book *Dynamic Manufacturing*:

> We have seen a number of companies that have been able to build a powerful competitive advantage around their internal capabilities and teamwork, even though their plants and equipment were not exceptional; but we have never seen one that was able to build a sustainable competitive advantage around superior hardware (technology) alone. . .It is almost impossible for a company to "spend" its way out of competitive difficulty.

Benefits rarely accrue when continuous improvements in both structural and infrastructural elements are not performed in tandem.

CASE STUDY

General Managers in Manufacturing and Innovation

It has been more than 10 years since the now well-known article entitled "Managing Our Way to Economic Decline" by Robert Hayes and William Abernathy appeared in the *Harvard Business Review* (July-August, 1980). Although everyone assumed the thesis in that article was that corporate presidents from law and finance were the cause of economic decline, little has ever appeared to substantiate that hypothesis. In fact, the *Wall Street Journal* (October 18, 1988) reported that CEOs from finance declined from 21.8 percent to 17.3 percent between 1984 and 1987, whereas CEOs from production and operations increased from 33.1 percent to 38.9 percent during the same period. What has been the impact of this changing profile of top management in manufacturing?

A study of more than three dozen companies that were in the process of modernizing their production processes and upgrading their products in 1987 shows that Hayes and Abernathy were only partially correct in their hypothesis. It is true that CEOs who have manufacturing experience do make an important difference in these North American manufacturing companies. In those companies that had manufacturing-experienced CEOs, there was a significantly higher likelihood that the company would take *calculated risks* and adopt new processing technologies such as flexible manufacturing and flexible assembly. These companies were characterized by four important differences:

- A reputation of being first to try new methods and equipment.
- An active campaign to recruit the best-qualified technical talent available.
- A commitment to technological forecasting.
- Awareness of new technological capabilities.

A company's commitment to *training* during modernization was much greater when senior vice-presidents and divisional managers had manufacturing experience. This suggests that training and development is still very much a strategic concern in North American manufacturing—although not as strategic as manufacturing technology policy. This commitment to training and development was reflected not only in plans and practices for training but in budgets as well. Training budgets for modernization that do not reach 10 percent of the total project cost cast serious doubt on the company's commitment.

Traditional senior managers were significantly more likely to emphasize *direct labor savings* from modernization and automation of assembly operations. Divisional managers, on the other hand, were far less focused on labor savings as an end-all of modernization. Further, divisional managers were significantly more likely than other general managers in the study to support the adoption of *administrative experiments*. These included the use of technology agreements in union contracts, flatter organizational structures in plants, and adoption of new, strategic plant charters for the future of the firm.

Not surprisingly, when divisional managers had manufacturing experience, the new system that was installed saw significantly higher *utilization* than when the divisional manager did not have manufacturing experience. In divisions where the general manager had manufacturing experience, average utilization was 80 percent as opposed to 61 percent.

Case prepared by J.E. Ettlie. Based in part on J.E. Ettlie, "Why Some Manufacturing Firms Are More Innovative," *Academy of Management Executives*, November 1990.

A Framework for Success

Several characteristics distinguish those companies that are the most likely to attempt to utilize advanced process technologies (see Figure 10-6). While these characteristics differentiate fertile from barren environments, the use of technology can grow anywhere.

Benchmarks for Implementing Advanced Technology Successfully

The characteristics of companies that automate and the processes that allow them to do so successfully vary from company to company and situation to situation. Research, however, has yielded a framework for manufacturers to use in implementing advanced manufacturing process technology. The framework is not a cookbook for success; the creation of competitive advantage through the application of advanced manufacturing technology does not lend itself to a recipe.

The five areas that seem to have the greatest impact on the success of any technology implementation effort are shown in Figure 10-7 and detailed below.

Top Management Viewpoint

Top management has broadened its perspective on technology to include more than just technical and equipment-related considerations. Management views technology not as individual components but as part of a total system of resources and capabilities. Top management acknowledges and understands the vital linkages between process technology, product design, suppliers, manufacturing engineering, human resources, and the shop floor. Technology decisions cannot be delegated completely to internal or external

FIGURE 10-6 Characteristics of Companies Most Likely to Automate

- Top management has a strong vision of the future and is able to communicate that vision effectively to the entire organization.
- Risk of investment tends to be a smaller percentage of the total cash flow. These are large companies or divisions of large companies with a significant capital base.
- A positive correlation exists between profitability and manufacturing process innovation.
- Historic earnings are depressed, future profitability is uncertain, and significant competitive pressures are mounting.
- Risk taking is encouraged and innovation is rewarded.
- Planning is a way of life.
- The financial function plays a complementary and nurturing role; "bean counting" is discouraged.
- Positive labor-management relations exist. Top management is valued by the workforce for their leadership abilities.

experts; top management maintains responsibility for the fundamental soundness of technology decisions.

Nature of Technology Decisions

Technology development follows a path rather than discrete steps. CIM, for example, is a step in the organization's development of technology, not the achievement of a certain level of technological development. Every technology decision can and should have a strategic impact. The investment in knowledge of and experience with technology is more important than investments in hardware and software. Technology decisions are viewed as *capability enhancing*. Each technology decision has the potential to both open future opportunities to the firm and close off other opportunities.

Focus of Technology Development

Technology development focuses on processes that are central to competitive advantage. Where management chooses to develop process technology

FIGURE 10-7 Areas that Have the Greatest Impact on the Success of Technology Implementation

1. **Top management's viewpoint:** The way in which top management thinks about technology and its place within the strategic framework of the organization.
2. **The nature of technology decisions:** The way in which specific technology decisions are regarded and the organization's understanding of the implications of those decisions.
3. **The technology development focus:** How the organization channels its vision of strategic advantage into technology development projects; how it decides where to expend resources and how to obtain technology.
4. **The justification process:** The process through which technology projects are evaluated and justified; the factors considered and the way in which cost and benefits are analyzed.
5. **The implementation process:** The mechanisms by which the organization implements its technology vision; the structure of the process and management's support of the process.

internally, the creation of trade secrets is the objective. In areas that are not central to the company's plans for competitive advantage, the best off-the-shelf technology available is used. The organization cannot tolerate a "not invented here" mentality. It must recognize that competitive advantage can be created through superior integration of purchased technology. Continual development and experimentation are encouraged to reduce the requirement to initiate high risk, great-leap-forward projects to catch up to competitors.

Justification Process

The technology justification process is based on the company's strategic understanding of the marketplace and its current competitive position. The response of competitors to specific technology investments by the firm is considered explicitly. The cost of not implementing new technology is factored into the justification and planning process. The costs of obsolescence and catching up with competitors are considered. Sunk costs are ignored in assessing relevant cash flows. Realistic hurdle rates that reflect the true cost of capital are used. The planning horizon used for evaluating the appropriateness of the investment is sufficiently long to reflect the technology life cycle. Net present value methods of justification evaluation are used only as a logic check for the soundness of the strategic decision, rather than as a hurdle to overcome. Chapter 5 details current justification policies.

Implementation Process

The company fosters an atmosphere supportive of innovation; constant experimentation and piloting of new ideas are encouraged. Incentives reward innovation. Fundamental elements of innovative behavior are tied to basic compensation. Projects phase in the new technology, supporting both the learning process and the developmental migration path.

Outrageous goals are established for technology implementation projects to force all aspects of current operations and plans to be challenged. Aggressive milestones are established for the project. At each milestone, success is measured against previously defined criteria.

A team approach has been adopted; multi-disciplinary teams are drawn from every area of the organization. The teams are comprised of the organization's brightest and most dedicated personnel. Management commits to keeping the teams together through project completion, and provides the financial support required, regardless of short-term earnings implications.

Can Technology Implementation Succeed?

Can success be achieved without using this framework? Possibly. But the real question U.S. manufacturers must ask themselves is why? Why attempt to implement technology within a different or modified framework? This framework is largely behavioral; it requires commitment and dedication. Every factor is within the control of management, and every factor significantly enhances the organization's probability of success. Adherence to this framework can result in a process that is both repeatable and sustainable. The alternative is failure.

The Essence of Failure

What most often goes wrong? This is an important question. The framework for establishing the ideal advanced manufacturing development environment is complex and difficult to create. Many problem areas can and will be encountered when a company attempts to implement advanced manufacturing technologies. The five most common are:

1. Lack of top management commitment and middle management initiative.
2. Inadequate education and training.
3. Plans that do not address strategic variables.
4. Settling for less than what's possible.
5. Not invented here (NIH) and the technical adventure.

These problems are described below and represent a formula for failure.

Lack of Top Management Commitment, Middle Management Initiative

You've heard it over and over again. Top management is to blame for not being committed to the project, the plan, the vision. Certainly the importance of top management's participation and leadership in the effort cannot be overemphasized. Anything but unswerving dedication to the cause results in short-lived and expensive one-time efforts that end in failure.

However, discussions with executive and middle managers have identified another, equally significant problem: lack of middle management initiative. For every advanced technology zealot found within the ranks of middle management there are 10 managers who display total apathy and such a lack of initiative that all the top management commitment and support in the world can never make advanced technology and, ultimately, CIM a reality in that organization. Lack of initiative for whatever reason—fear, inadequate financial incentives, lack of a sense of ownership—causes any automation effort to end in failure. While CIM fanaticism is hardly required or even preferred, middle managers must be motivated toward accomplishing significant goals and providing the fundamental energy that drives the advanced technology engine. Top management must ensure that the middle management team is prepared to act, not only in executing the vision, but also in expanding it.

Inadequate Education and Training and Development

Human resistance to change and lack of knowledge have been identified by numerous studies as the key obstacles to success in implementing advanced manufacturing process technology. Companies that have successfully implemented advanced manufacturing technologies like CIM—IBM, Motorola, Xerox, and Lockheed, for example—spend thousands of dollars per worker per year on education and training and development.

These expenditures are directed at not only the hourly worker but also the engineers, supervisors, and managers responsible for guiding the planning and designing of the implementation efforts. Those companies that fail in implementing advanced manufacturing technologies typically spend less than $100 per employee per year on worker training and education.

While a company cannot buy its way into a successful advanced manufacturing technologies implementation, most companies underestimate the importance of education and training and development.

- The creation and maintenance of a technical knowledge base is essential.
- Most companies' technical experts have less expertise than required to successfully implement advanced technology solutions.
- There are almost always fewer of these experts than required.
- The importance of overcoming basic ignorance and fear and developing a common understanding of capabilities, limitations, and implications is critical to quelling worker and management fears.

Putting advanced manufacturing technology into the hands of the uneducated and untrained will generate worse business performance than if the technology had not been introduced in the first place.

Plans Do Not Address Strategic Variables

Companies often do not understand why they are implementing advanced manufacturing technology. Their reasons may not have been derived from careful thought about the strategic variables of their business. If nothing else, U.S. manufacturers should have learned over the last decade that technology is not a quick fix to their competitiveness woes. Even General Motors had to rethink how its investments were contributing to the realization of its strategic success.

Technology plans must be integrated with business needs and critical success factors, including higher quality, lower cost, faster deliveries, and reliable delivery schedules. Competitors' responses must be gauged. While the strategic plans of successful companies incorporate technological solutions, unsuccessful companies often do not have strategic plans to start with. *Unless a company's technology policy is integrated into its strategic business plan, it runs the risk of failure from applying technological solutions to problems that do not exist or are not critical to the overall success of the business.*

Settling for Less than What's Possible

Few companies have attempted to automate their manufacturing environments for competitive advantage. They see the simple solutions the Japanese have applied to achieve victories in so many markets and the many moderate, short-term successes achieved by U.S. manufacturers in applying technology. Executives say: "It's good enough" and "don't fix what's not broken."

Smart companies understand that it never really is good enough. The competitive tide keeps pounding upon our beaches. If you think you have implemented enough advanced manufacturing technology, think again. Technology must be subject to the principle of continuous improvement.

The companies that fail are often the companies that stop along the way. Jodie Ray, senior vice president for automation at Texas Instruments, was asked if companies that have achieved 80 percent of the benefits of world-class manufacturing with only moderate investments in advanced manufacturing technology should bother to continue to invest in technology. He replied, "If all you're worried about is making it through 1990 and retiring, that's probably good enough. But if you're worried about an ongoing business, you must have a road map for continuous improvement."[7]

Not Invented Here and the Technical Adventure

The not-invented-here (NIH) syndrome is a common trap for companies implementing advanced process technology. Companies that have proud and competent engineering and information systems departments are those most often affected. MIT's Commission on Industrial Productivity concluded that, "Parochialism has blinded Americans to the growing strength of scientific and technological innovation abroad and, hence, the possibility of adapting the (new) discoveries for use in the United States."[8]

The NIH syndrome paves the way for the technical adventure. These same companies are so confident of their ability to create a better mouse trap that the creation process takes priority over all else, including return on investment, implementation time frame, and the necessity of creating something new. Unfortunately, this is all too often encouraged by a top management team that believes it can save a few dollars by "doing it ourselves."

Fortunately, leading companies have awakened to the costly and often destructive impact of NIH and the technical adventure. Motorola is a prime example of a company that overcame these problems with tremendous success.

Motorola's Paging Products Division, located in Boynton Beach, Florida, developed state-of-the-art CIM capabilities—capabilities that were created through a project called Bandit. The name Bandit grew out of a top management mandate for the project effort: *do not create what already exists somewhere else.*

The project team is proud of the fact that "we stole an idea from everyone." What did it steal? Japanese robots from Seiko. Computers from Hewlett-Packard, IBM, and Apple. Software from Automated Technology Associates and Bradly Ward. The Bandit team pieced together ideas and technologies from each of these vendors, concentrating its creative efforts on only those items necessary to differentiate competitive capabilities. Motorola now enjoys the leading market share for pagers in both the U.S. and Japan.

By watching out for and avoiding these common attributes of unsuccessful implementations, companies can achieve significant results.

The Implications of Success

Despite the gaps, leading companies are experiencing success with advanced manufacturing technology. Examples of benefits which have been achieved by a variety of companies utilizing advanced manufacturing technologies and techniques include the following:

- Flexible manufacturing systems can reduce production lead times by 30 to 60 percent, according to a study by Rockwell International. Other experiences with flexible manufacturing systems reveal benefits such as labor reductions of 50 percent, component price reductions of 25 percent, reduction in the number of separate machines by 10 to 50 percent, and decreased machine setup times.[9]
- Cellular production techniques have been responsible for results such as: work-in-process savings of 25 percent, floor space savings of 15 percent, production-output increases of 150 percent, and setup-time reductions of 35 percent.[10]
- Group technology, when applied with proper user education and computer processing support, can produce the following: direct labor savings of 15 percent, indirect labor savings of 15 percent, flow-time reductions of 70 percent, throughput-time reductions of 70 percent, industrial engineering time savings of 60 percent, process planning productivity increases of 50 percent, setup-time

reductions of 50 percent, and part design and retrieval savings of 50 percent.[11]

- CIM installations have been associated with these benefits: lead-time reductions of 50 to 70 percent, raw material inventory cutbacks of 60 percent, WIP reductions of 40 percent, material handling cost reductions of 40 percent, annual operating cost reductions of 25 percent, scrap reductions of 10 percent, output per worker increases of 10 percent, shortage reductions of 80 percent, manufacturing personnel reductions of 5 to 20 percent, overall manufacturing cost savings of 10 to 15 percent, and late shipment reductions of 80 percent.[12]

In addition to the measurable benefits attributable to the application and integration of advanced manufacturing technologies, there are a number of other benefits that are just as real but do not lend themselves to easy quantification. Such benefits can be seen in flexibility, cycle time, quality, asset integration, and customer satisfaction.

Conclusion

While technology has not been a key ingredient in the strategic agendas of most U.S. manufacturers through the 1980s, technology can and should be a key component of the competitive solution for world-class manufacturers throughout the 1990s and into the twenty-first century. Technological solutions still far outpace most companies' ability to understand and implement them. Consequently, technology transfer, not technology creation, should continue to be a primary objective for U.S. manufacturers.

Manufacturers are slowly overcoming their fears regarding their failures in implementing technology. Their lessons have been painful and costly. The formula suggested by those further along is:

- Develop a clear **vision** and make a tremendous **commitment** to it.
- Create **outrageous goals** and **experiment** constantly.
- Combine the best, brightest and most motivated **people** into autonomous **teams.**
- **Reward innovation** and thoughtful **risk taking** in all involved.

The world-class manufacturers of the twenty-first century will surely be leaders in the application of advanced manufacturing technology. In addition, they will be leaders in a number of other attributes of manufacturing, including the development of strategy, the management of human resources, and the application of the very best management techniques. In the last chapter, the attributes of world-class manufacturers are discussed and the operating principles for competitive success into the twenty-first century are defined.

CHAPTER 11

ATTRIBUTES OF WORLD-CLASS MANUFACTURING

Manufacturers today can no longer look at just the domestic markets for competitors. National boundaries have disappeared. Manufacturers from every industrialized nation on earth are expanding their markets. Companies aspiring to compete successfully are striving to become world-class manufacturers – able to compete with the very best the world has to offer.

This book has described the strategies, practices, and operations of world-class manufacturers within the United States and around the world. Throughout this book, the attributes that differentiate world-class competitors have been discussed and the principles that these world-class companies adhere to have been implied. As a result of this study, a common framework emerged. This framework (see Figure 11-1), first introduced in Chapter 1, underlies the success of these world-class competitors.

At the heart of the framework are quality and the customer. Far and away the most critical attribute of world-class competitors is the way in which they approach the development of quality and their focus on customer service. The definition of quality at these world-class companies is variable rather than absolute. This is because world-class manufacturers define quality from their customers' perspective, and customers' perspectives are ever-changing.

Interacting with quality and the customer are these elements of the manufacturing organization:

- Management Approach
- Manufacturing Strategy
- Manufacturing Capabilities
- Performance Measurement
- Organization

- Human Assets
- Technology

As this framework has emerged, so has a common set of operating principles. Within each of these elements in the framework is a set of operating principles that drives manufacturing excellence and sustains world-class performance. Companies that have achieved world-class performance have adopted these principles in total. Those that have missed the world-class mark have treated them merely as a *menu* of operating principles, selecting only those that can be implemented easily. While this book has focused on the approaches, techniques, practices, and strategies used by the best manufacturing organizations, it is the operating principles to which they adhere that provide world-class manufacturers with the constant and unifying force required to drive their quest for excellence.

This chapter reviews these fundamental operating principles and provides a view of those attributes which will differentiate the world-class competitors of the twenty-first century.

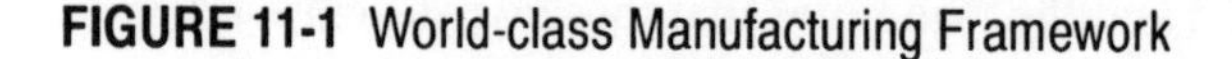

FIGURE 11-1 World-class Manufacturing Framework

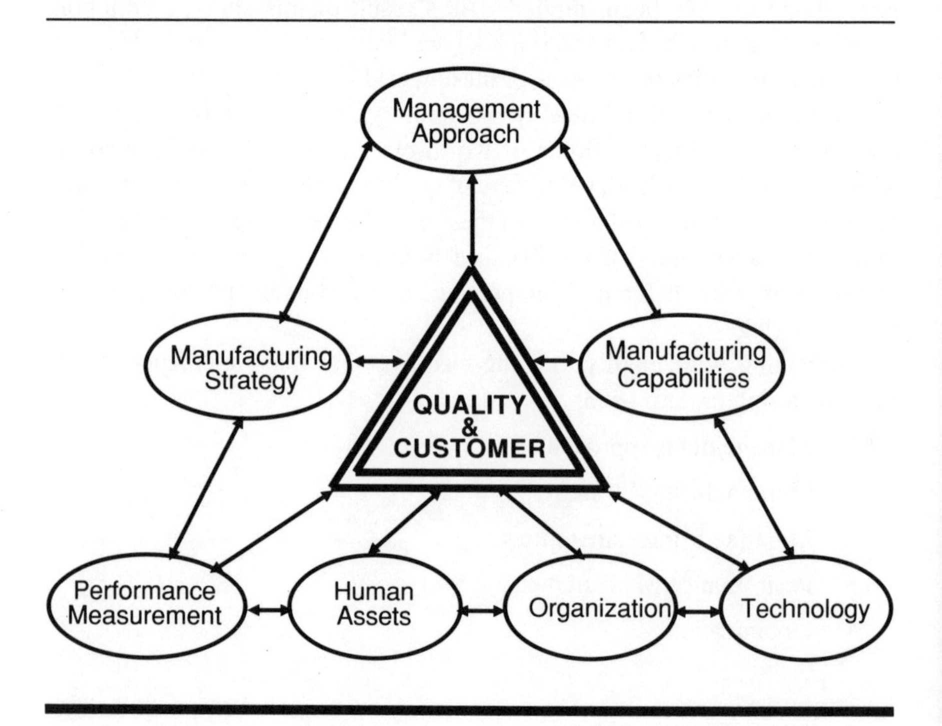

Operating Principles of World-class Organizations

The operating principles described in this chapter represent the critical success factors for manufacturers now and into the twenty-first century. World-class manufacturers will be those manufacturers that not only understand the significance of these principles but make them an integral part of their daily operations.

Executed in concert, these operating principles will increase the competitiveness of any manufacturer in a quest to be a world-class performer. The key, however, is to employ *all* the principles as a way of life. Since they do not represent a menu of options for improving performance, adherence to some while ignoring others will result in marginal overall improvement at best.

Becoming a world-class manufacturer is not an easy task; it requires a clear vision of the future and a tremendous commitment to the operating principles defined here. The growing number of excellent manufacturing organizations in the world today proves that the development of world-class manufacturing capabilities is not only possible, but required if the company is to remain competitive in the 1990s and beyond.

Quality and Customer

Superior quality no longer differentiates competitors; instead, it validates a company's worthiness to compete. While improving product quality is the first step toward regaining U.S. competitiveness in the 1990s, to move forward to the twenty-first century, U.S. manufacturers must expand their definition of quality to include all elements of both product and process. Attempting to compete on the basis of superior product quality alone will lead to competitive problems as surely as poor quality products unraveled the competitive fiber of the U.S. manufacturing community in the 1980s. In the twenty-first century, product quality will not be in question. But, to remain a serious global competitor, quality of process will be necessary. To achieve competitive cost structures and response times, a manufacturer's process must be superior. It must be efficient, effective, flexible, and in tune with customer needs.

Quality in the twenty-first century will become indistinguishable from customer service. Already in consumers' minds the concepts of quality of service and product are becoming intermingled. The Japanese assault on the U.S. auto industry, for example, has relied as much on exploitation of the poor service offered by too many U.S. auto dealerships, as it has on new and sophisticated high-quality products. It will soon be impossible for a product to be perceived as high quality if the service and support structure are not of equal stature. While this will remain more pronounced in consumer-related markets, all manufacturers will be subject to this broadening of the quality definition in the twenty-first century.

With regard to quality, these principles will differentiate world-class companies:

- *Define quality in terms of the customers' needs. Make customer closeness the number one priority.*
- *Integrate the concept of customer closeness into the organization so that everyone in the organization has a customer, and everyone's goal is to provide quality product and service to his or her customer.*
- *View quality from a global perspective. Achieving quality of product should be no more and no less important than achieving quality of process and service.*

The world-class competitor of the twenty-first century will expand the definition of quality along all dimensions of product, process, and service.

Management Approach

The development of manufacturing excellence requires management direction that is firm, innovative, and aggressive. Excellent manufacturing executives have developed a vision of the future that provides consistent guidance in their leadership. From top to bottom, world-class manufacturing organizations share the values and vision of their top manufacturing executives. Top management leadership, through shared vision and values, will continue to differentiate the world-class competitors of the twenty-first century.

It is management that develops and promotes the superior manufacturing strategies. Management defines the customer as the most important element of success; has the vision to adopt the appropriate technologies at the right time; reorganizes to promote learning, growing, team-oriented work environments; and develops the rewards that promote the never-ending cycle of continuous improvement so critical to success. To accomplish these goals, management must monitor the indicators that will help them identify problems, formulate solutions, and implement the appropriate actions.

In the twenty-first century, it will be management that provides the vision to redefine the concept of manufacturing from that of the isolated factory to that of an important element being merely a stage in the process of creating national economic value. The factory and the product as the output of the factory are narrow definitions of manufacturing that will be redefined. Peter Drucker describes manufacturing as "the process that converts things into economic satisfaction."[1] In the evolution of manufacturing in the next century, it will become clear that production does not stop when the product leaves the factory. As Drucker states: "physical distribution and product service are still part of the production process and should be integrated with it, coordinated with it, managed together with it."

To create this new manufacturing system of the twenty-first century, managers will be required to work across functional boundaries, integrating the organization with both its internal elements and its external relationships, including customers and suppliers. As described in Chapter 3, boundary management will become a critical element of world-class management in the twenty-first century. The manufacturing manager of the future will be a boundary manager, integrating people, skills, equipment, requirements, time, and capital in a system focused on achieving the economic objectives of the organization.

The operating principles for world-class companies defined below will lead to success only when executed by world-class executives who can create and share their vision of manufacturing excellence and who are capable of leading their organizations through the continual improvements required to become and remain world class. To do this, they must:

- *Develop firm yet open management direction, strategic in thought and effective in the implementation of innovation.*
- *Constantly establish "stretch goals" for the organization that require continual, incremental improvements.*
- *Foster an environment in which sensible risk taking toward innovation in product, process, and service is considered a fundamental necessity for long-term viability and is rewarded accordingly.*
- *Develop a thorough understanding of the products that are being produced and the critical manufacturing process capabilities required.*
- *Develop a systems perspective on manufacturing, one which treats manufacturing as a process extending from customer need to customer satisfaction.*
- *Manage the organization across boundaries. Establish transparent borders between customers, suppliers, and departments within the manufacturing system.*

Manufacturing Strategy

Companies that have risen to global leadership positions have done so with efforts that were beyond their original capabilities and resources. This notion of achieving global dominance through long-term adherence to a set of simple but consistent concepts of winning has been defined as *strategic intent.* The development of superior strategic plans is achievable only as organizations fully define their strategic intentions.

Following from their strategic intentions, world-class competitors have clearly defined their manufacturing strategies and plans. These strategies are congruent with their overall business goals and objectives and flexible enough to adapt to change. The real winners of the twenty-first century will be skilled in defining strategic intent and implementing the resultant strategy. The winners will not necessarily be the best in the art of production *per se.* As Kim Clark points out, "It is nearly impossible to build a lasting edge through a unique device developed by R&D or through an innovative, computer-driven process. There is not a theoretical limit to getting good at making things, no permanent advantage to being first."[2]

Customer-driven strategies, the highest evolution of strategy, allow a company to pull customer requirements and customer needs through the entire organization. Often these needs have not yet been articulated by the customer, yet they form the essence of the opportunity to achieve competitive advantage.

The operating principles that will underlie the development of manufacturing strategy in the twenty-first century include:

- *Establish a clearly defined strategic intent; define success in terms of winning for the long term.*
- *Establish a strategy consistent with the potential for developing needed manufacturing capabilities. Base the strategy on a realistic assessment of capabilities and priorities, but always look to push beyond the present.*
- *Develop a global perspective on competition, responding to international competition with at least the same intensity as to domestic challenges.*
- *Make the strategy more than a formal statement of policy. Make it a blueprint for action, a pattern of decisions to be executed over time. Deal with structural (brick and mortar) elements, infrastructural ("soft") issues, and integration elements.*

- *Put the strategy in writing. Manufacturers that have written strategies have been more forceful in the change process than those that have no written strategy.*
- *Develop the strategy through a participative approach and freely share it with all employees in the organization.*
- *Develop a strategy that is flexible and adaptive over time as the competitive environment changes. Review the strategy on a scheduled, periodic basis to ensure congruence with current and future goals and capabilities. Make modifications and enhancements as required.*
- *Implement the strategy to effect change as a gradual process of small, continuous steps over time. Allow your strategic intentions to govern the size of your steps.*

Manufacturing Capabilities

To achieve superior manufacturing capabilities, many priorities, such as adherence to quality, dependability, cost, and flexibility, are required. While the timing and staging of these priorities may vary from company to company, manufacturing excellence in the future will require the organization to achieve multiple capabilities that set it apart from its competition.

Indeed, competition appears unlikely to stop at dependable deliveries or delivery speed. After reduced cycle time comes cost effectiveness. Flexibility, in terms of both design and volume, is the ultimate competitive capability in the time-based race.

Lowering the breakeven point, i.e., reducing the fixed cost of the business, implies greater flexibility. When this is accomplished, the volume of output required to break even can be dramatically reduced, giving the company more maneuverability in tight markets and greater profitability in expanding markets.

In the twenty-first century, it is likely that world-class companies will have merged the concepts of quality function deployment and simultaneous engineering to achieve the flexibility and cost advantages required to command markets. Speed, dependability, cost advantage, and flexibility are linked to this manufacturing capability. Superior manufacturing capabilities in the future will be judged by the degree to which QFD and simultaneous engineering can be combined to deliver superior processes and products and satisfied customers. This will require a restructuring of both the product and process engineering functions and the intimate linking of these functions with those traditionally defined as quality control.

Leanness of operations, achieved through techniques like Just-in-Time and focused factories, will also be fundamental to the world-class environment of the twenty-first century. No other attribute was more closely linked to manufacturing success in the 1980s than efficiency of operation. As a result, JIT became synonymous with success. While success in the future will depend on much more than a lean operation achieved through JIT, leanness will be a common component of the foundation of successful companies.

Development of manufacturing capabilities and processes in the twenty-first century will have to take into consideration environmental and product safety requirements. This activity will be the result less of a desire for corporate "good citizenship" than of a pragmatic approach to reducing product development and delivery times. A common element of the simultaneous engineering methods of the twenty-first century will be identifying product safety and building specifications needed to secure approval from environmental watch dog organizations. Working closely with these organizations during product and process design, across traditional boundaries, will allow the post-process and product development time, often spent waiting for approvals, to be reduced. Products will arrive at the market sooner and at less cost. While this practice has already been incorporated by many manufacturers today, particularly those in highly regulated industries, it will become common for all companies in the future.

These principles will underlie the development of world-class manufacturing capabilities in the future:

- *Make dependability and consistency in quality, delivery, and service to your customers the goal of all operations.*
- *Develop manufacturing operations that are flexible and able to respond rapidly to changes in products and markets. Value leanness of operation above all other manufacturing performance indicators.*
- *Restructure engineering operations to reduce waste and inefficiency and improve quality. Develop engineering capabilities characterized by adherence to concurrent development techniques that integrate customer quality requirements.*
- *Consider the environmental impact of all products and processes, making environmental considerations an integral component of the design and production processes.*

Performance Measurement

Performance measurement systems may be transformed more significantly than any single element of the manufacturing system. This transformation is underway today, and it must be accomplished if U.S. companies are to compete globally. Traditional systems, focusing on outdated and inaccurate measures of financial performance, are already beginning their long overdue change. The critical activities of the manufacturing system are being defined, and drivers of the associated costs are being identified. Operational measures that drive customer satisfaction are rapidly becoming the key metrics by which success is determined in world-class companies. The twenty-first century may see the long battle between Wall Street (the short-term financial focus advocates) and the customer (the quality, dependability, and value advocates) reach a logical end. Their goals have always been the same, but their measures of success have always been different. In the twenty-first century, customer-valued metrics will prevail, as more and more companies find that superior performance in the area of customer metrics yields financial success.

With varying global exchange rates making it difficult to compare performance among geographic regions and with accounting systems changing on a global scale, the twenty-first century competitor will find customer satisfaction, market share growth, and network relationships the only measures by which true performance can be judged.

At an operational level, time will be the measurement by which world-class competitors will differentiate themselves in the future. Time-based competitive capabilities, such as dependable delivery time, delivery speed, and flexibility, will be the measures of success for the factory of the future, much as quality is the current measure of success. It is not that quality will no longer be important, but superior quality will be assumed.

The principles that will guide the performance measurement methods of world-class companies in the twenty-first century will include:

- *Focus on the competitive variables the customer requires. Allow customer success to drive financial success.*
- *Promote and measure knowledge and skill development. Develop measurement systems that encourage continual learning.*
- *Increase the vitality of the entire business by focusing attention on integrated business management. Measure the effectiveness of the boundary management that occurs.*

- *Tailor the performance measurement system to the company's strategic action programs. Manufacturing strategy is defined explicitly in terms of performance measurement. As strategy changes, so too should the critical measures of success.*

Organization

World-class manufacturers have streamlined the operations of their companies. They now operate from organizational structures that have been made lean and flexible. They have adopted a philosophy of waste reduction, and they constantly monitor and evaluate ongoing operations from a value-added perspective. These companies have elected to regain and sharpen their competitive positions by removing the traditional organizational structures which prevented management sensitivity to factors critical to success.

Organizations of the twenty-first century will look very different from those of today. While focused manufacturing units will be common, the organizational boundaries between units will be fuzzy. This fuzziness will be the result of two trends that are reshaping our traditional understanding of organizational entities. First, joint ventures will grow in importance, number, and success. These ventures will evolve toward true partnerships, in which common, long-term economic objectives will be satisfied. As a result, the ability to clearly delineate between "our" organization and "their" organization will diminish. As ownership, strategies, tactics, and marketplace success become intertwined, ours versus theirs will become a trivial and even harmful distinction.

Second, the power and utility of the R&D consortium will produce a similar effect on the critical research activities necessary to compete on a global basis. Consortia, such as NCMS, will become more common, more effective, and more open. The knowledge transfer, sharing, and synergy that result will further blur the lines between traditionally separate organizations. These blurred boundaries will become the symbols of the successful organizations of the twenty-first century.

The operating principles for organizing the manufacturing system in the twenty-first century will include the following:

- *Focus factories into organizationally flatter structures built around strategic business units.*
- *Dissolve the boundaries between management and worker and between functionally segregated staff units, to create dynamic cross-functional teams charged with resolving both strategic and operational issues.*

- *Embrace the advantages of joining forces with suppliers, customers, and even competitors. Identify the common ground and organize the manufacturing system around those elements that yield common and mutual measures of success.*

Human Assets

World-class status in the 1990s and beyond is not dependent on technology or manufacturing techniques. It is dependent on the talent of the people who will implement and utilize the best of the technologies and techniques. The need to attract and retain the best and brightest is clear. Filling this need is not an option; it is a requirement for those who wish to be world class.

Training programs for world-class manufacturers in the twenty-first century will be tailor-made to the specific requirements of their products, manufacturing processes, and customers. Training and education will be a significant component of every manufacturer's operating budget. World-class manufacturers will recognize the significant challenge presented by training and education requirements and respond without waiting for government relief or incentives. They will work closely with academicians to ensure the best possible products from secondary and technical schools and universities.

Teams of workers will be empowered to operate the manufacturing system. Without the supervision of nonvalue-added foremen, these teams will be rewarded for their measurable success on parameters that directly affect the overall success of the company. Workers will be valued for their knowledge, skills, and ability to learn. Managers will be valued for their ability to innovate, lead, challenge, and facilitate the work of teams. The new role of management is to create and nurture the project teams whose capabilities provide competitive advantage. What will be managed are intellectual and skill capabilities, not equipment.[3]

The human resource principles under which world-class companies will operate in the twenty-first century will include the following:

- *Invest in people; develop a pattern for updating workforce skills and capabilities consistent with the evolution of technology within the organization.*
- *Empower teams of workers to carry out the mission of the organization. Seek ways to liberate the teams from traditional organizational controls, and reward and motivate, based upon ability to achieve meaningful goals.*
- *Eliminate the terms supervisors and supervision. Develop leaders who can create and execute the strategic vision through the teams.*

- *Evaluate the success of your human assets on the basis of their ability to learn, adapt to change, and improve performance within their areas of responsibility.*
- *Develop accelerated and integrative learning programs.*

Technology

Implementing new technology is neither the problem nor the entire answer. Infrastructural components of manufacturing are the critical foundation upon which to build a lasting competitive architecture. Leapfrogging the competition by employing technology without first having developed a sound infrastructure is a strategy for only those unwilling to accept the realities and challenges of the real competitive issues.

In the twenty-first century, technology will still be viewed as a tool to be applied to create competitive advantage. However, the adoption of new technology will be part of a continuous improvement process. It will play a major role in the renewal process and be a rallying point around which organizations grow and develop their knowledge base. Technology will be valued not so much for what it can do as for the knowledge it creates within the organization. Competition will be waged on the basis of knowledge, and technology will be the catalyst that drives the knowledge base to ever-higher levels.

The U.S. government will play a major role in the success or failure of domestic companies attempting to adopt advanced manufacturing technology in the twenty-first century. The degree to which the government develops an environment that supports research and development, the adoption of standards, and the production of skilled and knowledgeable workers will have a dramatic impact on the success of North American companies. While the final determinant of success will be the ability of the individual company to seize opportunities and properly implement the appropriate technologies at the right time, the influence of government will determine the extent of that success.

The operating principles relating to the adoption and use of advanced manufacturing technology in the twenty-first century will include the following:

- *Develop an investment strategy for the continual enhancement of technology throughout the organization, based on a clearly defined vision of future competitive requirements.*
- *Identify the competitive advantage of the knowledge base that advanced technology can create; simultaneously implement new technology and develop the new knowledge base.*

- *Carefully plan technological upgrades to be consistent with infrastructural upgrades. Benefits can be achieved only when the infrastructure is capable of integrating and exploiting the technology advantage offered.*

The Challenge for the Future

The challenge facing North American manufacturers is clear. The ability to respond to this challenge now and in the future will determine not only the continuing viability of manufacturing in North America, but, in many ways, the welfare of the nations as well. Lasting, sustained competitiveness cannot be created overnight and will never be reached if manufacturers focus on only some of the elements in the manufacturing equation. U.S. manufacturers today must have a clear vision and a strong commitment to face up to the hard work and tough decisions required to become and remain world class.

The elements necessary for becoming a world-class manufacturer are within the grasp of all competitors. Giants like General Motors, GE, and Toyota have no more or less opportunity to achieve world-class capabilities than does the small $10 million-a-year parts manufacturer that serves as a link in the large OEM's supply chain.

Being world class is more about an approach to competition than about company size or resources. World-class companies realize they must approach manufacturing from a balanced perspective, placing equal emphasis on structural, infrastructural, and integration elements of strategy.

- *They regard customers as the core of their existence.*
- *They invest judiciously in technology and in their employees, eager not to become complacent about their capability advantages relative to their competitors.*
- *World-class companies are nimble, responsive, and innovative. They closely monitor performance indicators in a continual search for incremental improvement, careful to reward all contributors to continued success.*

The final and, perhaps, most important component of developing world-class capabilities is managerial leadership. World-class companies are managed by leaders who have a clear vision of the future and who share this vision with all employees. These leaders, through their own convictions regarding manufacturing excellence, are the catalysts that created and now sustain world-class environments.

American manufacturers have both the capability and the resolve to respond to these competitive challenges. The drive of many U.S. companies to become competitive at a world-class level has already begun. The future is bright with promise for manufacturing in the U.S. and Canada. However, the stakes are also very high; there is much to gain and much to lose. Success in meeting this challenge will result in lasting competitiveness and nations of world-class manufacturers.

Notes

Two studies are referred to throughout this book:

Manufacturing Futures Project

The Manufacturing Futures Project (MFP) was initially developed at Boston University in 1981 by Professor J.G. Miller. In 1983, the project expanded internationally. The European MFP was sponsored by INSEAD University. The Japanese MFP is run by Waseda University. Each year, manufacturing executives from about 500 business units in the U.S., Europe, and Japan complete a survey. The questions pertain to manufacturing strategy, goals, structure, performance, and action plans.

The responses cover large manufacturing enterprises in each country and are biased toward industry leaders. Major industrial sectors covered include electronics, basic industries, machinery, industrial goods, and consumer products. Complete details of the study can be found in A.V. Roth and J.G.Miller, *1988 Manufacturing Futures Factbook*, Boston University Manufacturing Roundtable Research Series, Boston, MA, 1988.

Survey of North American Technology

The North American Manufacturing Technology Survey, sponsored by Deloitte & Touche Manufacturing Services, has been conducted annually for the past three years. It gathers data on the current and future uses of manufacturing technology, the role of technology and manufacturing innovation in developing critical success factor capabilities, and manufacturing performance.

The 759 North American manufacturing executives responding to the 1989 survey are representative of a wide variety of manufacturing settings, including electronics, food, basic industries, fabricated metals, machinery/machine tools, aircraft/aerospace, and transportation/automotive. No single industry dominates the responses. Respondents are market leaders with a market share for their primary products exceeding 25 percent. Over one quarter of the respondents claim market dominance; their primary products exceed 50 percent market share. Over 42 percent of the manufacturing units were part of a Fortune 500-sized company, with annual sales exceeding $500 million. Details of this study can be found in C. Giffi and A.V. Roth, "Making the Grade in the 1990's." Deloitte & Touche Manufacturing Services, Cleveland, OH, 1990.

Chapter 1: The World Surpasses American Manufacturing

1. U.S. Office of Technology Assessment, "Making Things Better—Competing in Manufacturing," 1989, p. 5. Second report in a series of three in OTA's assessment of Technology, Innovation, and U.S. Trade. The assessment was requested by the Senate Committee on Finance; the Senate Committee on Banking, Housing and Urban Affairs; and the House Committee on Banking, Finance and Urban Affairs.

2. Michael L. Dertouzos, Richard K. Lester, and Robert M. Solow, *Made in America—Regaining the Productive Edge* (Boston: MIT, 1989), pp. 12-13.

3. Michael E. Porter, *The Competitive Advantage of Nations* (New York: The Free Press, 1990).

4. Robert H. Hayes and Steven C. Wheelwright, *Restoring Our Competitive Edge; Competing Through Manufacturing* (New York: John Wiley & Sons, 1984), p. 3.

5. Robert H. Hayes, Steven C. Wheelwright and Kim B. Clark, *Dynamic Manufacturing* (New York: The Free Press, 1988), pp. 24-25.

6. Ernest C. Huge and Alan D. Anderson, *The Spirit of Manufacturing Excellence: An Executive's Guide to the New Mind Set* (Homewood, IL: Dow Jones-Irwin, 1988), pp. 24-25.

7. Richard J. Schonberger, *World-class Manufacturing: The Lessons of Simplicity Applied* (New York: The Free Press, 1986), p. 2.

8. Robert W. Hall, *Attaining Manufacturing Excellence* (Homewood, IL: Dow Jones-Irwin, 1987), p. 280.

9. J.G. Miller, A. Amano, A. DeMeyer, K. Ferdows, J. Nachane, and A.V. Roth, "Closing the Competitive Gaps," *Managing International Manufacturing* (K. Ferdows, ed.) (Amsterdam: Elsevier Science Publishers, 1989), p. 159.

10. Amy Borrus, "Beating America to the Punch," *International Management* 43 (March 1988), p. 31.

11. U.S. Department of Defense, *Bolstering Defense Industrial Competitiveness*, Report to the Secretary of Defense by the Under-Secretary of Defense, July 1988, p. 23.

12. *Ibid.*, p. 22.

Chapter 2: The Quality Revolution

1. M.S. Flynn and R.E. Cole, *Automotive Suppliers: Customer Relationships, Technology, and Competition: Report of the Supplier Change Project*, Industrial Technology Institute, July 1986, p. 27.

2. A.V. Roth, "A Vision for the 1990's: Gearing Up for the 21st Century through Intelligent Manufacturing," Duke University, 1990.

3. A. De Meyer, "Quality Up, Technology Down," INSEAD Working Paper No. 88/65 (Fontainbleau, France).

4. K. Ferdows and A. De Meyer, "Lasting Improvement in Manufacturing Performance: In Search of a New Theory," INSEAD Working Paper (Fontainbleau, France), January 1989.

5. Organizational Dynamics, Inc. (ODI) Executive Opinion Survey. Boston, Ma.

6. Robert Hall, "Quality Essential: Capability to Improve Continuously," *Target* (publication of the Association for Manufacturing Excellence) 3.1 (Spring, 1987), pp. 5-11.

7. Patricia P. Mishne, "A New Attitude Toward Quality," *Manufacturing Engineering*, October 1988, pp. 51-52.

8. Stanley J. Modic, "What Makes Deming Run?" *Industry Week*, 20 June 1988, p. 84.

9. Robert A. Dumas, Nancy Cushing and Carol Laughlin, "Making Quality Control Theories Workable," *Training & Development Journal*, February 1987, p. 310.

10. Jeremy Main, "Under the Spell of the Quality Gurus," *Fortune* 114 (18 August 1986), p. 34.

11. Carl Kirkland, "Taguchi Methods Increase Quality and Cut Costs," *Plastics World,* February 1988, p. 45.

12. L. A. Ealey, *Quality by Design: Taguchi Methods and U.S. Industry* (Dearborn, MI: ASI Press, 1988), pp. 61-62.

13. G. Taguchi and D. Clausing, "Robust Quality," *Harvard Business Review*, January-February 1990, pp. 65-75.

14. "BBN Software Products Deliver New Releases of Advanced Analysis Software," *PRNewswire* (Cambridge, MA: 25 October).

15. P. Noaker, "Variation Research: Debugging the Manufacturing Process," *Tooling & Production*, June 1987, pp. 53-56.

16. J. Main, "Under the Spell ...," p. 34.

17. Dumas, et al., "Making Quality Control...," pp. 32-33.

18. I.I. Mitroff, *Break-away Thinking* (New York: John Wiley & Sons, 1988), p. 21.

19. Thomas Li-Ping Tang, Peggy Smith Tollison, and Harold D. Whiteside, "The Effect of Quality Circle Initiation on Motivation to Attend Quality Circle Meetings and on Task Performance," *Personnel Psychology* 40 (Winter 1987), p. 800.

20. Michael W. Piczak, "Quality Circles Come Home," *Quality Progress* 21 (December 1988), p. 37.

21. W.D. Hyde, "How Small Groups Can Solve Problems and Reduce Costs," *Industrial Engineering* 18 (December 1986), p. 42.

22. B. Dumaine, "Who Needs a Boss?," *Fortune*, 7 May 1990, p. 52.

23. *Ibid.*, p. 53.

24. R. Cole, "Target Information for Competitive Performance," *Harvard Business Review*, May-June 1985, pp. 100-109.

25. R. Cole, "The Macropolitics of Organizational Change: A Comparative Analysis of the Spread of Small Group Activities," *Administrative Science Quarterly* 30 (1985), pp. 560-585.

26. W. Morse and H.P. Harold, "Why Quality Costs Are Important," *Management Accounting* 69 (November 1987), p. 43.

27. Tom Hughes, "Make it Well; Make it Good; Don't Make-Good," *Automation* 34 (December 1987), p. 17.

28. Flynn and Cole, "Automotive Suppliers....," p. 27.

29. W.E. Eureka and N.E. Ryan, *The Customer Driven Company* (Dearborn, MI: ASI Press, 1988).

30. The Ernst & Young Quality Improvement Consulting Group, *Total Quality* (Dow Jones-Irwin/APICS Series, 1990).

31. George E.P. Box, Kackar N. Raghu, Vijay N. Nair, Madhav Phadke, Anne C. Shoemaker, and Jeff C.F. Wu, "Quality Practices in Japan," *Quality Progress*, March 1988, pp. 37-41.

32. Pickzak, "Quality Circles Come Home," p. 38.

33. "Beyond Customer Satisfaction Through Quality Improvement," Reprinted by permission from a paid advertisement appearing in the September 26, 1988 issue of *Fortune* Magazine.

34. J.R. Houghton, "Quality: The Competitive Advantage," *Quality Progress*, February 1988.

35. *Ibid.*

36. R. Pascale, *Managing on the Edge* (New York: Simon and Schuster, 1990), p. 214.

37. "Beyond Customer Satisfaction Through Quality Improvement," Reprinted by permission from a paid advertisement appearing in the September 26, 1988 issue of *Fortune* Magazine.

38. *Ibid.*

39. R. Cole, *U.S. Quality Improvement: Close But No Cigar* (University of Michigan, 1989).

Chapter 3: Manufacturing Strategies and Agendas

1. R.S. Eckley, "Caterpillar's Ordeal: Foreign Competition in Capital Goods," *Business Horizons*, March-April 1989, pp. 80-86.

2. R. Rose, "Caterpillar Sees Gains in Efficiency Imperiled by Strength of Dollar," *Wall Street Journal*, 6 April 1990, pp. A1, A10.

3. F. Miller, "Meeting Customer Needs with a Focused Factory," *Manufacturing Systems*, November 1987, pp. 34-39.

4. W. Skinner, "What Matters to Manufacturing," *Harvard Business Review*, January-February 1988, p. 16.

5. "Order-winning Capabilities" is a concept developed by Terry Hill in his 1988 book, *Manufacturing Strategy: Text and Cases* (Homewood, IL: Dow Jones-Irwin, 1989), p. 17.

6. R.B. Chase and D. Garvin, "The Service Factory," *Harvard Business Review*, July-August 1989, p. 61.

7. A.V. Roth "Boundary Management: Manufacturing's New Imperative," Duke University Working Paper, 1990.

8. Wickham Skinner, "The Focused Factory," *Harvard Business Review*, May-June, 1974, pp. 113-121.

9. J.G. Miller, A. Amano, A. DeMeyer, K. Ferdows, J. Nachane, and A.V. Roth, "Closing the Competitive Gaps: International Report of the Manufacturing Futures Project" in *Managing International Manufacturing* (K. Ferdows, ed.) (Amsterdam: Elsevier Science Publishers, 1989), pp. 153-168.

10. *Ibid.*

11. K. Ferdows, "Mapping International Factory Networks," *Managing International Manufacturing* (K. Ferdows, ed.) (Amsterdam: Elsevier Science Publishers, 1989), pp. 3-11.

12. G. Stalk and T.M. Hout, *Competing Against Time* (New York: The Free Press, 1990).

13. Brian Dumaine, "How Managers Can Succeed Through Speed," *Fortune*, 13 February 1989, pp. 54-59.

14. "GE Refrigerator Woes Illustrate the Hazards in Changing a Product," *Wall Street Journal*, 7 May 1990, p. 1.

15. George Stalk, Jr., "Time—The Next Source of Competitive Advantage," *Harvard Business Review*, July-August 1988, p. 44.

16. Miller, et. al., "Closing the Competitive Gaps . . .," pp. 153-168.

17. Wickham Skinner, *Manufacturing: The Formidable Competitive Weapon* (New York: John Wiley & Sons, 1985), pp. 53-68.

18. R.G. Van Der Hooning, "Manufacturing Strategy and the Implementation Problem," *AutoFact '87 Conference Proceedings*, November 9-12, 1987, Dearborn, MI, pp. 7:85-7:97.

19. R.H. Hayes, "Strategic Planning — Forward or Reverse?" *Harvard Business Review*, November-December, 1985, pp. 115-116.

20. Gary Hamel and C.K. Prahalad, "Strategic Intent," *Harvard Business Review* 67.3 (May-June 1989), pp. 64-68.

21. *Ibid.*

Chapter 4: The Power of Manufacturing

1. Aleda V. Roth, Arnoud DeMeyer, and Akio Amano, "International Manufacturing Strategies: A Competitive Analysis," *Managing International Manufacturing* (K. Ferdows, ed.) (Amsterdam: Elsevier Science Publishers, 1989), pp. 187-211.

And:

Aleda V. Roth, "Differentiated Manufacturing Strategies for the Competitive Advantage: An Empirical Investigation," (Boston University School of Management, February 1987), abstract.

2. A.V. Roth and J.G. Miller, "Manufacturing Strategy, Manufacturing Strength, Managerial Success, and Economic Outcomes," in *Manufacturing Strategy*, (J.E. Ettlie, M.C. Burstein, and A. Feigenbaum, editors) (Boston: Kluwer Academic Publishers, 1990), pp. 92-108.

3. Thomas Moore, "Old-Line Industry Shapes Up," *Fortune*, 27 April 1987, p. 23.

4. Roth and Miller, "Manufacturing Strategy...," pp. 92-108.

5. Robert D. Buzzell and Bradley T. Gale, *The PIMS Principles* (New York: The Free Press, 1987).

6. G. Stalk and T.M. Host, *Competing Against Time* (New York: The Free Press, 1990), p. 31.

7. Roger W. Schmenner, "The Merit of Making Things Fast," *Sloan Management Review* 30 (Fall 1988), p. 13.

8. Stalk and Hout, *Competing Against Time*, pp. 153-154.

9. Michael A. Cusumano, "Manufacturing Innovations: Lessons from the Japanese Auto Industry," *Sloan Management Review* 30 (Fall 1988), p. 31.

10. Schmenner, "The Merit of Making Things Fast," pp. 11-17.

11. John F. Kratick, "Triumph of the Lean Production System," *Sloan Management Review* 30 (Fall 1988), pp. 41-52.

12. Moore, "Old-Line Industry...," p. 23.

13. Kim B. Clark, Robert H. Hayes, and Christopher Lorenz, *The Uneasy Alliance: Managing the Productivity-Technology Dilemma* (Boston: Harvard University Press, 1985), p. 163.

Chapter 5: Reconciling Accounting and Manufacturing

1. Duncan C. McDougall, "Effective Manufacturing Performance Measurement Systems: How to Tell When You've Found One" (Boston University School of Management, February 1988), p. 12.

2. A.V. Roth, "A Vision for the 1990s: Gearing Up for the 21st Century through Intelligence Manufacturing," Duke University, 1990, p. 39.

3. Robert S. Kaplan, "Accounting Lag: The Obsolescence of Cost Accounting Systems" (Harvard Business School 75th Anniversary Colloquium on Productivity and Technology, March 28-29, 1984), pp. 27-30.

4. Robin Cooper and Robert S. Kaplan, "How Cost Accounting Distorts Product Costs," *Management Accounting,* April 1988, p. 22.

5. P. Drucker, "The Emerging Theory of Manufacturing," *Harvard Business Review*, May-June 1990, p. 97.

6. Toshiro Hiromoto, "Another Hidden Edge—Japanese Management Accounting," *Harvard Business Review,* July-August 1988, pp. 22-26.

7. Robert S. Kaplan, "One Cost System Isn't Enough," *Harvard Business Review 66 (January-February 1988),* p. 62.

8. Robert H. Hayes, Steven C. Wheelwright, and Kim B. Clark, *Dynamic Manufacturing* (New York: The Free Press, 1988), p. 139.

9. Robert A. Howell and Stephen R. Soucy, "Management Reporting in the New Manufacturing Environment," *Management Accounting,* February 1988, p. 22.

10. Hayes et al., *Dynamic Manufacturing,* p. 140.

11. Howell and Soucy, "Management Reporting...," p. 22.

12. Cooper and Kaplan, "How Cost Accounting...," p. 22.

13. John Miller, *Productivity's New Math* (American Productivity & Quality Center, Houston, TX, October 1988).

14. Cooper and Kaplan, "How Cost Accounting ...," p. 20.

15. Robert D. McIlhattan, "How Cost Management Systems Can Support the JIT Philosophy," *Management Accounting* 69 (September 1987), p. 26.

16. "Manufacturing Technology Trends: North American Manufacturing Survey," Touche Ross & Company, 1988, p. 12.

17. David Callahan, "Pitfalls of GT Cells & Flowlines," Presented at the 4th Biennial International Machine Tool Technology Conference, Chicago, IL, September 7-14, 1988, pp. 11-75.

18. Carol A. Beatty and John R. M. Gordon, "Barriers to the Implementation of CAD/CAM Systems," *Sloan Management Review* 29 (Summer 1988), p. 29.

19. Stephen R. Rosenthal, "Progress Toward the Factory of the Future," *Journal of Operations Management* 4 (May 1984), p. 213.

20. M. Burnstein and P. Graham, "Market, Manufacturing Strategy, and Technology Acquisition: An Integrative Development Approach," in *Manufacturing Strategy* (J.E. Ettlie, M.C. Burstein, and A. Fiegenbaum) (Boston: Kluwer Academic Publishers, 1990) p. 109.

21. A.V. Roth, C. Gaimon, and L. Krajewski, "Optimal Acquisition of FMS Technology Subject to Technological Progress," *Decision Sciences*, to appear.

Chapter 6: Value-Added Performance Measurement

1. "Manufacturing Technology Trends: North American Manufacturing Survey," Touche Ross & Company, 1988.

2. Robert McNulty, "Management Decision Making in World-class Manufacturing," *American Production and Inventory Control Society, 1987 Conference Proceedings*, pp. 472-474.

3. Robert H. Hayes, Steven C. Wheelwright, and Kim B. Clark, *Dynamic Manufacturing* (New York: The Free Press, 1988), p. 378.

4. Walter E. Goddard, "How to Measure Performance Beyond Class A," *American Production and Inventory Control Society, 1987 Conference Proceedings,* p. 490.

5. Donald G. Reinerstein, "Who Dunit? The Search for the New-product Killers," *Electronic Business,* 9(8), July 1983, pp. 62-64.

Chapter 7: Adopting Best Manufacturing Practices

1. The surveys and database available are elements of the Best Manufacturing practices (BMP) Program, initiated in 1985 by the Office of the Assistant Secretary of the Navy for Shipbuilding and Logistics. For more information, contact Mr. Ernie Renner, director, Best Manufacturing Practices Program, Office of the Assistant Secretary of the Navy, Washington, DC, 20360.

2. This section is based upon a Deloitte & Touche research report entitled "Restructuring the Engineering Process for Competitive Advantage," written by Tom Captain and Craig Giffi.

3. Bro Uttal, "Speeding New Ideas to Market," *Fortune*, 115 (2 March 1987), p. 62.

4. C. Voss and D. Clutterbuck, *Just-in-Time: A Global Report,* Springer-Verlag, 1989, p. 8.

5. William A. Wheeler III, "Social and Strategic Implications of JIT in the Future," Presented at the Total Manufacturing Performance Seminar, APICS, 1987. Unpublished paper.

6. "Just in (the Wrong) Time," *Industry Week*, August 15, 1988, p. 28.

7. Peter C. Reid, *Well Made in America –Lessons From Harley-Davidson on Being The Best* (McGraw Hill: New York, 1990), p. 11.

8. Ernest Raia, "Journey to World Class; For American Industry the Time Has Come to Shape Up or Ship Out," *Purchasing* 103, September 24, 1987, p. 69.

9. *Ibid.*, p. 63.

10. C.R. O'Neal, "The Buyer-Seller Linkage in a Just-In-Time Environment," *Journal of Purchasing and Materials Management,* Vol. 23, Spring 1987, p. 11.

11. U.S. Department of Commerce, Bureau of the Census, *Manufacturing Technology 1988*, p. 4.

12. Raia, "Journey to World...," p. 49.

13. Frank Barton, Surendra P. Agrawal, and L. Mason Rockwell, Jr., "Meeting the Challenge of Japanese Management Concepts," *Management Accounting* 70, September 1988, pp. 49-53.

14. Fred McGrail and Nancy McSharry, "Manufacturing Managers on Labor and Automation," *Electronic Business* 12, October 1, 1986, p. 56.

15. Touche Ross Logistics Consulting Services, "Implementing Just In Time Logistics," 1988 National Survey on Progress, Obstacles and Results, p. 5.

16. "Just in (the Wrong) Time," p. 28.

17. *Ibid.*, p. 28.

18. "Implementing Just In Time....", p. 5.

19. Vincent G. Reuter, "Becoming Competitive with Value Engineering/Value Analysis, *Journal of Systems Management*, October 1985, p. 30.

20. Lea Tonkin, "Value Analysis: Diversity in Application and Benefits," *Datapro* 2, October 1988, pp. 6-8.

21. Terry Wireman, "Integrating Maintenance and CIM," *SME/CASE Autofact 1987 Conference Proceedings*, Dearborn, MI, p. 15-2.

22. Theodore Armstrong, "World of Plant Maintenance Management," *American Production and Inventory Control Society, 1987 Conference Proceedings*, p. 117.

23. Kishan Bagadia, "Microcomputer Aided Maintenance Management and Inventory Control," *SME/CASE Autofact, 1987 Conference Proceedings*, Dearborn, MI, pp. 15-10.

24. Stephen Macaulay, "Amazing Things Can Happen If You Keep It Clean," *Production*, May 1988, pp. 72-74.

Chapter 8: Restructuring the American Workplace

1. Thomas Moore, "Old-Line Industry Shapes Up," *Fortune*, 27 April 1987, p. 23.

2. Peter Austin, "Critical Factors in Assembly Systems Performance," 4th Biennial International Machine Tool Technology Conference, September 7-14, 1988, pp. 5-173 – 5-190.

3. Wickham Skinner, "The Focused Factory," *Harvard Business Review*, May-June 1974, p. 114.

4. R. Hayes and R. Schmenner, "How Should You Organize Manufacturing?" *Harvard Business Review*, January-February 1978, p. 105.

5. Susan Lee and Christie Brown, "The Pro-team Corporation," *Forbes* 140 (24 August 1987), pp. 77-78.

6. William M. Evan and Paul Olk, "R&D Consortia: A New U.S. Organizational Form," *Sloan Management Review*, Spring 1990, pp. 37-45.

7. *Ibid.*, pp. 37-40.

8. *Ibid.*, pp. 37-45.

9. Perry Pascarella, "Tom Peters Invites Chaos for Survival," *Industry Week* 235 (19 October 1987), p. 48.

10. Ronald Henkoff, "Cost Cutting: How To Do It Right," *Fortune*, 9 April 1990, p. 40.

11. William Bridges, "How to Manage Organizational Transition," *Training: The Magazine of Human Resources Development* 22 (September 1985), pp. 29-30.

12. David W. Rhodes, "Employees – Strategy Makers or Breakers," *Journal of Business Strategy* 9 (July-August 1988), pp. 55-56.

13. Terrence E. Deal and Allan A. Kennedy, *Corporate Cultures – The Rites and Rituals of Corporate Life,* (Reading, MA: Addison-Wesley Publishing Company, 1982), pp. 175-176.

Chapter 9: Developing Human Resources

1. Michael Brody, "Helping Workers to Work Smarter," *Fortune* 115 (8 June 1987), pp. 86-89.

2. Dale Fuer and Chris Lee, "The Kaizen Connection," *Training* 25 (May 1988), p. 24.

3. *Ibid.*, p. 31.

4. Patricia Galagan, "Here's the Situation: A Quick Scan of the Trends that Experts Think Will Affect You Most," *Training & Development Journal* 41 (July 1987), p. 21.

5. Joel Dreyfuss, "The Three R's on the Shop Floor," *Fortune*, Spring 1990 Special Issue, p. 88.

6. Alan C. Greenspan, "Our Human Capital: the Power of Investment," *Directors & Boards* 12 (Summer 1988), p. 9.

7. Brian Dumaine, "Making Education Work," *Fortune*, Spring 1990 Special Issue, p. 13.

8. Judith F. Vogt and Bradley D. Hunt, "What Really Goes Wrong With Participative Work Groups?" *Training & Development Journal* 42 (May 1988), p. 96.

9. John Ryan, "Labor/Management Participation: The A.O. Smith Experience," *Quality Progress* 21 (April 1988), p. 39.

10. William H. Wagel, "Working (and Managing) Without Supervisors," *Personnel* 64 (September 1987), p. 8.

11. Jack C. Bailes and Robert L. Edwards, "Productivity Boost: Treating Employees as Independent Contractors," *Management Accounting* 69 (October 1987), p. 50.

12. Marc Bassin, "Teamwork at General Foods: New and Improved," *Personnel* 67 (1987), p. 67.

13. Patricia Galagan, "Donald E. Petersen: Chairman of Ford and Champion of its People," *Training and Development Journal* 24 (August 1988), p. 21.

14. Wagel, "Working (and Managing)...," p. 10.

15. Frank Hull, Koya Azumi, and Robert Wharton, "Suggestion Rates and Sociotechnical Systems in Japanese Versus American Factories: Beyond Quality Circles," *IEEE Transactions on Engineering Management* 35 (February 1988), p. 20.

16. Jack C. Horn, "Bigger Pay for Better Work," *Psychology Today*, July 1987, p. 56.

17. William Fechter and Renee B. Horowitz, "The Role of the Industrial Supervisor in the 1990s," *Industrial Management* 30 (May-June 1988), p. 18.

18. *Ibid.*, p. 19.

19. J.A. Klein, "Good Supervisors Are Good Supervisors—Anywhere," *Harvard Business Review*, November-December 1986, p. 125.

20. "Countdown to the Future: The Manufacturing Engineer in the 21st Century," (Profile 21 Executive Summary, Society of Manufacturing Engineers, 1988).

21. Drew Winter, "Technical Outsourcing—Reach Out and Touch Someone," *Ward's Autoworld*, March 1988, p. 45.

22. Donald C. Reeve, "Seeing the Light in Manufacturing," *High Technology* 6 (October 1986), p. 13.

23. John F. Sullivan, "The Future of Merit Pay Programs," *Compensation and Benefits Review* 20 (May-June 1988), p. 23.

24. Carla Dell and C. Jackson Grayson, Jr., "Flex Your Pay Muscle," *Across The Board* 25 (July-August 1988), p. 46.

25. Christopher W. Musselwhite, "Knowledge, Pay, and Performance," *Training & Development Journal* 42 (January 1988), p. 64.

26. Musselwhite, "Knowledge, Pay, and Performance," p. 34.

27. Dell and Grayson, "Flex Your Pay Muscle," p. 46.

28. Edward J. Ost, "Team-Based Pay: New Wave Incentives," *Sloan Management Review*, Spring 1990, p. 19.

Chapter 10: Technology Investment and Implementation

1. R.B. Reich, "The Quiet Path to Technological Preeminence," *Scientific American*, 261 (4), October 1989, pp. 41-47.

2. J. Ettlie and J.D. Penner-Hahn, "Focus, Modernization, and Manufacturing Technology Policy," *Manufacturing Strategy* (J.E. Ettlie, M.C. Burstein, and A. Feigenbaum, eds) (Norwell, MA: Kluwer Academic Publishers, 1989), p. 155.

3. R. Jalkumar, "Postindustrial Manufacturing," *Harvard Business Review*, November-December 1986, pp. 69-76.

4. U.S. Department of Commerce, "Current Industrial Reports: Manufacturing Technologies: 1988," 1989.

5. Wickham Skinner, *Manufacturing: The Formidable Competitive Weapon* (New York: John Wiley & Sons, 1985), pp. 27-28.

6. Robert A. Howell, James D. Brown, Stephen R. Soucy, and Allen H. Seed, "Management Accounting in the New Manufacturing Environment," National Association of Accountants, 1987.

7. John H. Sheridan, "Toward the CIM Solution," *Industry Week*, 16 October 1989, p. 50.

8. Michael L. Dertouzos, Richard K. Lester, and Robert M. Solow, *Made in America—Regaining the Productivity Edge* (Boston: MIT Press, 1989), p. 50.

9. "Volvo Accelerates its Drive to Lead Europe in Automation," *Production*, October 1986, p. 85.

10. David Callahan, "Pitfalls of GT Cells & Flowlines," 4th Biennial International Machine Tool Technology Conference, September 7-14, 1988, p. 11-53.

11. *Ibid.*, pp. 11-72.

12. Jack Scrimgeour, "Don't Ignore CIM Any Longer," *Canadian Machinery and Metalworking*, January 1988, p. 34.

Chapter 11: Attributes of World-Class Manufacturing

1. Peter Drucker, "The Emerging Theory of Manufacturing," *Harvard Business Review*, May-June 1990, p. 101.

2. Kim B. Clark, "What Strategy Can Do for Technology," *Harvard Business Review*, November-December 1989, pp. 94-95.

3. R. Jaikumar, "Postindustrial Manufacturing," *Harvard Business Review*, November-December 1986, pp. 69-76.

BIBLIOGRAPHY

"1988 Capital Spending Survey: Summary Findings 1988-1989 Capital Spending Survey and Forecast." Conducted by *Production Magazine* Research, 6600 Clough Pike, Cincinnati, OH, 1988.

"A New Challenge for Managers." *Advertising Age* 58 (31 August 1987): p. 16.

"AGVs Proceed With Caution." *Assembly Engineering* 31 (April 1988): pp. 20-22.

"Air Force Takes Initiative in Industrial Modernization Incentive Program for DOD." *Industrial Engineering*, November 1986, pp. 32-39.

"An Assembly Line on Wheels." *Industry Week*, 21 March 1988, pp. 64, 66.

"Anything Japan Can Do." *The Economist*, 20 February 1988, pp. 19-22.

"Applications: Material Handling Costs Cut By One-Third With AS/AR System." *Industrial Engineering* 15 (March 1983): p. 103.

"BBN Software Products Deliver New Releases of Advanced Analysis Software." *PRNewswire*, Cambridge, MA, October 25.

"Beyond Customer Satisfaction Through Quality Improvement." *Fortune*, 26 September 1988.

"Can Small Companies Compete?" *Production* 98 (December 1986): pp. 51-57.

"Career Development Programs: Planning for the Company's Future." *Small Business Report* 12 (November 1987): pp. 30-36.

"Caterpillar Introduces Free-Ranging AGV That Scans." *American Machinist & Automated Manufacturing*, October 1987, p. 41.

"Computer Aided Inspection in the Foundry." *Modern Casting*, September 1987, pp. 38-39.

"Computer-Controlled Manufacturing." *Machine Design* 60 (16 June 1988): pp. 99-109.

"Consumers' Perceptions Concerning the Quality of American Products and Services." Gallup Survey conducted for the American Society for Quality Control, Milwaukee, WI, October, 1988, pp. 1-62.

"Cost Accounting for the 90's." *Management Accounting* 68 (July 1986): pp. 58-60.

"Countdown to the Future: The Manufacturing Engineer in the 21st Century." *Profile 21 Executive Summary*, Society of Manufacturing Engineers, 1988.

"Cutting Tool Manufacturers' Role Becomes More Sophisticated." *Production* 94 (August 1984): pp. 23-25.

"Delco Products: Using VA to Gain a Competitive Edge." *Purchasing*, 11 June 1987, pp. 55-75.

"Deming's Point Four: A Study." *Quality Progress* 21 (December 1988): pp. 31-35.

"Do U.S. Schools Make the Grade?" *Fortune*, Spring 1990 Special Issue, pp. 50-51.

"Eastern Europe." *Production* 99 (January 1987): pp. 47-53.

"Employee-Owned Companies Are More Productive, Survey Says." *Journal of Accountancy*, June 1987, pp. 20-21.

"Establishing Competitive Advantage Through Employee Involvement." *Zero Inventory Philosophy and Practices, APICS Seminar Proceedings*, 1984, pp. 231-235.

"European Companies Forge New Multinational Links." *Aviation Week & Space Technology*, 5 September 1988, pp. 44-50.

"Explaining Internation Difference in the Cost of Capital." *FRBNY Quarterly Review*, Summer 1989, p. 7.

"FMS Market Growth Forecast." *Electrical Review*, 22 March 1988, p. 18.

"FMS Masters the Factory." *Production*, July 1985, p. 51.

"Factory of the Future: A Survey." *The Economist* 303 (30 May 1987): pp. s3-s15.

"Flat Out for Flexibility." *Engineering* 227 (October 1987): pp. 588-589.

"Flexible Laser Marketing Comes of Age." *Machine and Tool Blue Book*, July 1988, p. 47.

"GE Refrigerator Woes Illustrate the Hazards in Changing a Product." *Wall Street Journal*, 7 May 1990, p. 1.

"GM to Cut Half of its Suppliers." *The Cleveland Plain Dealer*, 8 November 1988, sec. D, p. 6.

"Group Technology: Its Time is Now." *Production* 99 (December 1987): pp. 66-67.

"Hard Times Are No Time to Forget People." *Wall Street Journal*, 15 November 1988, sec. B, p. 7.

"High Style and High Tech." *Small Business Report* 12 (October 1987): pp. 32-35.

"High-Powered Workstations Top Manufacturing Show." *Computerworld* 21 (16 November 1987): p. 16.

"Industry Outlook." *Machine and Tool Blue Book* 83 (August 1988): pp. 20-22.

"Insight: Computer-Integrated Manufacturing." *Insight* 8 (October 1988): Special Section pp. S2-S11.

"Integrate or Evaporate: The Coming Evolution in Plant Floor Systems." *IBM*, 6 October 1988.

"Internationalization Study Analyzes U.S. Competitive Position in World Market." *Aerospace Industries Association Newsletter*, June 1988.

"Internationalization: What is it? Why is it? Will it Go Away?" *Aerospace Industries Association Newsletter* 7 (July 1988).

"Japan Hasn't Lost Lead." *Industry Week*, 7 September 1987, p. 28.

"JIT, TQC, CIM Cure Many Ills." *Modern Materials Handling*, May 1988, p. 5.

"Just in (the Wrong) Time." *Industry Week*, 15 August 1988, p. 28.

"Keep U.S. Competitive, R&D 2000 Leaders Urge." *Research & Development* 30 (July 1988): pp. 15-18.

Key Technologies for the 1990's: An Industry Study of High-Leverage, Enabling Aerospace Technologies and Roadmaps to Attain Them. Aerospace Industries Association of America, November 1987.

"Linking Pay to Productivity." *Management Review* 76 (September 1987): pp. 9-11.

"Look at Foreseeable Trends in Parts Handling Offers Strategies for System Planners." *Industrial Engineering* 16 (March 1984): pp. 46-48, 50.

"MIG Welding Robot Increased Productivity Three to Five Times." *Industrial Engineering*, January 1986, pp. 96-97.

Machine Vision Activity in Europe: 1986. Dearborn, MI: Society of Manufacturing Engineers, 1987.

"Machine Vision Eyes its Future." *Production* 98 (August 1986): pp. 7-8.

"Management Incentives—Motivating Managers to Achieve Goals." *Small Business Report*, May 1988, pp. 48-51.

"Management Patterns Must Shift if U.S. is to Compete on Quality." *Marketing News* 22 (20 June 1988): pp. 13-15.

"Manufacturing Cells: Blue Print for Versatility." *Production*, December 1984, pp. 38-41.

Manufacturing Strategies—Sweden in an International Perspective. Institute of Management of Innovation and Technology and the Chalmers University of Technology, 1987.

"Material Handling Market to Nearly Double, Study Says." *Industrial Engineering*, January 1985, p. 6.

"MRP-ABCD Checklist," Essex Junction, UT: Oliver Wright Companies, 1988.

"New P&W Plant Designed as 'High-Tech Factory of the Future.'" *Industrial Engineering* 16 (September 1984): pp. 4, 6.

"Off-Wire AGVs Use LEDs to Travel Without Interference." *Industrial Engineering* 17 (April 1985): p. 105.

"PC-Based CAD/CAM Systems Pace Automation Drives at Many US Firms." *Metalworking News*, 20 June 1988, pp. 21-22.

"Partnering Is A Strategy — Not a Gizmo for Selling MRO." *Purchasing* 104 (28 April 1988): pp. 60-71.

"Peak Performance — It Can Be Learned and Taught." *Management Solutions* 31 (June 1986): pp. 26-28.

"Planning Profit from Advanced Manufacturing Technology." *Personnel Management*, December 1987, pp. 49-53.

"Plugging the Gaps in the U.S. Labor Market." *Electronic Business*, 15 August 1988, pp. 14-15.

"Preliminary Report: Second International Mathematics Study." University of Illinois, 1984.

"Purchasing's 12th Value Analysis Contest." *Purchasing*, 16 June 1988, pp. 103-105.

"Quality Controls 'Powerful Leverage.'" *Metalworking News*, 13 July 1987, pp. 27-28.

"Robotics: No One Said it Would Be Easy." *Production*, April 1988, pp. 38-49.

"Robots Unpinned." *Industrial Management & Data Systems*, July-August 1984, pp. 27-28.

"Seeking the Right Blend: Panel on Corporate Restructuring." *Mergers & Acquisitions*, November-December 1986, pp. 48-50.

"Service Quality." In *Council on Financial Competition, Retail Excellence* 7 (1987).

"Shaping of Things to Come." *Engineering* 228 (April 1988): pp. 201-202.

"Sheetmetal Shop Leans to Lasers." *Machine and Tool Blue Book* 83.7 (July 1988): pp. 51-52.

"Simulation: The Last Frontier." *Electronics*, 9 June 1986, pp. 30-32.

"Special Report: The Five Best Managed Companies." *Business Month*, December 1988, pp. 30-58.

"Study Sees An Emphasis on Competing in the 1990s." *Wall Street Journal*, 15 November 1988, sec. B, p. 1.

"Test and Inspection Equipment." *Production Engineering* 31 (January 1984): pp. 54-59.

"The American Gleam?" *Industry Week*, 20 June 1988, pp. 41-45.

"The National Quality Awards: Cutting the Mustard Wasn't Easy." *Business Week*, 12 December 1988, p. 82.

"The Networked Organization." *Industry Week*, 18 April 1988, pp. 46-47.

"The Tough Human Challenge in Automation." *Industry Week*, 20 June 1988, pp. 58-64.

"Three Cheers for Product Quality." *Business Week*, 12 December 1988, p. 136.

"Tort Woes Cast Global Shadow." *National Underwriter* 92 (20 June 1988).

"Toyota Awarded Deming Prize for New Generation of Machine Tools." *Iron Age*, 20 June 1986, p. 514.

"Training Vital for Automation." *Modern Materials Handling*, November 1988, p. 5.

"Trends on the Horizon." *Production* 98 (August 1988): pp. 80-82.

"U.S. Industrial Automation: A View to the Future." *Automation Forum*, March 1987.

"VA Case Histories." *Purchasing*, 16 June 1988, pp. 119-134.

"Vision Allows Robot System to Track Weld Joint Variances." *Industrial Engineering*, 19.8 (August 1987): p. 95.

"Volvo Accelerates its Drive to Lead Europe in Automation." *Production*, October 1986, pp. 84-85.

"Warehousing Trends: Service Moves to Center Stage." *Traffic Management* 23 (October 1984): pp. 88-90.

"What's New in Robotics and Machine Vision." *Tooling & Production*, September 1988, pp. 73-75.

"Where Have All the Foremen Gone?" *Production*, August 1988, p. 24.

"Will Another Round of Surgery Help Baxter?" *Business Week*, 30 April 1990, p. 92.

"Worldwide Executive Mobility." *Harvard Business Review* 66 (July-August 1988): pp. 105-123.

Abbott, Jan. "In Practice." *Training & Development Journal*, August 1988, pp. 12-13.

Abegglen, J.C., and Stalk, G. Jr., *Kaisha: The Japanese Corporation.* New York, NY: Basic Books Inc., 1985.

Abernathy, William J.; Clark, Kim B.; and Kantrow, Alan M. "The New Industrial Competition." In *Survival Strategies for American Industry*, pp. 72-95. New York: John Wiley & Sons, 1983.

Abraham, Richard. *Computer Integrated Manufacturing.* Revised By Warren Shrensker, 1986, pp. 1-7

Adkins, Lynn. "Phillips Industries' Cost Cutting Mavens." *Dun's Business Month*, November 1986, pp. 47-48.

Adler, Paul S. "Managing Flexible Automation." *California Management Review* 30 (Spring 1988): pp. 34-56.

Ahrens, Roger. "Productivity Improvements Using Inventory Turns as a Barometer." *APICS 1987 Total Manufacturing Performance Seminar Proceedings*, Washington, DC, July 13-15, 1987.

Aiman, Patricia S. "Employees Must Gear Up to New Technologies." *Iron Age* 228 (20 September 1985): p. 23.

Allen, Mike and Hodgkinson, Robert. "It's What's Beyond the Figures That Really Counts." *Accountancy*, November 1986, pp. 110-111.

Allio, Robert J. "Flexibility and the Global Shop Floor." *Across the Board*, December 1986, pp. 56-57.

Almoodovar, Angel R. "Manufacturing Simulators Save Time, Provide Good Data for Layout Evaluation." *Industrial Engineering* 20 (June 1988): pp. 28-33.

Alspatch, Philip H. "Building People in Manufacturing." *NMBTA 3rd Biennial International Machine Tool Technical Conference Proceedings,* Chicago, IL, September 3-11, 1986, pp. 5-130—5-137.

Alster, Norm. "Strategies in Automotive Electronics: The Race to Build the Car of the Future." *Electronic Business*, 14.11 (1 June 1988): p. 33.

Alsup, Fred. "SQC in Integrated Manufacturing Environments." *NMBTA 4th Biennial International Manufacturing Technology Conference Proceedings,* Chicago, IL, September 7-14, 1988, pp. 5-127—5-136.

Anderson, D. Alan. "Cost Management Impact of JIT." *American Production and Inventory Control Society, 1987 Conference Proceedings*, St. Louis, MO, October 19-23, 1987, pp. 614-617.

Anderson, D. Alan and Ostrenga, Michael R. "MRP II and Cost Management: A Match Made in Theory?" *CIM Review* 3.4 (Summer 1987): pp. 24-28.

Andrews, Charles G. "The Drive for Productivity Excellence: Gaining Senior Management Leadership." *APICS 1984 Zero Inventory Philosophy and Practice Seminar Proceedings*, Washington, DC, July 13-15, 1984, pp. 345-354.

Ansari, Ahsanuddin. "Survey Identifies Critical Factors in Successful Implementation of Just-In-Time Purchasing Techniques." *Industrial Engineering* 18 (October 1986): pp. 44-50.

Ansari, Ahsanuddin and Modarress, B. "The Potential Benefits of Just-In-Time Purchasing for U.S. Manufacturing." *Production and Inventory Management* 28 (Second Quarter 1987): pp. 30-35.

Appleton, David S. "The CIM Database." in *SME Technical Article Series*, originally presented to SME/CASA Technical Council, 1982, pp. 1-9.

Arbose, Jules. "'Fortress Europe' Finally Takes Shape." *International Management* 43 (January 1988): pp. 48-52.

Arbose, Jules. "How Countries Compare in Their Efforts to Think Internationally." *International Management*, November 1986, pp. 24-31.

Arcuri, Gerald. "Chart a New MAP." *Computerworld* 22 (30 May 1988): p. 52.

Armitage, Howard M., and Langdon, William E. "Management Accounting: The New Relevance." *The Management Accounting Magazine* 62 (June 1988): pp. 52-56.

Armstrong, Theodore. "World of Plant Maintenance Management." *American Production and Inventory Control Society, 1987 Conference Proceedings*, St. Louis, MO, October 19-23, 1987, pp. 116-119.

Asfahl, C. Ray. "Many Subtle Pitfalls of Automating Production Lines Can be Predicted." *Industrial Engineering* 18 (May 1986): pp. 34-42.

Atkinson, Anthony A. "Choosing a Future Role for Management Accounting." *CMA Magazine*, July-August 1987, pp. 29-35.

Austin, Peter. "Critical Factors in Assembly Systems Performance." *NMBTA 4th Biennial International Manufacturing Technology Conference Proceedings*, Chicago, IL, September 7-14, 1988, pp. 5-173—5-190.

Bacigalupo, Paul. "Master Production Planning and Scheduling for the 80s." c. 1982, *APICS 1987 Total Manufacturing Performance Seminar Proceedings*, Washington, DC, July 13-15, 1987, pp. 2-5.

Bagadia, Kishan. "Computerized Maintenance Management and Inventory Control." *NMBTA 4th Biennial International Manufacturing Technology Conference Proceedings*, Chicago, IL, September 7-14, 1988, pp. 3-113—3-126.

Bagadia, Kishan. "Micro-Computer Aided Maintenance Management Systems." *NMBTA 3rd Biennial International Machine Tool Technical Conference Proceedings*, Chicago, IL, September 3-11, 1986, pp. 6-20—6-25.

Bagadia, Kishan. "Microcomputer Aided Maintenance Management and Inventory Control." *SME/CASA Autofact, 1987 Conference Proceedings*, Dearborn, MI, November 9-12, 1987, pp. 15-7—15-22.

Bagadia, Kishan. "Selecting A Manufacturing Software System." *NMBTA 4th Biennial International Manufacturing Technology Conference Proceedings*, Chicago, IL, September 7-14, 1988, pp. 2-35—2-50.

Bailes, Jack C. and Edwards, Robert L. "Productivity Boost: Treating Employees as Independent Contractors." *Management Accounting* 69 (October 1987): pp. 48-51.

Bajaria, Hans J. "SPC Provides Strategic Guidance for Automation Investments." *Quality Progress* 21 (December 1988): pp. 22-24.

Baker, George P.; Jensen, Michael C.; and Murphy, Kevin J. "Compensation and Incentives: Practice Versus Theory." *Journal of Finance* 43 (July 1988): pp. 593-617.

Bakerjian, Ramon and Mishne, Patricia P. "A Whole New Ball Game." *Manufacturing Engineering*, October 1988, pp. 61-62.

Balakrishnan, Srinivason and Wernerfelt, Birger. "Technical Change, Competition and Vertical Integration." *Strategic Management Journal* 7 (1986): pp. 347-359.

Balkin, David B. and Logan, A. James W. "Reward Policies that Support Entrepreneurship." *Compensation & Benefits Review* 20 (January-February 1988): pp. 18-25.

Bancroft, Carol E. "Design for Manufacturability: Half Speed Ahead." *Manufacturing Engineering* 101 (September 1988): pp. 67-69.

Bangs, Edmund. "Automated Inspection on Machine Tools." *NMBTA 4th Biennial International Manufacturing Technology Conference Proceedings*, Chicago, IL, September 7-14, 1988, pp. 10-221—10-234.

Barckhoff, Jack. "A Management Plan for Technology Transfer." *NMBTA 4th Biennial International Manufacturing Technology Conference Proceedings*, Chicago, IL, September 7-14, 1988, pp. 1-117—1-131.

Barker, William E. "It's Time to Make America Competitive Again." *APICS 1984 Zero Inventory Philosophy and Practices Seminar Proceedings*, St. Louis, MO, October 29-31, 1984, pp. 444-451.

Barkman, Donald F. "Team Discipline: Put Performance on the Line." *Personnel Journal* 66 (March 1987): pp. 58-64.

Barrack, John B. and Edwards, Don. "A New Method of Inventory Accounting." *Management Accounting* 69 (November 1987): pp. 49-58.

Barrett, F.D. "Managing Robots Strategically." *Industrial Management & Data Systems*, March-April 1985, pp. 20-23.

Bartczak, Gene. "Machine Vision Keeps a Careful Eye on GM Forgings." *Machine and Tool Blue Book* 81 (December 1986): pp. Q-2—Q-7.

Bartlett, Christopher A. and Ghosbal, Somantra. "Managing Across Borders: New Organizational Responses." *Sloan Management Review*, Fall 1987, pp. 43-53.

Barton, Frank; Agrawal, Surendra P.; and Rockwell, L. Mason, Jr. "Meeting the Challenge of Japanese Management Concepts." *Management Accounting* 70 (September 1988): pp. 49-53.

Bassin, Marc. "Teamwork at General Foods: New and Improved." *Personnel* 67 (1987): pp. 62-71.

Baxter, John D. "U.S. Manufacturers Set to Counterattack." *Iron Age* 229 (21 November 1986): pp. 32-39.

Beach, David W. "Computer-Aided Prototyping and Design Education." *SME/CASA Autofact, 1987 Conference Proceedings*, Dearborn, MI, November 9-12, 1987, pp. 10-1 — 10-4.

Beatty, Carol A. and Gordon, John R. M. "Barriers to the Implementation of CAD/CAM Systems." *Sloan Management Review* 29 (Summer 1988): pp. 25-33.

Bechtel, Tom. "MRP is JIT: Use it Don't Abuse it." *APICS 1987 Total Manufacturing Performance Seminar Proceedings*, Washington, DC, July 13-15, 1987.

Beck, Larry. "AS/RS Delivers—Accurately and on Time." *Modern Materials Handling* 43 (April 1988): pp. 52-55.

Beck, Larry. "Cut Handling Costs With the Right Container." *Modern Materials Handling* 43 (April 1988): pp. 60-62.

Becker, Franklin J. and Boyd, P. Michael. "CIM Case Study: John Deere Manufacturing Systems." *APICS 1987 Total Manufacturing Performance Seminar Proceedings*, Washington, DC, July 13-15, 1987.

Beckmen, Jeff. "Integrating Modeling into a CAD/CAM System." *NMBTA 4th Biennial International Manufacturing Technology Conference Proceedings*, Chicago, IL, September 7-14, 1988, pp. 1-157 — 1-166.

Bell, Chip; Thurn, Walt; McLagan, Patricia; et al. "Looking Ahead." *Training: the Magazine of Human Resources Development* 25 (July 1988): pp. 79-84.

Bell, D.A. "Technology and the Human Factor." *IEEE* 135 (May 1988): pp. 309-312.

Ben Daniel, Elisabeth. "Using Statistical Process Control With Robotic Testing Improves Quality Level." *Industrial Engineering* 20 (February 1988): pp. 26-31.

Benjamin, Alan and Benson, Neil. "Why Ignore the Value of the People?" *Accountancy*, February 1986, pp. 81-84.

Bennet, Robert E. and Hendricks, James A. "Justifying the Acquisition of Automated Equipment." *Management Accounting* 69 (July 1987): pp. 39-48.

Bennett, Robert E.; Hendricks, James A.; Keys, David E.; and Rudnicki, Edward J. *Cost Accounting for Factory Automation*. Montvale, NJ: National Association of Accountants, 1987.

Bentley, David A. "Emphasis on Total Quality Management for JIT Success." *American Production and Inventory Control Society, 1987 Conference Proceedings*, St. Louis, MO, October 19-23, 1987, pp. 20-23.

Berger, Gus. "Lead Time Reduction: Management's Best Competitive Strategy." *APICS 1987 Total Manufacturing Performance Seminar Proceedings*, Washington, DC, July 13-15, 1987.

Berger, Michael. "Japan's Energetic New Search for Creativity." *International Management* 42 (October 1987): pp. 71-75.

Berger, Michael. "Restoring the U.S. Spirit after Years of Bludgeoning." *International Management*, November 1987, pp. 77-87.

Berkwitt, George. "The Future Years: 1986-2000." *Industrial Distribution* 75 (October 1986): pp. 38-52.

Bernek, Bretislav; Boillot, Jean-Paul; and Ferrie, F.P. "Laser Sensor for Adaptive Welding." *SPIE, Optical Techniques for Industrial Inspection* 665 (1986): pp. 195-199.

Bhote, K.R. "DOE—The High Road to Quality." *Management Review*, January 1988.

Bieber, Owen. "U.S. Labor and Global Manufacturing the Case for Accountability." *Vital Speeches of the Day*, 22 March 1988, pp. 456-459.

Bihun, Thomas. "The Impact of Just-In-Time on MRP II." *NMBTA 3rd Biennial International Machine Tool Technical Conference Proceedings*, Chicago, IL, September 3-11, 1986, pp. 1-192 — 1-200.

Bihun, Thomas. "The Impact of Just-In-Time on Manufacturing Resource Planning." *NMBTA 4th Biennial International Manufacturing Technology Conference Proceedings*, Chicago, IL, September 7-14, 1988, pp. 3-81 — 3-89.

Billett, Tony. "Current Trends in CAD and Integrated CAD/CAM for Sheet Metal Punching and Profiling." *Sheet Metal Industries* 65 (February 1988): pp. 85-86.

Birchmore, Sue. "Trimming off the Fat." *New Scientist*, 7 January 1988, p. 72.

Blevins, Preston W. "Implementing JIT: Opportunities, Concerns, and a Strategy for Success." *APICS 1987 Total Manufacturing Performance Seminar Proceedings*, Washington, DC, July 13-15, 1987, pp. 318-325.

Blevins, Preston W. "MPS: What it is and Why You Need it, Without the Jargon and Buzzwords!" c. 1984, *APICS 1987 Total Manufacturing Performance Seminar Proceedings*, Washington, DC, July 13-15, 1987, pp. 13-15.

Boker, David W. "Ten Steps to Class A MRP II." *American Production and Inventory Control Society, 1987 Conference Proceedings*, St. Louis, MO, October 19-23, 1987, pp. 120-124.

Bonsack, Robert A. "Direct Costing and the Factory of the Future." *Journal of Accounting & EDP* 4 (Spring 1988): pp. 60-62.

Bonsack, Robert A. "How to Justify Investments in Factories of the Future." *Management Review*, January 1988, pp. 38-40.

Bonsack, Robert A. "How to Justify the Investment in Factory Automation and CIM." *American Production and Inventory Control Society, 1987 Conference Proceedings*, St. Louis, MO, October 19-23, 1987, pp. 546-547.

Booth, Patricia. "Employee Involvement and Corporate Performance." *Canadian Business Review*, Spring 1988, pp. 14-16.

Borg, Dwayne. "Software for Discrete Parts Manufacturing." *NMBTA 3rd Biennial International Machine Tool Technical Conference Proceedings*, Chicago, IL, September 3-11, 1986, pp. 10-2 — 10-9.

Borrus, Amy. "Beating America to the Punch." *International Management* 43 (March 1988): pp. 31-33.

Bose, Bimal K. "Technology Trends In Microcomputer Control of Electrical Machines." *IEEE Transactions On Industrial Electronics* 35 (February 1988): pp. 160-176.

Boulton, A.H. *Business Consortia*. London: Eastern Press, 1961.

Box, George E.P.; Raghu, N. Kackar; Nair, Vijay N.; Phadke, Madhav; Shoemaker, Anne C.; and Wu, Jeff C.F. "Quality Practices in Japan." *Quality Progress*, March 1988, pp. 37-41.

Boyd, Joseph A. "International Trade Competition: The Declining Position of the United States and its Effect on the American Way of Life," Paper presented at Competing in a Global Economy Seminar, Melbourne, FL, December 11, 1986.

Boyd, P. Michael. "CIM at Deere & Co.: The 1981 LEAD Award and Beyond." *APICS 1987 Total Manufacturing Performance Seminar Proceedings*, Washington, DC, July 13-15, 1987, pp. 11-17 — 11-23.

Bradley, Keith. "Employee-Ownership and Economic Decline in Western Industrial Democracies." *Journal of Management Studies*, January 1986, pp. 51-71.

Bradley, Keith and Gelb, Alan. "Employee Buyouts of Troubled Companies." *Harvard Business Review*, September-October 1985, pp. 121-130.

Branam, James W. "Flexible Manufacturing Systems Eliminate the Need for Work Orders." *CASA/SME Flexible Manufacturing Systems '85 Conference Proceedings*, Dallas, TX, March 1985, pp. 37-42.

Branam, James W. "JIT vs. FMS — Which One Will Top Management Buy?" *SME/CASA Autofact, 1987 Conference Proceedings*, Dearborn, MI, November 9-12, 1987, pp. 30-36.

Branam, James W. "MRP and Schedule Execution in the Job Shop of the Future." *SME/CASA Autofact, 1987 Conference Proceedings*, Dearborn, MI, November 9-12, 1987, pp. 9-85 — 9-97.

Brandt, Richard. "At the Starting Gate of Machine Vision." *Business Week*, 18 March 1985, pp. 67, 70.

Brandt, Richard and Port, Otis. "How Automation Could Save the Day." *Business Week*, 3 March 1986, pp. 72-74.

Braun, Carlton. "Total Customer Satisfaction: the Quality Challenge of the 90s." Presented at the 2nd Annual Quality Conference, School of Business Administration, The University of Michigan, Ann Arbor, MI, April 27, 1989.

Brauninger, Paul. "Computer-Aided Manufacturing in Small Companies." *NMBTA 4th Biennial International Manufacturing Technology Conference Proceedings*, Chicago, IL, September 7-14, 1988, pp. 1-3 – 1-6.

Brennan, Paul J. "Managing Emerging Technologies." Paper presented at NMTBA Manufacturing Conference, Cambridge, Massachusetts, 24 May 1984.

Brew, John. "Practical FMS." *NMBTA 3rd Biennial International Machine Tool Conference Proceedings*, Chicago, IL, September 3-11, 1986, pp. 3-50 – 3-62.

Bridges, William. "How to Manage Organizational Transition." *Training: The Magazine of Human Resources Development* 22 (September 1985): pp. 28-33.

Brimson, James A. "Cost Management for Advanced Manufacturing." c. 1987, *APICS 1987 Total Manufacturing Performance Seminar Proceedings*, Washington, DC, July 13-15, 1987, pp. 1-18.

Brimson, James A. "How Advanced Manufacturing Technologies Are Replacing Cost Management." *Management Accounting*, March 1986, pp. 25-29.

Brimson, James A. and Frescola, Leonard D. "Technology Accounting – the Value Added Approach to Capital Depreciation." *CIM Review*, Fall 1986, pp. 44-52.

Bristow, Clinton Jr. "Management Strategies for Implementing Just-In-Time Production/Zero Inventory." *APICS 1984 Zero Inventory Philosophy and Practices Seminar Proceedings*, St. Louis, MO, October 29-31, 1984, pp. 354-363.

Brock, William C. "Preparing U.S. Workers for a Change." *Business America* 10 (27 April 1987): pp. 6-8.

Brody, E.W. "How to Reward Employee Performance." *Public Relations Quarterly*, Spring 1988, p. 18.

Brody, Herb. "CAD Meets CAM." *High Technology* 7 (May 1987): pp. 12-18.

Brody, Herb. "Robots: Hands Across Japan." *High Technology* 6 (August 1986): pp. 22-24.

Brody, Herb. "U.S. Robot Makers Try to Bounce Back." *High Technology* 7 (October 1987): pp. 18-25.

Brody, Michael. "Helping Workers to Work Smarter." *Fortune* 115 (8 June 1987): pp. 86-89.

Brooks, Brian J. "Long-Term Incentives: International Executives Need Them Too." *Personnel* 65 (August 1988): pp. 40-43.

Brooks, George and Linkletter, J.R. "Statistical Thinking and W. Edwards Deming's Teachings in the Administrative Environment." *National Productivity Review*, Summer 1986, pp. 271-280.

Brooks, Sandra Lee. "Machining Centers: a Real World Technology That Makes Sense." *Production* 98 (December 1986): pp. 36-43.

Brown, James F. Jr. "How U.S. Firms Conduct Strategic Planning." *Management Accounting*, February 1986, pp. 38-55.

Brown, Paul B.; Hyatt, Joshua; Lammers, Teri; Posner, Bruce G.; and Solomon, Stephen D. "Motivation – Make the Carrot Count." *Inc.*, March 1988, p. 112.

Brown, Ronald. "Why Apply SPC to Automatic Assembly Systems." *NMBTA 4th Biennial International Manufacturing Technology Conference Proceedings*, Chicago, IL, September 7-14, 1988, pp. 5-149 – 5-153.

Brownfield, Hal. "Automated Automobile Assembly." *NMBTA 4th Biennial International Manufacturing Technology Conference Proceedings*, Chicago, IL, September 7-14, 1988, pp. 11-79 – 11-85.

Brownfield, Hal. "Techno-Economics – A Modern Analytical Tool." *NMBTA 4th Biennial International Manufacturing Technology Conference Proceedings*, Chicago, IL, September 7-14, 1988, pp. 7-87 – 7-92.

Bruggeman, John J. "Zero Inventory's Role in Competitive Strategy." *APICS 1984 Zero Inventory Philosophy and Practices Seminar Proceedings*, Washington, DC, July 13-15, 1984, pp. 326-329.

Brunn, Richard J. "Excess Capacity Management: The Forest and the Trees." c. 1986, *APICS 1987 Total Manufacturing Performance Seminar Proceedings*, Washington, DC, July 13-15, 1987, pp. 270-272.

Bruns, T. and Schuneman, T.M. "Forecasting the Diffusion of Industrial Robots." *The Industrial Robot* 14 (December 1987): pp. 219-228.

Buffa, Elwood. *Meeting the Competitive Challenge: Manufacturing Strategy for U.S. Companies*. Homewood, IL: Richard D. Irwin, 1984.

Burt, David N. "Managing Suppliers Up To Speed." *Harvard Business Review*, July-August 1989, pp. 127-135.

Bullinger, H.J.; Warnecke, H.J.; and Lentes, H. "Toward the Factory of the Future." *International Journal of Production Research* 24 (July-August 1986): pp. 697-741.

Bullinger, H.J. and Sauer, H. "Planning and Implementing a Flexible Assembly System Supported By Simulation." *International Journal of Production Research* 25 (1987): pp. 1625-1634.

Burnham, John M. "Just-In-Time in a Major Process Industry." *APICS 1986 Zero Inventory/JIT Seminar Proceedings*, Hilton Head, SC., July 21-23, 1986, pp. 1-42.

Burnstein, M. and Graham, P. "Market, Manufacturing Strategy, and Technology Acquisition: An Integrative Development Approach." in *Manufacturing Strategy* (J.E. Ettie, M.C. Burstein, and A. Fiegenbaum, eds.), Boston: Kluwer Academic Publishers, pp. 109-116.

Buzzell, Robert D. and Gale, Bradley T. *The PIMS Principles*. New York: The Free Press, 1987.

Byham, W.C. "HRD and the Steel Collar Worker." *Training: The Magazine of Human Resources Development* 21 (January 1984): pp. 59-64.

Byrne, John A.; Zellner, Wendy; and Ticer, Scott. "Caught in the Middle." *Business Week*, 12 September 1988, pp. 80-88.

Callahan, David. "Pitfalls of GT Cells & Flowlines." *NMBTA 4th Biennial International Manufacturing Technology Conference Proceedings*, Chicago, IL, September 7-14, 1988, pp. 11-53 – 11-75.

Campanella, J. and Corcoran, F.J. "Principles of Quality Costs." *Quality Progress*, April 1983, p. 17.

Cantwell, John. "The Reorganization of European Industries After Integration: Selected Evidence on the Role of Multinational Enterprise Activities." *Journal of Common Market Studies* 26 (December 1987): pp. 127-151.

Captain, Tom and Giffi, Craig. "Restructuring the Engineering Process for Competitive Advantage." Deloitte & Touche research report, 1990.

Carcone, Joseph; Jensen, John; and Johnson, Dave. "Machine Vision Explosion in Sight." *IC&S-The Industrial and Process Control Magazine*, December 1984, pp. 29-33.

Carlopio, Jim and Roitman, David. "Assessing Work Environment and Organization Characteristics." *NMBTA 4th Biennial International Manufacturing Technology Conference Proceedings*, Chicago, IL, September 7-14, 1988, pp. 3-165 – 3-172.

Carlyle, Ralph Emmett. "Sins of Omission." *Datamation*, 1 January 1988, pp. 48-54.

Carr, Clay. "Making the Human-Computer Marriage Work." *Training & Development Journal*, May 1988, pp. 65-74.

Case, John. "Every Other Worker an Owner?" *Inc.*, May 1987, pp. 14-15.

Case, John. "Zero Defect Management." *Inc.*, February 1987, pp. 17-18.

Cavinato, Joseph. "What Does Your Inventory Really Cost?" *Distribution* 87 (March 1988): pp. 68-72.

Chaimberlain, Woodrow. "Our Strategy is Flawed in World Competition." *American Production and Inventory Control Society, 1987 Conference Proceedings*, St. Louis, MO, October 19-23, 1987, pp. 514-516.

Chalofsky, Neal E. and Reinhart, Carlene. "Your New Role in the Organizational Drama: Measuring Effectiveness." *Training & Development Journal* 42 (August 1988): pp. 30-38.

Chan, Kim W. "Industry Competition, Corporate Variables, and Host Government Intervention in Developing Nations." *Management International Review* 28 (April 1988): pp. 16-28.

Chao, Georgia T. and Kozlowski, Steve W.J. "Employee Perceptions on the Implementation of Robotic Manufacturing Technology." *Journal of Applied Psychology* 71 (February 1986): pp. 70-77.

Chase, R.B. and Garvin, D. "The Service Factory." *Harvard Business Review*, July-August 1989, pp. 61-69.

Chernik, Roger. "Flexible Machining Cells." *NMBTA 3rd Biennial International Machine Tool Technical Conference Proceedings*, Chicago, IL, September 3-11, 1986, pp. 5-38—5-44.

Chiantella, Nathan A. "Achieving Integration Automation through Computer Networks." *SME Technical Article Series*, 1987, pp. 1-6.

Child, John and Smith, Chris. "The Context and Process of Organizational Transformation." *Journal of Management Studies* 24 (November 1987): pp. 565-594.

Chitaley, A. D. "Justifying High-Tech Manufacturing Solutions Through JIT-TQC-CIM." *Robotics Engineering* 8 (April 1986): pp. 12-14.

Choate, Pat and Linger, Juyne. "Tailored Trade: Dealing With the World as It Is." *Harvard Business Review* 66 (January-February 1988): pp. 86-93.

Cholakis, Peter N. "Sensors: Workable, Reliable Sensor Technology is Now in the Marketplace." *Robotics Today* 8 (April 1986): pp. 39-42.

Christy, David P. "Sole Source or Multiple Vendors? Costs and Benefits." *American Production and Inventory Control Society, 1987 Conference Proceedings*, St. Louis MO, October 19-23, 1987, pp. 605-607.

Chui, Kwok-Sang. "Preparations for Manufacturing in the 1990s." *NMBTA 4th Biennial International Manufacturing Technology Conference Proceedings*, Chicago, IL, September 7-14, 1988, pp. 3-139—3-152.

Chui, Kwok-Sang and Adams, David. "A Computerized Model for FMS Economics." *NMBTA 4th Biennial International Manufacturing Technology Conference Proceedings*, Chicago, IL, September 7-14, 1988, pp. 4-165—4-177.

Ciampa, Dan. "Integrated Manufacturing: The Only Answer for U.S. Companies." *Managing Automation*, January 1987, pp. 15-16.

Cissell, Michael J. "Designing Effective Reward Systems." *Compensation & Benefits Review* 19 (November-December 1987): pp. 49-55.

Civerolo, John. "MRP II and JIT Make a Tough Competitive Combination." c. 1986, *APICS 1987 Total Manufacturing Performance Seminar Proceedings*, Washington, DC, July 13-15, 1987 pp. 120-124.

Clark, James T. "Computer Integrated Manufacturing: A Management Perspective." *American Production and Inventory Control Society, 1987 Conference Proceedings*, St. Louis, MO, October 19-23, 1987, pp. 526-532.

Clark, Kim B. "What Strategy Can Do for Technology." *Harvard Business Review*, November-December 1986, pp. 94-95.

Clark, Kim B.; Hayes, Robert H.; and Lorenz, Christopher. *The Uneasy Alliance: Managing the Productivity-Technology Dilemma.* Boston: Harvard University Press, 1985.

Cloutier, Edward J. "Top Management Experiences in Applying CIM to the Enterprise." *SME/CASA Autofact, 1987 Conference Proceedings*, Dearborn, MI, November 9-12, 1987, pp. 1-57—1-78.

Coffman, Cathy. "Make Me a Match: Getting Design and Manufacturing Together—Simultaneously." *Automotive Industries* 167 (December 1987): pp. 62-64.

Cohen, Alan and Quarrey, Michael. "Performance of Employee Owned Small Companies: A Preliminary Study." *Journal of Small Business Management*, April 1986, pp. 58-63.

Cole, Raymond C. "Issues of Survival and Competitiveness Lie at Heart of Automation Debate." *Industrial Engineering*, November 1987, pp. 28-34.

Cole, Raymond C. "Target Information for Competitive Performance." *Harvard Business Review*, May-June 1985, pp. 100-109.

Cole, Raymond C. "The Macropolitics of Organizational Change: A Comparative Analysis of the Spread of Small Group Activities." *Administrative Science Quarterly* 30 (1985): pp. 560-585.

Cole, Raymond C. "U.S. Quality Improvement: Close But No Cigar." University of Michigan, 1989.

Collins, Joseph E. "New Techniques Vital to Economic Growth." *Data Management* 25 (October 1987): pp. 7-9.

Collins, Robert P. "MAP is Key to U.S. Manufacturing Comeback." *Industry Week*, 16 May 1988, p. 16.

Conner, Daryl. "Peopleware Skills for APICS Professionals." *American Production and Inventory Control Society, 1987 Conference Proceedings*, St. Louis, MO, October 19-23, 1987.

Connor, Susan and Thompson, Olin. "...and Select the Right Automation Software." *Process Engineering* 68 (March 1987): pp. 33-35.

Contino, Anthony V. "Check Out the Competition." *Chemical Engineering*, 20 June 1988, pp. 127-132.

Cook, James. "Back to Simplicity." *Forbes* 138 (25 August 1986): pp. 32-33.

Cook, Mary F. "What's Ahead in Human Resources?" *Management Review*, April 1988, pp. 41-45.

Cooper, Robin. "Implementing an Activity Based Cost System." *The President and Fellows of Harvard University*, January 1988, pp. 1-30.

Cooper, Robin and Kaplan, Robert S. "How Cost Accounting Distorts Product Costs." *Management Accounting*, April 1988, pp. 20-27.

Cooper, Robin and Kaplan, Robert S. "Measure Costs Right: Make the Right Decision." *Harvard Business Review*, September-October 1988, pp. 96-100.

Cope, Ken. "A New Concept in Manufacturing Centers." *NMBTA 4th Biennial International Manufacturing Technology Conference Proceedings*, Chicago, IL, September 7-14, 1988, pp. 11-5—11-17.

Cortes-Camera, Nhora. "The Custom Route to Flexible Manufacturing." *Mechanical Engineering*, August 1987, pp. 60-66.

Coshnitzke, James. "Robotic Welders: Matching Work and Workplace." *Automation* 35 (June 1988): pp. 12-20.

Coughlin, Neil. "Flexible Manufacturing for Medium to High Production Factories." *NMBTA 4th Biennial International Manufacturing Technology Conference Proceedings*, Chicago, IL, September 7-14, 1988, pp. 11-21—11-36.

Cox, Jeff and Goldratt, Eliyahu M. *The Goal: A Process of Ongoing Improvement*. Rev. Ed. Croton-on-Hudson, New York: North River Press Inc., 1986.

Craddock, Jack. "Changing Role of AS/RS Equipment." *Plant Engineering* 42 (18 August 1988): pp. 76-78.

Craig, Charles E. "Total Productivity Measurement at the Firm Level." *Sloan Management Review*, Spring 1973, pp. 13-29.

Crosby, Philip B. "Eight Ways the Quality Drive Changes Purchasing." from *Electronics Purchasing*, a special section in *Purchasing Magazine* 100 (8 May 1986): p. 66A37.

Crosby, Philip B. *Quality is Free*. New York: New American Library, 1979.

Crowley, Robert E. "Let's Discuss CAD/CAM Integration." *Modern Machine Shop* 57 (January 1985): pp. 86-95.

Cudworth, E.F. "Pratt & Whitney's $200 Million Factory Showcase." *Industrial Engineering* 16 (October 1984): pp. 90-97.

Curtis, Donald. *Management Rediscovered; How Companies Can Escape the Numbers Trap*, Dow Jones-Irwin, 1990.

Cusumano, Michael A. "Manufacturing Innovations: Lessons From the Japanese Auto Industry." *Sloan Management Review* 30 (Fall 1988): pp. 29-39.

Czaplicki, David. "Bar Code Applications in Manufacturing." *NMBTA 4th Biennial International Manufacturing Technology Conference Proceedings*, Chicago, IL, September 7-14, 1988, pp. 3-65 – 3-71.

D'Amore, Robert. "Changing Mfg's Outdated Accounting & Control Systems." *APICS 1984 Zero Inventory Philosophy and Practices Seminar Proceedings*, St. Louis, MO, October 29-31, 1984, pp. 363-371.

Dauch, Richard R.; Richardson, Marion S.; Geier, James A.D.; et al. "Managing for Success: What Needs to be Done." *Production*, August 1986, pp. 47-72.

Davidson, Jeffrey P. "Time-Saving Plan Can Benefit You and Your Staff." *Data Management*, August 1987, pp. 17-23.

Davis, Arthur G. "Competitiveness: Survival Strategies for the 1990's." *Machine and Tool Blue Book* 83 (October 1988): p. 71.

Deal, Terrence E. and Kennedy, Allan A. *Corporate Cultures – The Rites and Rituals of Corporate Life.* Reading, MA: Addison-Wesley Publishing Company, 1982.

Dean, Roxanne and Prior, Daniel W. "Your Company Could Benefit From a No-Layoff Policy." *Training & Development Journal* 40 (August 1986): pp. 38-42.

Debow, Yvette. "G.E.: Easing the Pain of Layoffs." *Management Review* 76 (September 1987): pp. 15-18.

Delaney, Patrick S. "Help Wanted: Plant Manager." *American Production & Inventory Control Society, 1987 Conference Proceedings*, St. Louis, MO, October 19-23, 1987, pp. 132-135.

Dell, Carla and Grayson, C. Jackson, Jr. "Flex Your Pay Muscle." *Across The Board* 25 (July-August 1988): pp. 43-48.

De Meyer, A. "Quality Up, Technology Down." *INSEAD* Working paper No. 88/65, Fountainbleau, France.

Deming, W. Edwards. "New Principles of Leadership." *Modern Materials Handling*, October 1987, p. 37.

Dentzer, Susan. "The Foibles of ESOP's." *Newsweek* 110 (19 October 1987): pp. 58-60.

Dertouzos, Michael L.; Lester, Richard K.; Solow, Robert M. *Made in America – Regaining the Productive Edge*. Boston: The Massachusetts Institute of Technology, 1989.

Deutsch, Steven. "Successful Worker Training Programs Help Ease the Impact of Technology." *Monthly Labor Review* 110 (November 1987): pp. 14-21.

Deveny, Kathleen; Brown, Corie; and Hampton, William J. "Going for the Lion's Share: The Time is Right For U.S. Companies to Reclaim Lost Markets." *Business Week*, 18 July 1988, pp. 70-72.

Diesch, K.H. and Malstrom, E.M. "Physical Simulator Analyzes Performance of Flexible Manufacturing System." *Industrial Engineering* 17 (June 1985): pp. 66-72.

Dilts, D.M. and Russell, G.W. "Accounting for the Factory of the Future." *Management Accounting* 66 (April 1985): pp. 34-40.

Dixon, J. Robb; Nanni, Alfred J.; and Vollmann, Thomas E. *"The Performance Challenge: Measuring Operations for World-class Competition."* Homewood, IL: Dow Jones Irwin, 1990.

Dodgson, Mark. "Small Firms, Advanced Manufacturing Technology, and Flexibility." *Journal of General Management* 12 (Spring 1987): pp. 58-76.

Donoghue, Ronald. "Bar Code Factory Data Collection Systems." *NMBTA 4th Biennial International Manufacturing Technology Conference Proceedings*, Chicago, IL, September 7-14, 1988, pp. 3-75 – 3-77.

Dornan, Sandra Brooks. "Cells and Systems: Justifying the Investment." *Production* 99 (February 1987): pp. 30-36.

Dornan, Sandra Brooks. "Just-in-time: The Home-Grown Technique Comes Home." *Production* 99 (August 1987): pp. 60-63.

Dornan, Sandra Brooks. "Planning for Manufacturing in Large Organizations – the Year 2000 and Beyond." *Production* 99 (March 1987): pp. 34-40.

Dornan, Sandra Brooks. "Sensing Systems for the Real-World Factory." *Production* 99 (September 1987): pp. 62-69.

Dornan, Sandra Brooks. "The Factory of the Future: Visions Taking Shape." *Production* 99 (May 1987): pp. 45-49.

Dowst, Somerby. "VA '86 Buyers Say VA is More Important Than Ever." *Purchasing*, 26 June 1986, pp. 64-83.

DoDoyle, Hoyt, and Riley, Thomas. "Considerations in Developing Incentive Plans." *Management Review* 76 (March 1987): pp. 34-38.

Dreyfuss, Joel. "Get Ready For The New Workforce." *Fortune*, 23 April 1990, pp. 165-181.

Dreyfuss, Joel. "The Three R's on the Shop Floor." *Fortune*, Spring 1990 Special Issue, p. 88.

Drucker, P. "The Emerging Theory of Manufacturing." *Harvard Business Review*, May-June 1990, p. 94-102.

Dumaine, Brian. "How Managers Can Succeed Through Speed." *Fortune*, 13 February1989, pp. 54-59.

Dumaine, Brian. "Making Education Work." *Fortune*, Spring 1990 Special Issue, p. 13.

Dumaine, Brian. "Who Needs a Boss?" *Fortune*, 7 May 1990, pp. 52-60.

Dumas, Roland A. "Shaky Foundations of Quality Circles." *Training*, April 1983, pp. 32-33.

Dumas, Roland A.; Cushing, Nancy; and Laughlin, Carol. "Making Quality Control Theories Workable." *Training & Development Journal*, February 1987, pp. 30-35.

Dunlap, Glenn C. "CIM – Communication Integrated Manufacturing." *Autofact 1987 Conference Proceedings*, pp. 7-99 – 7-112. Detroit, MI, 1987.

Dunlop, John T. "Have the 1980's Changed Industrial Relations?" *Monthly Labor Review* 111 (May 1988): pp. 29-34.

Dunn, Alan G. "How to Determine the Real Cost of Your Product." *American Production and Inventory Control Society, 1987 Conference Proceedings*, St. Louis, MO, October 19-23, 1987, pp. 644-647.

Duvall, Thomas J. "Realistic Cost Accounting Systems." *NMBTA 3rd Biennial International Machine Tool Technology Conference Proceedings*, Chicago, IL, September 3-11, 1986, pp. 11-1 – 11-9.

Dvorak, Karen S. "AGVs Blaze Path to Productivity." *Assembly Engineering* 30 (September 1987): pp. 32-34.

Dvorak, Karen S. and Kelly, Robert T. "Assembly Systems 2000." *Assembly Engineering* 30 (June 1987): pp. 48-50.

Dybeck, Martin and Bozenhardt, Herman, "Justifying Automation Projects." *Chemical Engineering*, 20 June 1988, pp. 113-116.

Ealey, L.A. *Quality by Design: Taguchi Methods and U.S. Industry*. Dearborn, MI: ASI Press, 1988.

Earl, Michael; Feeny, David; Lockett, Martin; and Runge, David. "Competitive Advantage Through Information Technology: Eight Maxims for Senior Managers." *Multinational Business*, Summer 1988, pp. 15-21.

Eckley, R.S. "Caterpillar's Ordeal: Foreign Competition in Capital Goods." *Business Horizons*, March-April 1989, pp. 80-86.

Edosomwan, Johnson Aimee. "Industrial Engineers' Roles are Numerous and Varied in High Technology Environment." *Industrial Engineering* 19 (December 1987): pp. 36-40.

Edwards, J. Nicholas. "Master Scheduling for a JIT Environment: A Case Study." *APICS 1987 Total Manufacturing Performance Seminar Proceedings*, Washington, DC, July 13-15, 1987, pp. 9-11.

Edwards, James B. and Heard, Julie A. "Is Cost Accounting the No. 1 Enemy of Productivity?" *Management Accounting* 65 (June 1984): pp. 44-49.

Eiler, Robert G.; Goletz, Walter K.; and Keegan, Daniel P. "Is Your Cost Accounting Up to Date?" *Harvard Business Review* 60 (July-August 1982): pp. 133-139.

Eisenstein, Paul. "MIS in Transition, Part III: Detroit's Counterattack Strategy." *Computer Decisions* 20 (May 1988): pp. 52-56.

Elliott, Paul H. "Integrating Human Resources into Advanced Manufacturing Systems." *Proceedings From the 1987 IEEE, International Conference on Systems, Man and Cybernetics*, Alexandria, VA, 1987, pp. 503-507.

Emmelheing, Margaret A. "Electronic Data Interchange: Does it Change the Purchasing Process." *Journal of Purchasing and Materials Management*, Winter 1987, pp. 2-8.

Emmett, Arielle. "Factories of the Future: The Hidden Element in CIM." *Communications Consultant*, November 1988, pp. 30-34.

Engel, Alan K. "How to Profit From Japanese Technology." *Directors & Boards* 12 (Summer 1988): pp. 14-17.

Engel, Alan K. "Number One in Competitor Intelligence." *Across the Board* 24 (December 1987): pp. 43-48.

Epner, Paul. "Defining General Purpose Automation Controls." *NMBTA 4th Biennial International Manufacturing Technology Conference Proceedings*, Chicago, IL, September 7-14, 1988, pp. 2-215 – 2-223.

The Ernst and Young Quality Improvement Consulting Group. *Total Quality*. Dow Jones-Irwin/APICS Series, 1990.

Ettlie, John E.; Klein, Janice A.; and Vossler, Marika L. "Robotics Training." *Training and Development Journal*, March 1988, pp. 54-56.

Ettlie, John E.; Bridges, W.; and O'Keefe, R. "Organization, Strategy and Structural Differences for Radical Versus Incremental Innovations." *Management Science* 30 (June 1984): pp. 682-695.

Ettlie, John E. and Penner-Hahn, J.D. "Focus, Modernization, and Manufacturing Technology Policy." *Manufacturing Strategy*, (J.E. Ettlie, M.C. Burnstein, and A. Feigenbaum, eds.), Norwell, MA: Kluwer Academic Publishers, 1989, pp. 153-164.

Ettlie, John E. and Stoll, Henry W. *Managing the Design-Manufacturing Process.* New York: McGraw-Hill, 1990.

Ettorre, John J. "Can Labor Work With Automation?" *Training and Development Journal*, July 1988, pp. 15-18.

Ettorre, John J. "Robotics: Search for New Markets in Warehousing." *Handling & Shipping Management* 28 (May 1987): pp. 20-24.

Eureka, W.E. and Ryan, N.E. *The Customer Driven Company*. Dearborn, MI: ASI Press, 1988.

Evan, William M. and Olk, Paul. "R&D Consortia: A New U.S. Organizational Form." *Sloan Management Review*, Spring 1990, pp. 37-45.

Evans, Bill. "Simultaneous Engineering." *Mechanical Engineering* 110 (February 1988): pp. 38-39.

Evans, Karen M. "On-the-Job Lotteries: A Low-cost Incentive that Sparks Higher Productivity." *Personnel*, April 1988, pp. 22-26.

Ewaldz, Donald B. "Crossing the Bridge to CIM." *NMTBA 1987 Southern Manufacturing Technology Conference Proceedings*, Charlotte, NC, January 26-29, 1987, pp. 2-3 – 2-29.

Ewaldz, Donald B. "The Wages of CIM." *NMBTA 4th Biennial International Manufacturing Technology Conference Proceedings*, Chicago, IL, September 7-14, 1988, pp. 1-23 – 1-42.

Faiola, Tony and DeBloois, Michael L. "Designing Visual Factors-Based Screen Display Interface: The New Role of the Graphic Technologist." *Educational Technology,* August 1988, pp. 12-21.

Farley, John U.; Kahn, Barbara; Lehmann, Donald R.; and Moore, William L. "Modeling the Choice is Automatic." *Sloan Management Review,* Winter 1987, pp. 5-15.

Fasci, Martha A.; Weiss, Timothy J.; and Worrall, Robert L. "Everyone Can Use This Cost/Benefit Analysis System." *Management Accounting,* January 1987, pp. 45-47.

Fechter, William F. "Achieving Manufacturing Excellence Through the Sociotechnical Aspects of Cycle Time Management." *SME/CASA Autofact, 1987 Conference Proceedings,* Dearborn, MI, November 9-12, 1987, pp. 8-1 – 8-11.

Fechter, William F. and Horowitz, Renee B. "The Role of the Industrial Supervisor in the 1990's." *Industrial Management* 30 (May-June 1988): pp. 18-19.

Feder, Barnaby J. "The Drive to Speed Automation." *The New York Times,* Wednesday 15 June 1988, sec. D, pp. 1, 8.

Feigenbaum, Armand V. *Total Quality Control.* New York: McGraw-Hill Book Company, 1983.

Ferdows, K. "Mapping International Factory Networks." *Managing International Manufacturing,* Amsterdam: Elsevier Science Publishers, 1989, pp. 3-11.

Ferdows, K. and De Meyer, A. "Lasting Improvement in Manufacturing Performance: In Search of a New Theory." *INSEAD* Working Paper, Fountainbleau, France, January 1989.

Ferdows, K. and Lindberg, P. "FMS as Indicator of the Strategic Role of Manufacturing." *International Journal of Production Research* 25 (1987): pp. 1563-1571.

Fernandez, John. "New Life for Old Stereotypes." *Across The Board,* July-August 1988, pp. 24-25.

Firschein, Oscar. "Machine Vision and Image Understanding." *IEEE Control Systems Magazine* 8 (June 1988): pp. 3-5.

Fisher, Sharon. "Connectivity the Byword at PC Expo Product Show." *Infoworld* 10 (27 June 1988): pp. 11-13.

Fitzgerald, Karen. "Special Report – The Global Automobile: Compressing the Design Cycle." *IEEE Spectrum* 24.10 (October 1987): pp. 39-43.

Fleck, James D., and D'Cruz, Joseph R. "The Globalization of Manufacturing." *Business Quarterly* 52 (Winter 1987): pp. 42-52.

Flick, Willi. "Material Handling in a Flexible Fabricating System." *NMBTA 4th Biennial International Manufacturing Technology Conference Proceedings,* Chicago, IL, September 7-14, 1988, pp. 3-23 – 3-25.

Flynn, M.S. and Cole, R.E. "Automotive Suppliers: Customer Relationships, Technology, and Competition." Report of the Supplier Change Project, Industrial Technology Institute, July 1986, p. 27.

Foley, Mary Jo. "How Computer Products Linked MRP to AS-RS." *Electronic Business* 13 (1 June 1987): pp. 80-82.

Foley, Mary Jo. "The Winds of Change Are Blowing for MRP II Users." *Datamation* 34 (1 February 1988): pp. 19-24.

Forbes, David. "A Consensus on Helping Idled Workers." *Business Month* (March 1987): pp. 55-57.

Forger, Gary. "Coming to Grips With Global Competition." *Modern Materials Handling* 43 (January 1988): pp. 64-67.

Foster Badi G.; Jackson, Gerald; Cross, William E.; Jackson, Bailey; and Hardiman, Rita. "Workforce Diversity and Business." *Training & Development Journal* 42 (April 1988): pp. 38-43.

Fox, Robert E., and Goldratt, Eliyahu M. *The Race.* Croton-on-Hudson, NY: North River Press Inc., 1986.

Francis, Tom. "FIS Ties it All Together for Chrysler." *Machine and Tool Blue Book* 80 (April 1985): pp. 79-80.

Franck, Gerald. "Flexible Inspection for the Factory Environment." *NMBTA 4th Biennial International Manufacturing Technology Conference Proceedings*, Chicago, IL, September 7-14, 1988, pp. 10-201 – 10-217.

Frank, Frederic and Taylor, Craig. "Assessment Centers in Japan." *Training & Development Journal*, February 1988, pp. 54-57.

Franklin, Jerry. "For Technical Professionals: Pay for Skills and Pay for Performance." *Personnel* 65 (May 1988): pp. 20-28.

Frost, Chad C.; McIntyre, Robert H.; and Nauta, Randal W. "Developing a Manufacturing Strategy for the Smaller Company... Tomorrow, the Day After, and Beyond." *Production*, March 1987, pp. 40-46.

Fruhan, William E. "Management, Labor, and the Golden Goose." *Harvard Business Review*, September-October 1985, pp. 131-141.

Fry, Timothy D. and Smith, Allen E. "A Procedure for Implementing Input/Output Control: A Case Study." *Production and Inventory Management Journal*, Fourth Quarter 1987, pp. 50-52.

Fuer, Dale and Lee, Chris. "The Kaizen Connection." *Training* 25 (May 1988): pp. 23-35.

Funk, Jeffrey L. "How Does Japan Do it?" *Production* 100 (August 1988): pp. 57-63.

Fuqua, Dan. "Key Technologies for America's Future." Keynote Speech to the Aviation/Space Writers Association Yearend Review & Forecast Luncheon, Washington, DC, 16 December, 1987.

Galagan, Patricia. "Donald E. Petersen: Chairman of Ford and Champion of its People." *Training and Development Journal* 24 (1987): pp. 20-25.

Galagan, Patricia. "Here's the Situation: A Quick Scan of the Trends that Experts Think Will Affect You Most." *Training & Development Journal* 41 (July 1987), pp. 20-23.

Gale, Bradley T. and Klavans, Richard. "Formulating a Quality Improvement Strategy." *Journal of Business Strategy*, pp. 21-32.

Garcia, Chris. "Quality Intelligence: The Missing Link in CAD/CAM." *NMBTA 4th Biennial International Manufacturing Technology Conference Proceedings*, Chicago, IL, September 7-14, 1988, pp. 6-175 – 6-189.

Garwood, Dave. "The Future of the Factory: Automation or Motivation?" *American Production & Inventory Control Society, 1987 Conference Proceedings*, St. Louis, MO, October 19-23, 1987, pp. 552-553.

Gaugler, Eduard. "HR Management: An International Comparison." *Personnel* 65 (August 1988): pp. 24-31.

Gauvin, Jacqueline. "Survey Rates Robot Vendors." *Robotics Today*, February 1986, p. 45.

Gayman, David J. "Carving Out a Legend." *Manufacturing Engineering*, March 1988.

Geber, Beverly. "The Hidden Agenda of Performance Appraisals." *Training*, June 1988, pp. 42-47.

Gee, Jack. "CIM Sales in Europe Send the Right Signals to U.S. Vendors." *Electronic Business* 14 (1 August 1988): pp. 68-69.

Gellerman, Saul W. and Hodgson, William G. "Cyanamid's New Take on Performance Appraisal." *Harvard Business Review* 66 (May-June 1988): pp. 36-41.

Gellman, Harvey. "Organization and Behavior – Keys to Successful Automation." *CMA – the Management Accounting Magazine* 61 (May-June 1987): p. 60.

Gersick, Connie J. G. "Time and Transition in Work Teams: Toward a New Model of Group Development." *Academy of Management Journal* 31 (March 1988): pp. 9-41.

Gerwin, Donald. "Do's and Don'ts of Computerized Manufacturing." *Harvard Business Review*, March-April 1982, pp. 107-116.

Gettleman, Ken. "Japan's TLC." *Modern Machine Shop* 60 (December 1987): pp. 78-87.

Gettleman, Ken and Czinski, Ron. "6 Steps to the Best." *Modern Machine Shop*, July 1988, pp. 85-92.

Giacomino, Don E. "The SAI Movement in Manufacturing." *The CPA Journal*, October 1986, pp. 64-73.

Gibble, D. "Data Links for Industrial Control." *Machine Design* 54 (12 August 1982): pp. 75-79.

Giffi, Craig and Roth, Aleda V. "Making the Grade in the 1990s," Deloitte & Touche Third Annual Survey of North American Manufacturing Technology, 1989.

Gill, M.S. "Stalking Six Sigma." *Business Month*, January 1990, pp. 421-45.

Giovannone, Dr. Ing. Marcello. "Measuring Robots in Manufacturing Systems." *The FMS Magazine*, January 1987, pp. 31-34.

Girard, Richard. "Is There a Need for Performance Appraisals?" *Personnel Journal*, August 1988, pp. 89-90.

Giunipero, Larry C. and O'Neal, Charles. "Obstacles to JIT Procurement." *Industrial Marketing Management* 17 (February 1988): pp. 35-42.

Glazer, Sarah. "High-Tech Storage for Storage Tech." *Electronic Business* 13 (1 February 1987): pp. 80-82.

Goddard, Robert W. "Post Employment: The Changing Current in Discrimination Charges." *Personnel Journal*, October 1986, pp. 34-40.

Goddard, Walter E. "How to Measure Performance Beyond Class A." *American Production and Inventory Control Society, 1987 Conference Proceedings*, St. Louis, MO, October 19-23, 1987, pp. 488-490.

Goddard, Walter E. *Just-In-Time Surviving By Breaking Tradition*. New York, NY: Oliver Wight Limited Publications, Inc., 1986.

Gold, Bela. "Charting a Course to Superior Technology Evaluation." *Sloan Management Review* 30 (Fall 1988): pp. 19-28.

Goldbogen, Geof and Abramson, Jerold. "A Practical and Realizable Definition of CIM Utilizing Existing Technology." *Autofact '87 Conference Proceedings*, pp. 4-99 – 4-116. Detroit, MI, 1987.

Goldman, Edward J.; Dickenson, Edward L.; and Captain, Khushroo M. "Planning Computerized On-Line Test and Inspection." *Assembly Engineering* 27.11 (November 1984): pp. 38-41.

Gorman, Christine. "The Literacy Gap." *Time*, 19 December 1988, pp. 56-57.

Gould, Lawrence. "What's the Technology in GT?" *Managing Automation*, July 1986, pp. 51-56.

Graff, Richard. "Material Control as a Manufacturing Process: JIT/TQC Synergies." *APICS 1987 Total Manufacturing Performance Seminar Proceedings*, Washington, DC, July 13-15, 1987, pp. 5-7.

Graham, J.H.; Meagher, J.F.; and Derby, S.J. "Safety and Collision Avoidance System for Industrial Robots." *IEEE Transactions on Industry Applications* IA-22 (1986): pp. 195-203.

Grant, Hank and Clapp, Christopher. "Making Production Scheduling More Efficient Helps Control Manufacturing Costs and Improve Productivity." *Industrial Engineering*, June 1988, pp. 54-62.

Grant, Philip C. "Rewards: the Pizzazz is the Package, Not the Prize." *Personnel Journal* 67 (March 1988): pp. 76-81.

Gray, James A. "Rustproof the Rust Belt." Unbound speech for the 85th Spring Meeting of the National Machine Tool Builders' Association, April 9, 1987.

Greenlaw, Paul S. "Cash Emerges as a Compensation Hero." *Personnel Journal*, July 1988, pp. 96-105.

Greenspan, Alan C. "Our Human Capital: the Power of Investment." *Directors & Boards* 12 (Summer 1988): pp. 4-9.

Gregory, Gene. "The Factory of Japan's Future." *Management Today*, April 1984, pp. 66-71.

Grieco, Peter L., Jr. "Competing at Home or Abroad." *APICS 1984 Zero Inventory Philosophy and Practices Seminar Proceedings*, St. Louis, MO, October 29-31, 1984, pp. 422-431.

Grieco, Peter L., Jr. "The Role TQC Plays in Our Business." *American Production and Inventory Control Society, 1987 Conference Proceedings*, St. Louis, MO, October 19-23, 1987, pp. 27-28.

Groocock, John. "A Quality Policy for the '80's." *Production*, January 1983, pp. 39-41.

Groover, Mikell P. and Wiginton, John C. "CIM and the Flexible Automated Factory of the Future." *Industrial Engineering* 18 (January 1986): pp. 75-86.

Gue, Frank. "Push, Pull JIT Production Methods: A Perspective." *APICS 1987 Total Manufacturing Performance Seminar Proceedings*, Washington, DC, July 13-15, 1987, pp. 116-119.

Guerrero, Hector, H. "Group Technology : II. The Implementation Process." *Production and Inventory Management*, Second Quarter, 1987, pp. 1-9.

Gunn, Thomas. "The CIM Connection." *Datamation*, 1 February 1986, pp. 50-58.

Gunn, Thomas. "The U.S. as Manufacturer: A Story of Pernicious Shortcomings." *Directors & Boards* 12 (Summer 1988): pp. 10-13.

Halbrecht, Herbert Z. and Nostrand, John W. "Is Management to Blame for the Unfulfilled Dream of CIM?" *Production Engineering*, July 1987, pp. 26-28.

Halcrow, Allan. "The Future is Made Now." *Personnel Journal* 67 (March 1988): pp. 12-14.

Hale, Carl D. "How to Professionally Qualify Your Suppliers." *American Production and Inventory Control Society, 1987 Conference Proceedings*, St. Louis, MO, October 19-23, 1987, pp. 590-592.

Hall, Robert W. *Attaining Manufacturing Excellence*. Homewood, IL: Dow Jones-Irwin, 1987.

Hall, Robert W. "Quality Essential: Capability to Improve Continuously." *Target*, publication of the Association for Manufacturing Excellence, 3.1 (Spring 1987), pp. 5-11.

Hall, Robert W. "The Maturing of JIT Manufacturing." Paper presented at APICS 1987 Total Manufacturing Performance Seminar, Washington, DC, July 13-15, 1987.

Hall, William K. "Survival Strategies in a Hostile Environment." In *Survival Strategies for American Industry*, pp. 52-71. New York: John Wiley & Sons 1983.

Halligan, Thomas M. and Rees, Tom. "Integrating Material Handling into the Automation Process." *Production* 98 (October 1986): pp. 62-69.

Hamel, Gary and Prahalad, C.K. "Strategic Intent." *Harvard Business Review* 67.3 (May-June 1989): pp. 63-76.

Hamerstone, James E. "How to Make Gain Sharing Pay Off." *Training & Development Journal*, April 1987, pp. 80-81.

Hampton, William J., "GM Bets an Arm and a Leg on a People-Free Plant." *Business Week*, 12 September 1988, pp. 72-73.

Hampton, William J. "Reality Has Hit General Motors—Hard." *Business Week*, 24 November 1986, p. 37.

Hannah, K.H. "Just-In-Time: Meeting the Competitive Challenge." *Production & Inventory Management 28* (Third Quarter 1987): pp. 1-3.

Hansen, Gary B. "Creating a Future for Workers Displaced by Foreign Competition." *Business Forum* 13 (Winter 1988): pp. 16-21.

Hanson, Russell G. "Managing the Human Aspects of CIM Implementation." *SME/CASA Autofact, 1987 Conference Proceedings*, Dearborn, MI, November 9-12, 1987, pp. 8-13—8-25.

Hardy, John W.; Orton, Bryce B.; and Moffit, J. Weldon. "Bonus Systems Do Motivate." *Management Accounting* 68 (November 1986): pp. 58-62.

Hartenstein, Annette. "Building Integrated HRM Systems." *Training & Development Journal* 42 (May 1988): pp. 90-94.

Hatch, James. "Training For the New High Tech Factories." *NMBTA 4th Biennial International Manufacturing Technology Conference Proceedings*, Chicago, IL, September 7-14, 1988, pp. 3-155—3-161.

Haught, David. "CIM Case Study: AT&T Technologies." *APICS 1987 Total Manufacturing Performance Seminar Proceedings*, Washington, DC, July 13-15, 1987, pp. 1-6.

Havlovic, S.J. "The Impact of a Worker Participation Process on Production and Human Resource Outcomes." Doctoral Dissertation, Ohio State University, 1987, pp. 146-150.

Hawbaker, Jim. "Standardization for Rapid CAM Implementation." *NMBTA 4th Biennial International Manufacturing Technology Conference Proceedings*, Chicago, IL, September 7-14, 1988, pp. 1-45 – 1-55.

Hawk, Martin R. "An Industry Perspective of IMIP." *Industrial Engineering* 18.11 (November 1986): p. 38(2).

Hay, Edward J. "Reduce any Setup by 75%." c. 1984, *APICS 1987 Total Manufacturing Performance Seminar Proceedings*, Washington, DC, July 13-15, 1987, pp. 173-178.

Hayes, Robert H. "Strategic Planning – Forward or Reverse?" *Harvard Business Review*, November-December 1985, pp. 111-120.

Hayes, Robert H. "Why Japanese Factories Work." *Survival Strategies for American Industry*, pp. 231-247. New York: John Wiley & Sons, 1983.

Hayes, Robert H. and Jaikumar, Ramchandran. "Manufacturing's Crisis: New Technologies, Obsolete Organizations." *Harvard Business Review*, September-October 1988, pp. 77-85.

Hayes, Robert H. and Schmenner, Roger W. "How Should You Organize Manufacturing?" *Harvard Business Review*, January-February 1978, pp. 105-117.

Hayes, Robert H. and Wheelwright, Steven C. *Restoring Our Competitive Edge; Competing Through Manufacturing*. New York: John Wiley & Sons, 1984.

Hayes, Robert H.; Wheelwright, Steven C.; and Clark, Kim B. *Dynamic Manufacturing*. New York: The Free Press, 1988.

Heard, Julie A. "JIT and Performance Measurement." *APICS 1984 Zero Inventory Philosophy and Practices Seminar Proceedings*, St. Louis, MO, October 29-31, 1984, pp. 372-379.

Heard, Julie A. "JIT for White Collar Workers: The Rest of the Story." *American Production and Inventory Control Society, 1987 Conference Proceedings*, St. Louis, MO, October 19-23, 1987, pp. 415-420.

Heard, Julie A. "JIT Implementation Structure Requirements." *APICS 1987 Total Manufacturing Performance Seminar Proceedings*, Washington, DC, July 13-15, 1987.

Hector, Gary. "Yes, You Can Manage Long Term." *Fortune*, 21 November 1988, pp. 64-76.

Hedgecock, C. "Testing for 'Worms.'" *Venture* 6 (August 1984): pp. 133-134.

Heenan, David A. "A Different Outlook for Multinational Companies." *Journal of Business Strategy* 9 (July-August 1988): pp. 51-55.

Heller, Russell S. "Performance Measurement for the Strategically Integrated Business." *American Production and Inventory Control Society, 1987 Conference Proceedings*, St. Louis, MO, October 19-23, 1987, pp. 460-463.

Hellwig, Helmut. "Technology and the Competitive Challenge." *Research & Development* 30 (July 1988): pp. 44-49.

Hemmen, Paul G. "Manufacturing Resource Planning – Making it Happen." *NMBTA 3rd Biennial International Machine Tool Technical Conference Proceedings*, Chicago, IL, September 3-11, 1986, pp. 1-168 – 1-176.

Hemmen, Paul G. "Inventory Control – Not By Itself." *NMBTA 4th Biennial International Manufacturing Technology Conference Proceedings*, Chicago, IL, September 7-14, 1988, pp. 3-105 – 3-110.

Henkoff, Ronald. "Cost Cutting: How To Do It Right." *Fortune*, 9 April 1990, p. 40.

Herzberg, Frederick. "One More Time: How Do You Motivate Employees?" *Harvard Business Review* 65 (September-October 1987): pp. 109-121.

Herzberg, Frederick. "Workers' Needs: the Same Around the World." *Industry Week*, 21 September 1987, pp. 29-32.

Hessler, Walt. "Simple Automation for Machine Loading." *NMBTA 4th Biennial International Manufacturing Technology Conference Proceedings*, Chicago, IL, September 7-14, 1988, pp. 5-193 – 5-204.

Heylin, Michael. "Competitive Moves By Industry Still Tough on Some Chemists." *C&EN*, 30 November 1987, pp. 35-36.

Hiebert, Murray B. and Smallwood, W. Norman. "Now for a Completely Different Look at Needs Analysis." *Training & Development Journal* 41 (May 1987): pp. 75-80.

Hill, John M. "The Changing Profile of Material Handling Systems & Controls." *Industrial Engineering* 18 (November 1986): pp. 68-73.

Hill, Rod. "Pick Your Poison or Select a Solution." *APICS 1987 Total Manufacturing Performance Seminar Proceedings*, Washington, DC, July 13-15, 1987.

Hill, T. *Manufacturing Strategy: Text and Cases*. Homewood, IL: Dow Jones-Irwin, 1989.

Hillkirk, John. "The Man Japanese Firms Follow." *USA Today*, 22 May 1988, sec. B, p. 7.

Hinson, Ray S. "Developing Tomorrow's End-Of-Arm Tooling." *Robotics Today*, April 1986, pp. 52-55.

Hiromoto, Toshiro. "Another Hidden Edge – Japanese Management Accounting." *Harvard Business Review*, July-August 1988, pp. 26.

Hoerr, John and Pollock, Michael A. "Management Discovers the Human Side of Automation." *Business Week*, 29 September 1986, pp. 70-76.

Hoffman, Donald. "Laser Cutting in the 90s." *Machine and Tool Blue Book* 83.7 (July 1988): pp. 54-56.

Hoffman, Richard C. "The General Management of Foreign Subsidiaries in the U.S.A: An Exploratory Study." *Management International Review* 28 (Second Quarter 1988): pp. 41-55.

Holbrook, William G. "Accounting Changes Required for Just-In-Time Production." *APICS 1984 Zero Inventory Philosophy and Practices Seminar Proceedings*, St. Louis, MO, October 29-31, 1984, pp. 366-393.

Holbrook, William G. "Management Accounting Changes Made Possible by JIT." *APICS 1987 Total Manufacturing Performance Seminar Proceedings*, Washington, DC, July 13-15, 1987.

Holland, Thomas E. and Rummel, Patricia A. "Human Factors are Crucial Component of CIM System Success." *Industrial Engineering*, April 1988, pp. 36-42.

Holmes, Geoffrey. "Accountants Who Thrive On Chaos." *Accountancy* 101 (May 1988): pp. 94-96.

Holmes, Geoffrey and Sugden, Alan. "Silk Purses Out of Sows' Ears." *Accountancy*, July 1986, pp. 81-82.

Hoover, William R. "Implementing CIM From the Top Down." *Chief Executive*, November-December 1987, pp. 28-32.

Horn, Jack C. "Bigger Pay for Better Work." *Psychology Today*, July 1987, pp. 54-57.

Horngren, Charles T. and Foster, George. "JIT: Cost Accounting and Cost Management Issues." *Management Accounting* 68 (June 1987): pp. 19-26.

Horton, Thomas R. "Competing Through Competition." *Management Review* 77 (February 1988): pp. 5-6.

Houghton, J.R. "Quality: The Competitive Advantage." *Quality Progress*, February 1988.

Houston, Jerry. "How SPC Smooths the Way for CIM." *NMBTA 4th Biennial International Manufacturing Technology Conference Proceedings*, Chicago, IL, September 7-14, 1988, pp. 5-139 – 5-146.

Howell, Robert A.; Brown, James D,; Soucy, Stephen R.; and Seed, Allen H. "Management Accounting in the New Manufacturing Environment." National Association of Accountants, 1987.

Howell, Robert A. and Soucy, Stephen R. "Capital Investment in the New Manufacturing Environment." *Management Accounting* 69 (November 1987): pp. 26-33.

Howell, Robert A. and Soucy, Stephen R. "Management Reporting in the New Manufacturing Environment." *Management Accounting*, February 1988, pp. 22-29.

Howell, Robert A. and Soucy, Stephen R. "Operating Controls in the New Manufacturing Environment." *Management Accounting* 69 (October 1987): pp. 25-32.

Howell, Robert A. and Soucy, Stephen R. "The New Manufacturing Environment: Major Trends for Management Accounting." *Management Accounting*, July 1987, pp. 21-27.

Howser, Kenneth J. "Vendors at Just-In-Time Partners." *APICS 1987 Total Manufacturing Performance Seminar Proceedings*, Washington, DC July 13-15, 1987, pp. 378-379.

Hsie, A. and Tu, X.Y. "Intelligent Control & Intelligent Robots." *NMBTA 4th Biennial International Manufacturing Technology Conference Proceedings*, Chicago, IL, September 7-14, 1988, pp. 9-3 – 9-20.

Huber, Robert F. "GM Invests to Learn." *Production* 99 (May 1987): pp. 52-56.

Huber, Robert F. "Machine Tools: A Global View." *Production* 100 (September 1988): pp. 40-45.

Huber, Robert F. "Simulate, Integrate, Innovate." *Production* 99.11 (November 1987): pp. 40(10).

Huber, Robert F. "Today it Takes More to Be Competitive." *Production* 99 (December 1987): pp. 36-44.

Huge, Ernest C. "Overcoming Cultural Barriers to JIT." c. 1986, *APICS 1987 Total Manufacturing Performance Seminar Proceedings*, Washington, DC, July 13-15, 1987, pp. 389-393.

Huge, Ernest C. and Anderson, Alan D. *The Spirit of Manufacturing Excellence: An Executive's Guide to the New Mind Set.* Homewood, IL: Dow Jones-Irwin, 1988.

Hughes, Tom. "Make it Well; Make it Good; Don't Make-Good." *Automation* 34 (December 1987): pp. 16-20.

Hull, Frank; Azumi, Koya; and Wharton, Robert. "Suggestion Rates and Sociotechnical Systems in Japanese Versus American Factories: Beyond Quality Circles." *IEEE Transactions on Engineering Management* 35 (February 1988): pp. 11-24.

Hultman, Kenneth E. "The Psychology of Performance Management." *Training & Development Journal* 42 (July 1988): pp. 34-40.

Hutchinson, G.K. and Holland, J.R. "The Economic Value of Flexible Automation." *Journal of Manufacturing Systems*, 1982, pp. 215-228.

Hutto, Todd. "Available-to-Promise: The Key to Demand Management's Payoff." c. 1986, *APICS 1987 Total Manufacturing Performance Seminar Proceedings*, Washington, DC, July 13-15, 1987, pp. 29-32.

Hyde, W.D. "How Small Groups Can Solve Problems and Reduce Costs." *Industrial Engineering* 18 (December 1986): pp. 42-48.

Iannacito, Alan C. "Valuing Assets to Expand Manufacturing." *NMBTA 4th Biennial International Manufacturing Technology Conference Proceedings*, Chicago, IL, September 7-14, 1988, pp. 9-209 – 9-232.

Iemmolo, George R. "Introduction of Change in a Manufacturing Environment." *American Production & Inventory Control Society, 1987 Conference Proceedings*, St. Louis, MO, October 19-23, 1987, pp. 192-195.

Imai, M. *KAIZEN*. New York: Random House, 1986.

Inglesby, Tom. "Vendors Speak Out: The Market Perception of CIM." *Manufacturing Systems*, October 1988, pp. 34-53.

International Association for the Evaluation of Educational Achievement. *Science Achievement in Seventeen Countries: A Preliminary Report.* New York: Pergamon Press, 1988.

Irie, A. and Nakao, M. "The AGVs Information System." *SME/CASA Autofact, 1987 Conference Proceedings*, Dearborn, MI, November 9-12, 1987, pp. 8-16.

Isaac, George A. *Creating a Competitive Advantage Through Implementing Just-In-Time Logistics Strategies.* Chicago, IL: Touche Ross & Co., 1988.

Iversen, Wesley R. "Is 1988 Finally the Year for MRP?" *Electronics*, 15 October 1987, p. 140.

Iwabuchi, A. "The Resistance of Toshiba." *Toshiba News*, Kondanshai, 1990.

Izari, Mahyar and Karbassioon, Ebrahim. "An Analysis of Industrial Investment Alternatives." *Commline*, Spring 1988, pp. 13-14.

Jacobs, James A. "TQC's Role in World-Class Manufacturing." *American Production and Inventory Control Society, 1987 Conference Proceedings*, St. Louis, MO, October 19-23, 1987, pp. 24-26.

Jalkumar, R. "Postindustrial Manufacturing." *Harvard Business Review*, November-December 1986, pp. 69-76.

Jasinowski, Jerry. "Results of the September 1988 NAM Board Survey." Report to the National Association of Manufacturers, Washington, DC, October 19, 1988.

Jayson, Susan. "Goldratt & Fox: Revolutionizing the Factory Floor." *Management Accounting*, May 1987, pp. 18-22.

Jelinek, M. and Goldhar, J.D. "The Strategic Implications of the Factory of the Future." *Sloan Management Review* 25 (Summer 1984): pp. 29-37.

Jensen, Christopher. "Putting Ideas to Good Use." *The Cleveland Plain Dealer*, 13 November 1988, sec. E, p. 1+.

Johnson, Carl L. "CIM Success Requires Integrated Planning Techniques." *SME/CASA Autofact, 1987 Conference Proceedings*, Dearborn, MI, November 9-12, 1987, pp. 7-122—7-123.

Johnson, H. Thomas. "Activity-Based Information: A Blueprint for World Class Management Accounting." *Management Accounting*, June 1988, pp. 23-31.

Johnson, H. Thomas. "Rediscovering Cost Management Just In Time." *APICS 1987 Total Manufacturing Performance Seminar Proceedings*, Washington, DC, July 13-15, 1987.

Johnson, H. Thomas and Kaplan, Robert S. *Relevance Lost; the Rise and Fall of Management Accounting*. Boston: Harvard University Press, 1987.

Johnson, H. Thomas and Kaplan, Robert S. "The Rise and Fall of Management Accounting." *Management Accounting*, January 1987, pp. 22-31.

Johnson, H. Thomas and Loewe, Dennis A. "How Weyerhaeuser Manages Corporate Overhead Costs." *Management Accounting* 69 (August 1987): pp. 20-27.

Johnson, Henry. "Bar Code Basics." *NMBTA 4th Biennial International Manufacturing Technology Conference Proceedings*, Chicago, IL, September 7-14, 1988, pp. 3-59—3-62.

Johnson, Richard Tanner and Ouchi, William G. "Made in America (Under Japanese Management)." *Harvard Business Review*, September-October 1974, pp. 61-69.

Johnston, Russell and Lawrence, Paul R. "Beyond Vertical Integration—the Rise of the Value-Adding Partnership." *Harvard Business Review*, July-August 1988, pp. 94-101.

Johnston, William B. and Parker, Arnold H. *Workforce 2000: Work and Workers for the 21st Century*. Indianapolis, IN: Hudson Institute, 1987.

Jones, Lou "Competitor Cost Analysis at Caterpillar." *Management Accounting*, October 1988, pp. 32-38.

Jones, M.S. and Pinkley D. "Upgrading an Automated Storage and Retrieval System." *Industrial Engineering* 19 (February 1987): pp. 37-38.

Jordan, Henry H. "Inventory Management in the JIT Age." *APICS 1987 Total Manufacturing Performance Seminar Proceedings*, Washington, DC, July 13-15, 1987.

Juran. J.M.; Gryna, Frank M.; and Bingham, R.S., Jr. *Quality Control Handbook.* New York: McGraw-Hill, 1974.

Kalb, Bill. "AGVs in the Auto Factory: Is the Honeymoon Over?" *Automotive Industries* 167 (April 1987): pp. 87-88.

Kandt, Kevin. "Simulation From Design Through Implementation." *NMBTA 4th Biennial International Manufacturing Technology Conference Proceedings*, Chicago, IL, September 7-14, 1988, pp. 11-89—11-97.

Kantner, Rosabeth Moss. "Increasing Competitiveness Without Restructuring." *Management Review* 76 (June 1987): pp. 41-47.

Kanter, Rosabeth Moss. "The Attack on Pay." *Harvard Business Review*, March-April 1987, pp. 60-67

Kantrow, Alan M. "Keeping Informed." *Harvard Business Review*, July-August 1980, pp. 6-21.

Kaplan, Robert S. "Accounting Lag: The Obsolescence of Cost Accounting Systems." Harvard Business School 75th Anniversary Colloquium on Productivity and Technology, Boston, MA, March 28-29, 1984, pp. 1-46.

Kaplan, Robert S. "Must CIM Be Justified By Faith Alone?" *Harvard Business Review*, March-April 1986, pp. 87-95.

Kaplan, Robert S. "One Cost System Isn't Enough." *Harvard Business Review* 66 (January-February 1988): pp. 61-66.

Kaplan, Robert S. "What You See (in Accounting Earnings) is Not What You Get." Harvard Business School Working Paper, May 1987, pp. 1-11.

Kaplan, Robert S. "Yesterday's Accounting Undermines Production." *Harvard Business Review*, July-August 1984, pp. 95-101.

Kapoor, Vinod K. "Focused Factories: a Simple Idea With a Big Payback." *American Production and Inventory Control Society, 1987 Conference Proceedings*, St. Louis, MO, October 19-23, 1987, pp. 468-471.

Karikari, John A. "International Competitiveness and Industry Pricing in Canadian Manufacturing." *Canadian Journal of Economics* 21 (May 1988): pp. 410-426.

Katz, Donald. "Coming Home." *Business Month*, October 1988, pp. 57-62.

Kaufman, Brian. "Manufacturing Rebounds." *Computerworld*, June 1988, pp. 21-24.

Kaul, Pamela A. "Motivation is More Than Pay." *Associate Management*, August 1988, pp. 241-244.

Kazanjian, Robert K. and Drazin, Robert. "Implementing Manufacturing Innovations: Critical Choices of Structure and Staffing Roles." *Human Resource Management* 25 (Fall 1986): pp. 385-404.

Kazuaki, Iwata and Moriwaki, Toshimichi. "The Direction of Future Development and the Role of Knowledge in Manufacturing Technology." *Robotics & Computer-Integrated Manufacturing* 1 (1984): pp. 25-34.

Kearns, David T. "National Quality Forum III: A Corporate Response." *Quality Progress* 21 (February 1988): pp. 28-30.

Keefe, Patricia. "EDI: Handle With Care." *Computerworld* 22 (23 May 1988): pp. 49-50.

Keefe, Patricia. "MAP, OSI Make the Grade at ENE." *Computerworld* 22 (13 June 1988): p. 6.

Keefe, Patricia. "Take the Bull By its Horns." *Computerworld* 22 (16 May 1988): pp. 47-52.

Kennedy, Paul. *The Rise and Fall of the Great Powers*. New York: Vintage Books, 1989.

Kerr, John. "Industrial Automation: the Reality Will Match the Dream." *Electronic Business*, 10 December 1985, pp. 104-106.

Key, Y.T. "1995 (A Scenario for Personnel Management Retraining in the Future)." *Business Quarterly* 50 (Autumn 1985): pp. 22-30.

Keys, Bernard and Wolfe, Joseph. "Management Education and Development: Current Issues and Emerging Trends." *Journal of Management* 14 (June 1988): pp. 205-230.

Kii, Shirley. "John Deere Improves Schedule, Tool Changeover Time." *Industrial Engineering* 19 (February 1987): pp. 30-32.

Kim, W. Chan. "Industry Competition, Corporate Variables, and Host Government Intervention in Developing Nations." *Management International Review* 28 (Second Quarter, 1988): pp. 16-27.

Kim, W. Chan. "The Effects of Competition and Corporate Political Responsiveness on Multinational Bargaining Power." *Strategic Management Journal* 9 (1988): pp. 289-295.

Kindel, Stephen. "The 10 Worst Managed Companies in America." *Financial World* 157 (26 July 1988): pp. 28-39.

Kindleberger, Charles. "Danger Zone: America's Industrial Problems are Eerily Similar to Those of the United Kingdom in the 19th Century." *Business Month* 132 (July-August 1988): pp. 72-74.

King, Alfred M. "Cost Accounting in the 1990's: Can Production Executives and Financial Executives Keep in Touch?" *FE: The Magazine for Financial Executives* 2 (November 1986): pp. 24-29.

King, Leslie. "Plant Switches to AGVs for Space Gain, System Improvement." *Industrial Engineering* 19 (February 1987): pp. 25-27.

Kinney, Hugh D., Jr. and McGuiness, Leon F. "Design and Control of Manufacturing Cells." *Industrial Engineering* 19 (October 1987): pp. 28-36.

Kinsey, John W. "Supplying JIT to NUMMI: A Case Study." *APICS 1987 Total Manufacturing Performance Seminar Proceedings*, Washington, DC, July 13-15, 1987, pp. 331-333.

Kiran, Ali S. and Krason, Richard J. "Automating Tooling in a Flexible Manufacturing System." *Industrial Engineering* 20 (April 1988): pp. 52-57.

Kirkland, Carl. "Quality Program Saves Big Bucks for Sheller-Globe." *Plastics World*, February 1988, pp. 48-51.

Kirkland, Carl. "Taguchi Methods Increase Quality and Cut Costs." *Plastics World*, February 1988, pp. 44-47.

Kirkland, Richard I., Jr. "Entering a New Age of Boundless Competition." *Fortune* 117 (14 March 1988): pp. 40-48.

Kirkpatrick, David. "Look Out, World, Here We Come." *Fortune* 117 (20 June 1988): pp. 54-55.

Klein, J.A. "Good Supervisors Are Good Supervisors—Anywhere." *Harvard Business Review*, November-December 1986, p. 125.

Klotz, Valentin. "Staff Suggestion Schemes." *International Labor Review* 127 (1988): pp. 335-353.

Knill, Bernie and Weimer, George. "Systems Integration II." *Material Handling Engineering*, Supplement, March 1988, pp. 2-23.

Knill, Bernie and Schwind, Gene. "Manufacturing '87: Selecting the Right Path is a Lonely Job." *Material Handling Engineering* 42 (January 1987): pp. 83-95.

Komanecky, A. Nicholas. "Developing New Managers at GE." *Training & Development Journal* 42 (June 1988): pp. 62-65.

Korbrin, Stephen J. "Expatriate Reduction and Strategic Control in American Multinational Corporations." *Human Resource Management* 27 (Spring 1988): pp. 63-76.

Koso, Gregor A. "When Material Handling Comes First: An Incremental Approach to Factory Automation." *SME/CASA Autofact, 1987 Conference Proceedings*, Dearborn, MI, November 9-12, 1987, pp. 4-157—4-170.

Kouth, Allan R. "Will the Real Class A MRP II Definition Please Stand Up?" *American Production and Inventory Control Society, 1987 Conference Proceedings*, St. Louis, MO, October 19-23, 1987, pp. 125-128.

Koziol, David S. "How the Constraint Theory Improved a Job-Shop Operation." *Management Accounting* 69 (May 1988): pp. 44-50.

Kozluk, Tad A., and Mathews, Vad T. "Managing Production for Export: Scope for and Limitations of Zero Inventory and Just-In-Time Manufacturing Strategy." *APICS 1984 Zero Inventory Philosophy and Practices Seminar Proceedings*, St. Louis, MO, October 29-31, 1984, pp. 466-473.

Krajewski, L.J.; King, B.E.; Ritzman, L.P.; and Wong, D.S. "Kanban, MRP, and Shaping the Manufacturing Environment." *Management Science* 33 (January 1987): pp. 39-57.

Kratick, John F. "Triumph of the Lean Production System." *Sloan Management Review*, 30 (Fall 1988): pp. 41-52.

Krechpin, Ira P. "Controls Tune Plants for Peak Performance." *Modern Materials Handling* 43 (July 1988): pp. 87-91.

Krepchin, Ira P. "Here, JIT and MRP II Work in Harmony." *Modern Materials Handling*, June 1988, pp. 87-89.

Krechpin, Ira P. "How Flexible Automation Delivers 1-Year Payback." *Modern Materials Handling* 42 (January 1987): pp. 91-95.

Kruger, Pamela and Scandura, Janette. "A Game Plan for the Future." *Working Woman*, 2 January 1990, p. 74.

Kuhlmann, Torsten M. "Adapting to Technical Change in the Workplace." *Personnel* 65 (August 1988): pp. 67-70.

Kuliviec, Ray. "Ten Obstacles to Competitiveness." *Modern Materials Handling*, November 1988, p. 9.

Kusiak, Andrew. "Artificial Intelligence Series, Part 8: An Expert System for Group Technology." *Industrial Engineering*, October 1987, pp. 56-61.

Kusiak, Andrew. "Designing Expert Systems for Scheduling of Automated Manufacturing." *Industrial Engineering* 19 (July 1987): pp. 42-47.

Kuzmits, Frank E. "Triage in the Organizational Emergency Room." *Training and Development Journal*, August 1988, pp. 38-40.

LaDuke, Bettie. "Manufacturing and Accountants: Friends or Foes?" *American Production and Inventory Control Society, 1987 Conference Proceedings*, St. Louis, MO, October 19-23, 1987, pp. 611-613.

Labs, Wayne and Kuhfeld, Ron. "Manufacturing Refrigerators with People Based Automation." *I&CS* 61 (February 1988): pp. 46-50.

Lademann, Don. "Simultaneous Engineering: Breakthrough in Machine Tool Buying." *Purchasing World* 31 (June 1987): pp. 73-74.

Lamb, John. "Gearing for Battle." *Datamation*, 1 September 1985, pp. 58-62.

Lamb, John. "Standing on the Sidelines." *Datamation*, 1 February 1986, pp. 39-40.

Landon, Wanda G. "Kanban and Deming's 14 Points." *Quality*, September 1988, pp. 50-51.

Landvater, Darryl. "How to Conquer the Most Difficult Part of JIT: Motivating People." *American Productivity & Inventory Control Society, 1987 Conference Proceedings*, St. Louis, MO, October 19-23, 1987, pp. 186-191.

Lankford, Ray. "The Beginning of the Post-MRP Era." *APICS 1987 Total Manufacturing Performance Seminar Proceedings*, Washington, DC, July 13-15, 1987, pp. 50-53.

Lankford, Ray and Schlegel, C. Frederic. "How to Schedule Aerospace Production: The State of the Art." *American Production & Inventory Control Society, 1987 Conference Proceedings*, St. Louis, MO, October 19-23, 1987, pp. 38-40.

Law, A.M. and McComas, M.G. "How Simulation Pays Off." *Manufacturing Engineering* 100 (February 1988): pp. 37-39.

Lawler, Edward E., III and Mohrman, Susan A. "Quality Circles: After the Honeymoon." *Organizational Dynamics* 15 (Spring 1987): pp. 42-55.

Lea, Dixie and Brostrom, Richard. "Managing the High-Tech Professional." *Personnel* 65 (June 1988): pp. 12-18.

Leach, John L. and Chakiris, B.J. "The Future of Jobs, Work, and Careers." *Training & Development Journal* 42 (April 1988): pp. 48-54.

Leahy, James A. "Management Issues in JIT and OPT Implementations." *APICS 1984 Zero Inventory Philosophy and Practices Seminar Proceedings*, St. Louis, MO, October 29-31, 1984, pp. 335-344.

Leavey, Richard A. "Production Activity Control in a JIT Environment." *APICS 1987 Total Manufacturing Performance Seminar Proceedings*, Washington, DC, July 13-15, 1987.

LeBoeuf, Michael. "Why All Executives Must Learn the Greatest Management Principle in the World." *Working Woman*, January 1988, pp. 70-74.

Lechler, Gerhard. "Tool Monitoring in Lathes and Machining Centers." *NMBTA 4th Biennial International Manufacturing Technology Conference Proceedings*, Chicago, IL, September 7-14, 1988, pp. 10-169 – 10-179.

Lee, Susan and Brown, Christie. "The Pro-team Corporation." *Forbes* 140 (24 August 1987): pp. 76-80.

Leibson, Steven H. "Developing CAE Tools Target Top-Down Design of Complex Systems." *EDN* 33 (17 March 1988): pp. 131-137.

Len, John J. "Inventory Value Engineering (IVE): A Proven Methodology to Implement the Just-in-time Inventory Concepts." *APICS 1984 Zero Inventory Philosophy and Practices Seminar Proceedings*, St. Louis, MO, October 29-31, 1987, pp. 220-225.

Lenk, Gerry. "Quality." *Supervision* 50 (June 1988): pp. 11-13.

Levin, Scott A. "Creativity and Innovation: The Key to Just-in-Time." c. 1986, *APICS 1987 Total Manufacturing Performance Seminar Proceedings*, Washington, DC, July 13-15, 1987, pp. 556-565.

Levitt, Ted. "The Chryslerization of America." *Harvard Business Review*, 65 (July-August 1987): p. 4.

Lewis, A.; Nagpal, B.K.; and Watts, P.L. "Robotics: Market Growth, Application Trends and Investment Analysis." *Proceedings from the Institute of Mechanical Engineering* 199 (1985): pp. 35-40.

Lewis, Geoffrey; Clark, John; and Moss, Bill. "BHP Reorganizes for Global Competition." *Long Range Planning* 21 (June 1988): pp. 18-26.

Liker, Jeffrey K. "Changing Everything All at Once." *Sloan Management Review* 28 (Summer 1987): pp. 29-47.

Lilly, Dick. "Software for Managing Capacity." *NMBTA 4th Biennial International Manufacturing Technology Conference Proceedings*, Chicago, IL, September 7-14, 1988.

Lindsay, Bruce. "Warehouse Business System and 60 AGVs Track and Route Products at Kodak Distribution Center." *Industrial Engineering* 20 (May 1988): pp. 50-54.

Lippa, Victor G. "The Emancipation of the Foreman." *APICS 1984 Zero Inventory Philosophy and Practices Seminar Proceedings*, St. Louis, MO, October 29-31, 1984, pp. 236-245.

Loebel, Alan. "JIT Manufacturing: Implications for Manufacturing Resources Planning." *APICS 1987 Total Manufacturing Performance Seminar Proceedings*, Washington, DC, July 13-15, 1987.

Lombardo, Richard. "Breaking the Barriers to Corporate Creativity." *Training & Development Journal* 42 (August 1988): pp. 63-66.

London, Manuel and MacDuffie, John Paul. "Technological Innovations Case Examples and Guidelines." *Personnel* 64 (November 1987): pp. 26-36.

Lopes, Peter F. "Simplification Before Automation: A Blueprint for Success." *American Production and Inventory Control Society, 1987 Conference Proceedings*, St. Louis, MO, October 19-23, 1987, pp. 563-567.

Lopes, Peter F. "The Factory with a Future Will Be the Factory of the Future." *NMBTA 4th Biennial International Manufacturing Technology Conference Proceedings*, Chicago, IL, September 7-14, 1988, pp. 7-13 – 7-37.

Lopez, Virginia C. and Yager, Loren. *The U.S. Aerospace Industry and the Trend Toward Internationalization*. Washington, DC: Aerospace Research Center, March 1988.

Lundy, Wilson T., Jr. "Computer Optimized Manufacturing." *NMBTA 4th Biennial International Manufacturing Technology Conference Proceedings*, Chicago, IL, September 7-14, 1988, pp. 2-3 – 2-26.

Lyons, John W. "Coming up with New Products that Beat the Odds." *Mechanical Engineering*, April 1988, pp. 38-41.

Macaulay, Stephen. "Amazing Things Can Happen if You Keep it Clean." *Production*, May 1988, pp. 72-74.

Macaulay, Stephen. "Complex Can Be Competitive." *Production* 100 (April 1988): pp. 74-76.

Macdonald, J. "The Japan of Europe: Which Country will Earn the Title?" *Management Decision* 25 (January 1987): p. 38.

Mackey, James T. "Eleven Key Issues in Manufacturing Accounting." *Management Accounting*, January 1987, pp. 32-38.

Maimon, Ozed and Gershwin, Stanley B. "Dynamic Scheduling and Routing for Flexible Manufacturing Systems That Have Unreliable Machines." *Operations Research* 36 (March-April 1988): pp. 279-293.

Main, Jeremy. "Under the Spell of the Quality Gurus." *Fortune* 114 (18 August 1986): pp. 30-34.

Majchrzak, Ann. "Sociotechnical Design of Advanced Manufacturing Technology." *SME/CASA Autofact, 1987 Conference Proceedings*, Dearborn, MI, November 9-12, 1987, pp. 8-27 – 8-39.

Maloney, William F. and McFillen, James M. "Influence of Foremen on Performance." *Journal of Construction and Engineering Management* 113 (September 1987): pp. 399-415.

Maloney, William F. and McFillen, James M. "Worker Perceptions of Contractor Behavior." *Journal of Construction & Engineering Management* 113 (September 1987): pp. 416-426.

Mammone, James L. "Productivity Measurement: A Conceptual Overview." *Management Accounting*, June 1980, pp. 36-42.

Mangin, Charles-Henri. "Component Insertion and Placement." *Assembly Engineering* 30 (May 1987): pp. 20-23.

Mangin, Charles-Henri. "SMT Snapshot: The U.S. Electronics Assembly Lags." *Electronic Business* 14 (15 June 1988): p. 9.

Manji, James F. "Building Cars With Mazda Quality." *Automation* 35 (June 1988): pp. 30-42.

Manji, James F. "Handling Systems Move Toward Full Integration." *Automation*, December 1987, pp. 36-40.

Manufacturing Studies Board Commission on Engineering and Technical Systems National Research Council. *The Future of Electronics Assembly.* Washington, DC, National Academy Press, 1988.

March, J.G. and Simon, H.A. *Organizations*. New York: John Wiley & Sons, 1958.

Margolis, Nell. "Report: Users to Shape AI's Future." *Computerworld* 22 (30 May 1988): pp. 19-21.

Marks, Mitchell Lee and Cutliffe, Joseph G. "Making Mergers Work." *Training & Development Journal* 42 (April 1988): pp. 30-37.

Marks, Peter. "AI on the Factory Floor." *Hardcopy* 7 (October 1987): pp. 100-103.

Marshal, Dan. "Managing Suppliers Up to Speed." *Harvard Business Review* (July-August 1989): p. 133.

Martin, John M. "The Final Piece of the Puzzle." *Manufacturing Engineering* 101 (September 1988): pp. 46-51.

Martin, Raymond L. "Understanding and Overcoming the Barriers to Becoming a World-Class Manufacturer." *American Production and Inventory Control Society, 1987 Conference Proceedings*, St. Louis, MO, October 19-23, 1987, pp. 484-487.

Mastenbroek, Willem. "A Dynamic Concept of Revitalization." *Organizational Dynamics* 16 (Spring 1988): pp. 52-61.

Materna, Anthony T. and Sepehri, Mehran. "Integrating Manufacturing Systems Using TRW's CIM Data Engine." *Industrial Engineering*, October 1986, pp. 70-80.

Mather, Hal. "Competing With the Best in the World in Your Own Backyard." *American Production and Inventory Control Society, 1987 Conference Proceedings*, St. Louis, MO, October 19-23, 1987, pp. 520-522.

May, Donald L. "How Much Backup for Process Control Computers?" *Chemical Engineering* 95 (25 April 1988): pp. 64-70.

Mayer, Richard J.; Phillips, Don T.; and Young, Robert E. "Artificial Intelligence—Applications in Manufacturing." *SME/CASA Autofact 6 Conference Proceedings*, Anaheim, CA, October 1-4, 1984, pp. 10-24.

McAdams, Jerry L. "Performance-Based Reward Systems." *Canadian Business Review* 15 (Spring 1988): pp. 17-19.

McArthur, C. Daniel. "Building on Quality: World-Class JIT." *APICS 1987 Total Manufacturing Performance Seminar Proceedings*, Washington, DC, July 13-15, 1987, pp. 14-15.

McCallum, Edward D., Jr. "Capital Spending: the Ascent Begins." *Production* 100 (January 1988): pp. 37-46.

McCallum, Edward D., Jr. "Capital Spending: Outlook & Trends for 1987 and Beyond." *Production* 99 (March 1987): pp. 49-56.

McClean, Tom. "Quality and Manufacturing: Issues and Information Systems." *The Accountant's Magazine* 92 (February 1988): pp. 49-51.

McClellan, Michael. "Computer Integrated Manufacturing With Real Time Shop Floor Control." *SME/CASA Autofact, 1987 Conference Proceedings*, Dearborn, MI, November 9-12, 1987, pp. 9-119—9-133.

McClenahen, John S. "Automation's Global Game: Who's Winning?" *Industry Week* 236 (23 June 1988): pp. 37-64.

McCluckie, John D. "Current Status of CIM Education in the United States." *SME/CASA Autofact, 1987 Conference Proceedings*. Dearborn, MI, November 9-12, 1987, pp. 10-5—10-18.

McDermott, Lynda C. "Keeping the Winning Edge: Strategies for Being a Business Partner." *Training & Development Journal* 41 (July 1987): pp. 16-20.

McDougall, Duncan C. "Effective Manufacturing Performance Measurement Systems: How to Tell When You've Found One." Working draft of a research paper for the Boston University School of Management, February 1988, pp. 1-22.

McEarchran, John R. "A Technology Assessment of Machine Vision Gaging and Inspection." SME Technical Paper, IQ87-906, pp. 1-13.

McGaughey, Nick W. "The Manufacturing Connection: A Competitive Requirement." *Industrial Management* 30 (January-February 1988): pp. 23-25.

McGill, Michael E. "American Business and the Quick Fix." New York: Henry Holt & Company, 1988.

McGrail, Fred and McSharry, Nancy. "Manufacturing Managers on Labor and Automation." *Electronic Business* 12 (1 October 1986): pp. 56-58.

McGuire, Kenneth J. "JIT: Providing Focus for the Converging Manufacturing Strategies of the 80s." c. 1984, *APICS 1987 Total Manufacturing Performance Seminar Proceedings*, Washington, DC, July 13-15, 1987, pp. 1-3.

McGuire, Kenneth J. "Zero Inventory: Providing Focus for the Converging Manufacturing Strategies of the 80s." *APICS 1984 Zero Inventory Philosophy and Practices Seminar Proceedings*, St. Louis, MO, October 29-31, 1987, pp. 318-325.

McIlhattan, Robert D. "How Cost Management Systems Can Supply the JIT Philosophy." *Management Accounting* 69 (September 1987): pp. 20-27.

McKee, Keith. "Beyond Machining." *NMBTA 4th Biennial International Manufacturing Technology Conference Proceedings*, Chicago, IL, September 7-14, 1988, pp. 1-135—1-143.

McKenna, Regis; Borrus, Michael; and Cohen, Stephen. "Industrial Policy and International Competition in High Technology—Part I Blocking Capital Formation." *California Management Review* 26 (Winter 1984): pp. 15-32.

McKlusky, Robert. "Boeing Develops a New Design to Cut Down on Corporate Drag." *International Management*, April 1987, pp. 57-58.

McLean, Tom. "Quality and Manufacturing: Issues and Information Systems." *The Accountant's Magazine* 92 (February 1988): pp. 49-51.

McLeod, Jonah. "Has CAE Lived Up to its Promise?" *Electronics* 59 (9 June 1986): pp. 25-27.

McLinn, James A. "Product Reliability: Extending Quality's Reach." *Manufacturing Engineering* 101 (September 1988): pp. 52-55.

McMahon, P.D. and Zoch, J.D. "Japanese are Better Managers." *APICS 1984 Zero Inventory Philosophy and Practices Seminar Proceedings*, St. Louis, MO, October 29-31, 1987, pp. 260-269.

McNair, C.J. and Mosconi, William. "Measuring Performance in an Advanced Manufacturing Environment." *Management Accounting* 69 (July 1987): pp. 28-32.

McNair, C.J.; Mosconi, William; and Norris, Thomas. *Meeting the Technology Challenge: Cost Accounting in a JIT Environment.* Montvale, NJ: National Association of Accountants, 1988.

McNulty, Robert. "Management Decision Making in World-Class Manufacturing." *American Production and Inventory Control Society, 1987 Conference Proceedings*, St. Louis, MO, October 19-23, 1987, pp. 472-474.

Means, Grady E. "The CFO, Information Flow, and Global Competition." *Financial Executive* 3 (September-October 1987): pp. 26-32.

Mehler, Mark. "The Crowded World of Robot System Integrators." *Electronic Business*, 15 July 1986, pp. 36-37.

Mehler, Mark. "Vision Systems: What You See is What You Get." *Electronic Business*, 1 June 1986, pp. 71-73.

Meredith, Jack R. "Acquiring New Manufacturing Technologies for Competitive Advantage." *Operations Management Review*, Winter 1987, pp. 9-13.

Meredith, Jack R. "Automating the Factory: Theory Versus Practice." *International Journal of Production Research* 25 (1987): pp. 1493-1510.

Meredith, Jack R. "Installation of Flexible Manufacturing System Teaches Management Lessons in Integration, Labor, Cost, Benefits." *Industrial Engineering* 20 (April 1988): pp. 18-27.

Meredith, Jack R. "Managing Factory Automation Projects." *Journal of Manufacturing Systems* 6 (1988): pp. 239-255.

Meredith, Jack R. "The Role of Manufacturing Technology in Competitiveness: Peerless Laser Processors." *IEEE Transactions on Engineering Management* 35 (February 1988): pp. 3-10.

Meyer, Ken. "The Pull of Just-In-Time Cost Issues." *NMBTA 4th Biennial International Manufacturing Technology Conference Proceedings*, Chicago, IL, September 7-14, 1988, pp. 9-271 — 9-284.

Meyer, Marc H. and Roberts, Edward B. "Focusing Product Technology for Corporate Growth." *Sloan Management Review* 29 (Summer 1988): pp. 7-17.

Miceli, Marcia P. and Near, Janet P. "Individual and Situational Correlates of Whistle-Blowing." *Personnel Psychology* 41 (Summer 1988): pp. 267-282.

Michaels, Lawrence T. "A Control Framework for Factory Automation." *Management Accounting* 69 (May 1988): pp. 37-43.

Michaels, Lawrence T. "Selecting and Justifying Advanced Manufacturing Technologies to Obtain a Competitive Advantage." *SME/CASA Autofact, 1987 Conference Proceedings*, Dearborn, MI, November 9-12, 1987, pp. 7-39 — 7-48.

Mickler, Mary Louise. "Merit Pay: Boon or Boondoggle?" *The Clearing House*, November 1987, pp. 137-141.

Middo, K.; Murdock, B.; and Karnosh, L. "Touch Panels Point the Way to Natural Data Entry." *Electronic Design* 35 (28 May 1987): pp. 123-126, 128-129.

Miller, Charles. "New Trends in Automated Welding & Soldering." *NMBTA 4th Biennial International Manufacturing Technology Conference Proceedings*, Chicago, IL, September 7-14, 1988, pp. 9-37 — 9-47.

Miller F. "Meeting Customer Needs with a Focused Factory." *Manufacturing Systems*, November 1987, pp. 34-39.

Miller, Jeffrey G. and Roth, Aleda V. "Manufacturing Strategies: Executive Summary of the 1987 North American Manufacturing Futures Survey." *Operations Management Review* 6 (January 1988): pp. 8-20.

Miller, Jeffery G. and Roth, Aleda V. "Manufacturing Strategies: Executive Summary of the 1988 North American Manufacturing Futures Survey." Boston University School of Management Manufacturing Roundtable, 1988.

Miller, Jeffrey G. and Vollman, Thomas E. "The Hidden Factory." *Harvard Business Review*, September-October 1985, pp. 141-150.

Miller, Jeffery, G.; Amano, Akio; DeMeyer, Arnoud; Ferdows, Kasra; Nakane, Jinichiro; and Roth, Aleda. "Closing the Competitive Gaps: The International Report of the Manufacturing Futures Project." Draft to appear in *Managing International Manufacturing*, (K. Ferdows, ed.), Amsterdam: Elsevier Science Publishers, 1989, pp. 153-168.

Miller, John. "Productivity's New Math." American Productivity & Quality Center, Houston, TX, October 1988.

Miller, Richard K. "Artificial Intelligence: A New Tool for Manufacturing." *Manufacturing Engineering*, April 1985, pp. 3-9.

Miller, Richard K. *Automated Guided Vehicles and Automated Manufacturing*. Dearborn, MI: SME, 1987, pp. 147-152.

Miller, Rock. "Quality is a Business Issue." *Managing Automation*, March 19988, pp. 44-47.

Miller, William H. "U.S. Improving? Simplified 'Index' Offers a Yardstick." *Industry Week* 236 (20 June 1988): p. 22.

Mishne, Patricia P. "A New Attitude Toward Quality." *Manufacturing Engineering*, October 1988, pp. 50-55.

Mitroff, I.I. *Break-away Thinking*. New York: John Wiley & Sons, 1988, p. 21.

Moad, Jeff. "DuPont Seeks Global Communications Reach." *Datamation*, 15 January 1988, p. 72.

Modic, Stanley J. "CIM's Final Link-People!" *Industry Week*, 2 November 1987, p. 7.

Modic, Stanley J. "Evolution and Strategic Communications." *Public Relations Journal*, January 1988, pp. 29-30.

Modic, Stanley J. "What Makes Deming Run?" *Industry Week*, 20 June 1988, pp. 84-90.

Moldoveanu, Andrei. "Safety Interlocking in The Automated Factory." *NMBTA 4th Biennial International Manufacturing Technology Conference Proceedings*, Chicago, IL, September 7-14, 1988, pp. 12-81 – 12-91.

Monroe, Leo. "An Advanced Workholding/Work Transfer System." *NMBTA 4th Biennial International Manufacturing Technology Conference Proceedings*, Chicago, IL, September 7-14, 1988, pp. 12-23 – 12-25.

Montgomery, Douglas C. "Experiment Design and Product and Process Development." *Manufacturing Engineering* 101 (September 1988): pp. 57-63.

Moore, Thomas. "Old-Line Industry Shapes Up." *Fortune*, 27 April 1987, pp. 23-32.

Moosbrucker, Jane and Berger, Emanuel. "Know Your Customer." *Training & Development Journal* 42 (March 1988): pp. 30-34.

Morgan, James P. "Quality is Real, it isn't Goodness." *Purchasing*, 5 November 1987, pp. 38-43.

Morse, Wayne J. and Poston, Kay. "Accounting for Quality Costs – A Critical Component of CIM." *CIM Review* 3.1 (Fall 1986): pp. 53-60.

Morse, Wayne J. and Harold, H.P. "Why Quality Costs are Important." *Management Accounting* 69 (November 1987): pp. 42-44.

Moskal, B.S., "Just in (the Wrong) Time." *Industry Week*, 15 August 1988, p. 28.

Moskal, B.S. "Shoveling Out – Harnischfeger Counts on High Tech." *Industry Week* 222 (9 July 1984): pp. 16-19.

Mourdoukoutas, Pantos and Sohng, Soong N. "The Japanese Industrial System: A Study in Adjustment to Automation." *Management International Review* 27 (October 1987): pp. 46-56.

Mullins, Peter. "At the Core: 'Soft Automation.'" *Production*, August 1988, pp. 44-47.

Mullins, Peter. "Europe Looks at Flexible Welding Systems." *Production* 99 (June 1987): pp. 50-52.

Mullins, Peter. "Switching from the Old to the New," *Production* 98 (October 1986): pp. 70-73.

Muramatsu, Rintaro; Miyazki, Haruo; and Ishii, Kazuyoshi. "Successful Application of Job Enlargement/Enrichment at Toyota." *IIE Transactions* 19 (December 1987): pp. 451-459.

Murphy, Erin E. and Werner, Kenneth I. "Technology '88: Design Automation." *IEEE Spectrum* 25 (January 1988): pp. 35-37.

Murphy, Robert H. "Master Scheduling: Key to Success." c. 1986, *APICS 1987 Total Manufacturing Performance Seminar Proceedings*, Washington, DC, July 13-15, 1987, pp. 84-86.

Musselman, Kenneth. "Simulation: Designing in an Imperfect World." *NMBTA 4th Biennial International Manufacturing Technology Conference Proceedings*, Chicago, IL, September 7-14, 1988, pp. 7-3 – 7-14.

Musselwhite, Christopher W. "Knowledge, Pay and Performance." *Training & Development Journal* 42 (January 1988): pp. 62-65.

Musson, Thomas; Linton, Larry; and Hoffman, Dennis. "What Design Automation is Needed for R&M Workshop Discussion." *IEEE Transactions on Reliability* R-36, (December 1987): pp. 489-494.

Myklebust, Arvid. "Mechanical Computer-Aided Engineering." *IEEE Computer Graphics and Applications* 8 (March 1988): pp. 24-26.

Mytelka, Lynn Krieger and Delapierre, Michel. "The Alliance Strategies of European Firms in the Information Technology Industry and the Role of ESPIRIT." *Journal of Common Market Studies* 26 (December 1987): pp. 231-251.

Naidish, Norman. "Competitive Automated Assembly." *NMBTA 3rd Biennial International Machine Tool Technical Conference Proceedings*, Chicago, IL, September 3-11, 1986, pp. 9-105 – 9-118.

Nakane, Jinichiro. "Implementing Manufacturing Strategies: Breaking the Performance Measurement Barriers, Japan 2001." Paper presented at the International Conference Boston Manufacturing Roundtable School of Management, Boston University, October 21, 1988.

Nanni, Alfred, Jr. "Financial Versus Non-financial Measures of Performance: Barriers to Strategic Control." Working draft of paper, Boston University School of Management, February 1988.

Nathan, Jay. "PERT: A Production Activity Control Manager for JIT." c. 1986, *APICS 1987 Total Manufacturing Performance Seminar Proceedings*, Washington, DC, July 13-15, 1987, pp. 152-153.

National Electrical Manufacturers Association. *U.S. Manufacturer's Five-Year Industrial Automation Plans for Automation Machinery and Plant Communication Systems.* Washington, DC, 1985.

Navas, Sri. "Making MRP II A Reality." *NMBTA 4th Biennial International Manufacturing Technology Conference Proceedings*, Chicago, IL, September 7-14, 1988, pp. 3-93 – 3-101.

Nelson, Craig A. "A Review of New Justification Models." *NMBTA 3rd Biennial International Machine Tool Technical Conference Proceedings*, Chicago, IL, September 3-11, 1986, pp. 7-16 – 7-23.

Nelton, Sharon. "Motivating For Success." *Nation's Business*, March 1988, pp. 18-26.

Nemeth, Ernest. "Trends in Software and Hardware Control Technologies." *NMBTA 4th Biennial International Manufacturing Technology Conference Proceedings*, Chicago, IL, September 7-14, 1988, pp. 2-181 – 2-191.

Neumeir, Shelley. "Markets of the World Unite." *Fortune* 30 July 1990, p. 102. Reprinted with permission from FORTUNE Magazine, 1990

Newton, D. "Simulation Model Calculates How Many Automatic Guided Vehicles Are Needed." *Industrial Engineering* 17 (February 1985): pp. 68-75.

Nicoletti, Guy. "Using the Manufacturing Automation Protocol." *NMBTA 4th Biennial International Manufacturing Technology Conference Proceedings*, Chicago, IL, September 7-14, 1988, pp. 8-23 – 8-51.

Niehaus, Thomas L. "The Courtship of JIT and MRP: A Case Study." *APICS 1987 Total Manufacturing Performance Seminar Proceedings*, Washington, DC, July 13-15, 1987, pp. 420-422.

Nitkiewicz, Joseph. "Machine Tool Monitors." *NMBTA 3rd Biennial International Machine Tool Technical Conference Proceedings*, Chicago, IL, September 3-10, 1986, pp. 2-97 – 2-119.

Noaker, Paula M. "At the Cutting Edge: Rethinking Strategies." *Production* 100 (July 1988): pp. 53-57.

Noaker, Paula M. "Variation Research: Debugging the Manufacturing Process." *Tooling & Production*, June 1987, pp. 53-56.

Noaker, Paula M. "Workholding: Firm but Flexible?" *Production* 100 (August 1988): pp. 50-56.

Norman, Richard L. and Minerich, Jon T. "Partners in Profit: The Manufacturing/Vendor Relationship." *American Production and Inventory Control Society, 1987 Conference Proceedings*, St. Louis, MO, October 19-23, 1987, pp. 600-602.

O'Brien, Thomas. "Improving Performance Through Activity Analysis." Proceedings of the Third Annual Management Accounting Symposium, San Diego, CA, March 1989,(Peter B.B. Turney, Ed.), Copyright 1990 by the American Accounting Association.

O'Connor, Frances and Diesslin, Richard. "Management Planning and Control System for Small to Medium Sized Shops." *SME/CASA Autofact, 1987 Conference Proceedings*, Dearborn, MI, November 9-12, 1987, pp. 9-1 – 9-13.

O'Guin, Michael C. "Information Age Calls for New Methods of Financial Analysis in Implementing Manufacturing Technologies." *Industrial Engineering* 19 (November 1987): pp. 36-41

O'Neal, C.R. "The Buyer-Seller Linkage in a Just-In-Time Environment." *Journal of Purchasing and Materials Management* 23 (Spring 1987): pp. 7-13.

O'Neal, Kim Rogers. "Roundtable Participants Discuss AI Trends and Developments." *Industrial Engineering* 19 (February 1987): pp. 52-59.

O'Reilly, Phillip A. "Fight Foreign Government Cartels." *Industry Week* 22 (25 June 1984): p. 8.

Oden, Howard W. "VIM (Valve Integrated Manufacturing) A Prerequisite for CIM." *SME/CASA Autofact, 1987 Conference Proceedings*, Dearborn, MI, November 9-12, 1987, pp. 9-15 – 9-28.

Odiorne, George S. "Measuring the Unmeasurable: Setting Standards for Management Performance." *Business Horizons*, July-August 1987, pp. 69-75.

Ohno, Taiichi. *Toyota Production System Beyond Large Scale Production.* Cambridge, MA: Productivity Press, 1988.

Opalka, Daniel P. and Williams, James B. "Employee Obsolescence and Retraining: An Approach to Human Resource Restructuring." *Journal of Business Strategy* 7 (Spring 1987): pp. 90-97.

Opie, Roy. "Vision Bridges – the Feasibility Gap." *Control and Instrumentation* 18 (July 1986): pp. 51, 53.

Organizational Dynamics, Inc. (ODI) Executive Opinion Survey, Boston, MA.

Orr, G.B.; Sopher, S.M.; and Apple, J.M., Jr. "Material Handling Equipment Alternatives Examined for Progressive Build in Light Assembly Operations." *Industrial Engineering* 17 (April 1985): pp. 68-73.

Ost, Edward J. "Team-Based Pay: New Wave Incentives." *Sloan Management Review*, Spring 1990, pp. 19-27.

Osterberg, William. "The Japanese Edge in Investment: The Financial Side." *Economic Commentary (Federal Reserve Bank of Cleveland)*, March 1, 1987, pp. 1-4.

Packer, A.F. "Factories Can Meet Current and Future Needs By Phasing in Flexible Controls." *Industrial Engineering* 16 (March 1984): pp. 72-76.

Paine, Thomas H. "Benefits in the 1990's." *Personnel Journal*, March 1988, pp. 82-92.

Panisset, Brian D. "Measuring the Implementation." *P&IM Review With APICS News*, November 1988, pp. 40-42.

Paris, Michael L. "Good Planning Plus Good Information Equals Good Justification." *NMBTA 4th Biennial International Manufacturing Technology Conference Proceedings*, Chicago, IL, September 7-14, 1988, pp. 4-181 – 4-194.

Paris, Michael L. "Planning for CIM – The Case for a Strategic Approach." *CIMTECH 1987 Conference Proceedings*, Los Angeles, CA, March 24-16, 1987, pp. MM87-212-1 – MM87-212-15.

Parisian, Joseph. "Implementing CIM Elements." *NMBTA 4th Biennial International Manufacturing Technology Conference Proceedings*, Chicago, IL, September 7-14, 1988, pp.1-11 – 1-19.

Parrish, David J. "Opening a Dialogue Between FMS and CIM." *Mechanical Engineering* 110 (May 1988): pp. 70-76.

Pascale, R. *Managing on the Edge*. New York: Simon and Schuster, 1990.

Pascarella, Perry. "Tom Peters Invites Chaos for Survival." *Industry Week* 235 (19 October 1987): pp. 48-53.

Passmore, Steve. "Affordable Manufacturing Cells." *NMBTA 4th Biennial International Manufacturing Technology Conference Proceedings*, Chicago, IL, September 7-14, 1988, pp. 5-159 – 5-170.

Pattell, James M. "Cost Accounting, Process Control, and Product Design: a Case Study of the Hewlett-Packard Personal Office Computer Division." *Accounting Review* 62 (October 1987): pp. 808-840.

Patterson, William Pat. "The Costs of Complexity." *Industry Week* 236.11 (6 June 1988): pp. 63-68.

Patterson, William Pat and Teresko, John. "Managing in the 90s: Brainstorming With a Machine." *Industry Week* 236 (18 April 1988): pp. 61, 67-68.

Pavsidis, Constantine. "Total Quality Control: An Overview of Current Efforts." *Quality Progress* 17 (September 1984): pp. 28-29.

Pearlstein, Steven. "The Corpocracy's Last Gasp." *Inc.*, October 1987, pp. 31-33.

Pearson, Gordon. "Why the Japanese Dominate." *Accountancy*, February 1987, pp. 84-85.

Peirson, R.A. "Adopting Horizontal Material Handling Systems to Flexible Manufacturing Setups." *Industrial Engineering*, March 1984, pp. 62-64.

Pellerin, Cheryl. "Eyes to the Future, Industry Puts AI to Work." *Washington Technology* 3 (5-18 May 1988), p. 1.

Pendleton, William E. "Performance Management: A Key to Improved Productivity and Quality." *American Production and Inventory Control Society, 1987 Conference Proceedings*, St. Louis, MO, October 19-23, 1987, pp. 523-524.

Pentelow, Lindsay. "Can Anything Be Wrong With A Bonus With Tax Benefits?" *Accountancy*, March 1988, pp. 99-100.

Perry, Nancy J. "Saving the Schools: How Business Can Help." *Fortune*, 7 November 1988, pp. 42-49.

Peters, Thomas J. "Facing Up to the Need for a Management Revolution." *California Management Review* 30 (Winter 1988): pp. 7-39.

Peters, Thomas J. "There Are No Excellent Companies." *Fortune* 115 (27 April 1987): pp. 341-345.

Peters, Thomas J. and Waterman, Robert H., Jr. *In Search of Excellence*. New York: Harper and Row, 1982.

Petersen, Mark F. "PM Theory in Japan and China: What's in it for the United States?" *Organizational Dynamics* 16 (Spring 1988): pp. 22-39.

Petroff, John N. "Introduction to Statistical Process Control: On the Way to Zero Defects." *American Production and Inventory Control Society, 1987 Conference Proceedings*, St. Louis, MO, October 19-23, 1987, pp. 1-5.

Phillips, B.J. "1992: Gearing up for the New Europe." *Institutional Investor* 22 (July 1988): pp. 124-130.

Piciacchia, F. Roy. "How to Develop People Integrated Engineering in the Factory of the Future." *American Production and Inventory Control Society, 1987 Conference Proceedings*, St. Louis, MO, October 19-23, 1987, pp. 554-558.

Piczak, Michael W. "Quality Circles Come Home." *Quality Progress* 21 (December 1988): pp. 37-39.

Pierson, Robert A. "Adapting Horizontal Material Handling Systems to Flexible Manufacturing Setups." *Industrial Engineering* 16 (March 1984): pp. 62-64, 66-71.

Pierson, Robert A. "Automated Workpiece Transportation." *NMBTA 4th Biennial International Manufacturing Technology Conference Proceedings*, Chicago, IL, September 7-14, 1988, pp. 6-207 – 6-212.

Pierson, Robert A. "Trends in Material Handling for the Next Decade." *Automation* 35 (March 1988): pp. 10-16.

Plossl, George. "Manage By the Numbers." *American Production and Inventory Control Society, 1987 Conference Proceedings*, St. Louis, MO, October 19-23, 1987, pp. 499-503.

Plossl, Keith R. "Computer Integrated Manufacturing." *Production Engineering*, June 1987, pp. 38-50.

Plossl, Keith R. "The Vital Role of Integrated Engineering in JIT Success." *American Production and Inventory Control Society, 1987 Conference Proceedings*, St. Louis, MO, October 19-23, 1987, pp. 85-89.

Plotkin, A.S. "Labor Called Technology Aid." *Automotive News*, 24 November 1986, p. 12.

Poe, Robert. "American Automobile Makers Bet on CIM to Defend Against Japanese Inroads." *Datamation*, 1 March 1988, pp. 43-51.

Polakoff, Joel. "Inventory Accuracy: Getting Back to Basics." *CPA Journal* 58 (February 1988): pp. 75-78.

Poole, Michael. "Factors Affecting the Development of Employee Financial Participation in Contemporary Britain: Evidence from a National Survey." *British Journal of Industrial Relations (UK)* 26 (March 1988): pp. 21-36.

Port, Otis; King, Resa; and Hampton, William J. "How the New Math of Productivity Adds Up." *Business Week*, 6 June 1988, pp. 103-113.

Porter, Michael E. "Technology and Competitive Advantage." *Journal of Business Strategy*, Winter 1985, pp. 60-79.

Porter, Michael E. *The Competitive Advantage of Nations*. New York: The Free Press, 1990.

Porter, Robert W. "How to Manage Purchase Costs." *American Production and Inventory Control Society, 1987 Conference Proceedings*, St. Louis, MO, October 19-23, 1987, pp. 593-596.

Posner, Bruce G. "Strategic Alliances." *Inc.*, June 1985, pp. 74-80.

Potter-Brotman, Jennifer. "How to Keep Ideas Moving." *Training & Development Journal* 42 (May 1988): pp. 32-35.

Pottorf, Douglass L. "CAM Cuts NC Programming Time by 60%." *Manufacturing Engineering*, December 1987, pp. 63-64.

Potts, Gregory E. "SPC: A Necessary Part of a True JIT Program." Paper presented at the 1987 Total Manufacturing Performance Seminar, APICS, Washington, DC, July 13-15, 1987, pp. 50-51.

Prestowitz, Clyde. "Japanese Versus Western Economics." *Technology Review*, May-June 1988, pp. 27-30.

Pukanic, R. "Flexibility and Integration are Key Material Handling Concepts In Electronic Assembly Environment." *Industrial Engineering* 17 (April 1985): pp. 40-43.

Quick, Finan & Associates. "Contracting for Machine and Tooling: The Hidden Costs of Sourcing Abroad." Report for the National Tooling and Machining Association, Ft. Washington, MA, September 1987.

Quick, Perry. "The True Cost of Foreign Sourcing." *NMBTA 4th Biennial International Manufacturing Technology Conference Proceedings*, Chicago, IL, September 7-14, 1988, pp. 9-245 – 9-267.

Quinn, James Brian; Baruch, Jordan J.; and Paquette, Penny C. "Exploiting the Manufacturing-Services Interface." *Sloan Management Review* 29 (Summer 1988): pp. 45-56.

Quinn, Jane Bryant. "Slow Profits, Slow Pay." *Newsweek*, 22 December 1986, p. 52.

Rahnjat, H. "Simulating for 'Resource Optimization' in Robot Assisted Automatic Assembly." *Proceedings From the Institute of Mechanical Engineers* 200 (1986): pp. 181-186.

Raia, Ernest. "Journey to World Class; For American Industry the Time Has Come to Shape Up or Ship Out." *Purchasing* 103 (24 September 1987): pp. 48-52.

Raia, Ernest. "Leave Out the Baloney When Reporting VA Savings." *Purchasing* 103 (7 May 1987): p. 35.

Raia, Ernest. "The Many Faces of VA at Westinghouse." *Purchasing* 104 (16 June 1988): pp. 74-79.

Raia, Ernest. "VA Takes Off at Textron Lycoming." *Purchasing* 104 (16 June 1988): pp. 90-92.

Raia, Ernest and Dowst, Somerby. "Can VA Save the Empire?" *Purchasing* 103 (11 June 1987): pp. 52-54.

Raju, Venkitaswamy. "Justifying CIM Implications." *NMBTA 1987 Eastern Manufacturing Technology Conference Proceedings*, Springfield, MA, November 10-12, 1987, pp. 4-17—4-21.

Ramano, Patrick L. "Meeting the Technology Challenge." *Management Accounting*, August 1988, p. 64.

Ransom, William J. "The Impact of Customer Service." *NMBTA 4th Biennial International Manufacturing Technology Conference Proceedings*, Chicago, IL, September 7-14, 1988, pp. 1-169—1-186.

Ransom, William J. "Using Audits To Upgrade Your Warehousing." *NMBTA 4th Biennial International Manufacturing Technology Conference Proceedings*, Chicago, IL, September 7-14, 1988, pp. 4-199-4-212.

Rao, Ashok and Scheraga, David. "Moving From Manufacturing Resource Planning to Just-In-Time Manufacturing." *Production and Inventory Management Journal*, First Quarter 1988, pp. 44-49.

Rayner, Bruce. "Can JIT Revive U.S. High-Tech Manufacturing?" *Electronic Business* 12 (1 November 1986): pp. 140-143.

Rayner, Keith. "MRP II-The Financial Manager's Perspective." *Management Accounting*, March 1988, pp. 34-35.

Reeve, Donald C. "Seeing the Light in Manufacturing." *High Technology* 6 (October 1986): pp. 13-15.

Reich, R.B. "The Quiet Path to Technological Preeminence." *Scientific American* 261(4) (October 1989): pp. 41-47.

Reichert, Bruce. "Back to the Future in Manufacturing." *SME/CASA Autofact, 1987 Conference Proceedings*, Dearborn, MI, November 9-12, 1987, pp. 15-37—15-41.

Reid, Peter C. "Well Made in America—Lessons From Harley-Davidson on Being the Best." (McGraw-Hill Publishing, 1990).

Reinerstein, Donald G. "Who Dunit? The Search for the New-product Killers." *Electronic Business* 9(8) (July 1983): pp. 62-64.

Reip, Robert W. "Make the Most of Customer Complaints." *Quality Progress* 21 (March 1988): pp. 24-25.

Rembold, U. and Levi, P. "Factory of the 90s." *Computers in Mechanical Engineering* 6 (March-April 1988): pp. 26-31.

Reuter, Vincent G. "Becoming Competitive with Value Engineering/ Value Analysis." *Journal of Systems Management*, October 1985, pp. 24-31.

Reutersward, Anders. "Educating and Training Tomorrow's Workforce." *OECD Observer*, December-January 1987, pp. 22-25.

Rhodes, David W. "Employees—Strategy Makers or Breakers." *Journal of Business Strategy*, 9 (July-August 1988): pp. 55-58.

Richardson, Peter R. "Adopting a Strategic Approach to Costs." *Canadian Business Review* 13 (Spring 1986): pp. 27-31.

Rizzi, Joseph V. "What Restructuring Has to Offer." *The Journal of Business Strategy*, 8 (Fall 1987): pp. 38-42.

Robertson, James M. "Downsizing to Meet Strategic Objectives." *National Productivity Review*, Autumn 1987, pp. 324-330.

Robinson, James P. "MRP II Financial Justification." *American Production and Inventory Control Society, 1987 Conference Proceedings*, St. Louis, MO, October 19-23, 1987, pp. 622-625.

Roch, Arthur J., Jr. "Flexible Automation Holds Key to Competitive Advantage for Aerospace Manufacturer." *Industrial Engineering*, 18 (November 1986): pp. 52-60.

Rohan, Thomas M. "Blue Collar Skills." *Industry Week*, 6 June 1988, pp. 41-46.

Rohan, Thomas M. "Into the Next Generation." *Industry Week*, 21 March 1988, p. 36.

Rohan, Thomas M. "Justifying Your CIM Investment." *Industry Week*, 9 March 1987, pp. 33-35.

Rohan, Thomas M. "Sophisticated Controls: Small Shops Reach Out." *Industry Week* 235 (19 October 1987): pp. 45-47.

Rolland, William C. "Industrial Automation in the Year 2000." *NMBTA 3rd Biennial International Machine Tool Technical Conference Proceedings*, Chicago, IL, September 3-11, 1986, pp. 1-64—1-98.

Rollins, Thomas. "Pay for Performance: the Pros and the Cons." *Personnel Journal*, June 1987, pp. 104-111.

Rollins, Thomas and Bratkovich, Jerrold R. "Productivity's People Factor." *Personnel Administrator*, February 1988, pp. 50-57.

Romano, Patrick L. "Advanced Cost Management Systems: Part II." *Management Accounting* 69 (March 1988): pp. 61-64.

Romano, Patrick L. "Manufacturing in Transition: The Turning Point for Cost Management Practice—Part II." *Management Accounting*, October 1987, p. 58.

Rose, R. "Caterpillar Sees Gains in Efficiency Imperiled by Strength of Dollar." *Wall Street Journal*, 6 April 1990, pp. A1, A10.

Rosen, Corey and Quarrey, Michael. "How Well is Employee Ownership Working?" *Harvard Business Review*, September-October 1987, pp. 126-129.

Rosenthal, Stephen R. "Progress Toward the Factory of the Future." *Journal of Operations Management* 4 (May 1984): pp. 203-229.

Ross, Marc. "Capital Budgeting Practices of Twelve Large Manufactures." *Financial Management*, Winter 1986, pp. 15-22.

Roth, Aleda V. "Strategic Planning for the Optimal Acquisition of Flexible Manufacturing Systems Technology." Doctoral Dissertation, Ohio State University, School of Business Administration, 1986.

Roth, Aleda V. "Differentiated Manufacturing Strategies for the Competitive Advantage: An Empirical Investigation." Abstract, Boston University School of Management, February 1987.

Roth, Aleda V. "Matching Business Directions and Manufacturing Capabilities." Duke University, Working Paper, 1990.

Roth, Aleda V. "Escaping Traditional Paradigms: Effective Management Requires New Perspectives." Working Paper, School of Management, Boston University, October 1988.

Roth, Aleda V. "A Vision for the 1990s: Gearing up for the 21st Century through Intelligent Manufacturing." Duke University, Working Paper, 1990, p. 30.

Roth, Aleda V. "Boundary Management: Manufacturing's New Imperative." Duke University, Working Paper, 1990.

Roth, Aleda V.; DeMeyer, Arnoud; and Amano, Akio. "International Manufacturing Strategies: A Competitive Analysis." *Managing International Manufacturing*, (K. Ferdows, ed.), Amsterdam: Elsevier Science Publishers, 1989.

Roth, Aleda V.; Gaimon, C.; and Krajewski, L. "Optimal Acquisition of FMS Technology Subject to Technological Progress." *Decision Sciences*, to be published.

Roth, Aleda V. and Giffi, Craig. "Changing the Basis of Competition." Duke University Working Paper, 1990.

Roth, Aleda V. and Miller, Jeffrey G. "Manufacturing Futures Fact Book, The North American Manufacturing Futures Survey." Research Report of the Boston University School of Management Manufacturing Roundtable, Boston University, 1988.

Roth, Aleda V. and Miller, Jeffrey G. "Manufacturing Strategy, Manufacturing Strength, Managerial Success, and Economic Outcomes." in *Manufacturing Strategy* (J.E. Ettlie, M.C. Burstein, and A. Feigenbau, eds.), Boston: Kluwer Academic Publishers, 1990, pp. 92-108.

Roth, Aleda V. and Schneider, H. "Customer-Driven Manufacturing Strategy: A Paradigm Shift." Duke University, Working Paper, 1990.

Rowland, Daniel C. and Greene, Bob. "Incentive Pay: Productivity's Own Reward." *Personnel Journal*, March 1987, pp. 48-57.

Rummel, Patricia. "Human Factors are a Crucial Component of CIM Success." *Industrial Engineering* 20 (April 1988): pp. 36-41.

Russell, Grant W. and Dilts, David M. "Costing the Unknown Product." *CMA Magazine* 60 (March-April 1986): pp. 38-43.

Ryan, John. "Consumers See Little Change in Product Quality." *Quality Progress* 21 (December 1988): pp. 16-20.

Ryan, John. "Labor/Management Participation: The A.O. Smith Experience." *Quality Progress* 21 (April 1988): pp. 36-40.

Sage, Lee A. "Just-In-Time: A Philosophy in Manufacturing Excellence." *APICS 1984 Zero Inventory Philosophy and Practices Seminar Proceedings*, St. Louis, MO, October 29-31, 1984, pp. 3-9.

Samuelson, Robert J. "How Companies Grow Stale: The Peter Principle Applies to Corporations as Well as People." *Newsweek* 108 (8 September 1986): p. 45.

Sanderson, Susan. "Where the Excellence Is." *Across the Board* 24 (September 1987): pp. 24-32.

Sandora, D. "Cutting Tool Management: A Lever on Manufacturing Effectiveness." *Production* 94 (December 1984): pp. 29-33.

Sankar, Y. "Organizational Culture and New Technologies." *Journal of Systems Management* 39 (April 1988): pp. 10-17.

Sarlin, Ronald A. "CIMS Methodology is Applied to CAD-CAM Integration in Factory of Future." CIMS Series, Part 21. *Industrial Engineering* 15 (September 1985): pp. 58-63.

Sasaoka, H. "Automation of Body Assembly Operations." *International Journal of Vehicle Design* 8 (1987): pp. 356-365.

Sata, Toshio. "Development of Computer Integrated Manufacturing in Japan." *NMTBA Proceeding Papers*, pp. 14-2—14-12.

Savage, Dr. Charles M. "Preparing for the Factory of the Future." in *SME Technical Article Series* (1986) from *Modern Machine Shop*, October 1983, pp. 1-17.

Savage, Dr. Charles M. (ed.). "Fifth Generation Management for Fifth Generation Technology: Roundtable Discussion." Society of Manufacturing Engineers, Dearborn, MI, 1987.

Savage, Peter. "CPT Likes CIM." *Chemical Engineering*, 28 March 1988, pp. 20-21.

Schafer, Joachim. "Finding That Competitive Edge for Your Company at an International Trade Fair." *American Salesman* 33 (August 1988): pp. 16-21.

Scheuermann, Larry. "Employee Discipline—Who Would Do That?" *APICS 1984 Zero Inventory Philosophy and Practices Seminar Proceedings*, St. Louis, MO, October 29-31, 1984, pp. 309-317.

Schiff, Jonathan, B. and Schiff, Allen I. "High-Tech Cost Accounting for the F-16." *Management Accounting* 10 (September 1988): pp. 43-47.

Schmenner, Roger W. "The Merit of Making Things Fast." *Sloan Management Review* 30 (Fall 1988): pp. 11-17.

Schneible, Seth. "Workpiece Transportation and Storage in the Automated Factory." *NMBTA 3rd Biennial International Machine Tool Technical Conference Proceedings*, Chicago, IL, September 3-11, 1986, pp. 1-229—1-248.

Schneider, George, Jr. and Mishne, Patricia P. "Keeping Up in an Uphill Struggle." *Manufacturing Engineering*, October 1988, pp. 68-71.

Schneier, Craig Eric; Brown, Amy; and Burchman, Seymour. "Unlocking Employee Potential: Managing Performance (Part 1)." *Management Solutions* 33 (January 1988): pp. 14-20.

Schonberger, Richard J. *Japanese Manufacturing Techniques.* New York: The Free Press, 1982.

Schonberger, Richard J. *World Class Manufacturing Casebook: Implementing JIT and TQC.* New York: The Free Press, 1987.

Schonberger, Richard J. *World Class Manufacturing: The Lessons of Simplicity Applied.* New York: The Free Press, 1986.

Schoor, John E. "High Performance Purchasing." *American Production & Inventory Control Society, 1987 Conference Proceedings*, St. Louis, MO, October 19-23, 1987, pp. 608-610.

Schreiber, Rita R. "Machine Loading Applications at P&W's Factory of the Future." *Robotics Today* 6 (August 1984): pp. 26-27.

Schreiber, Rita R. "Robots 10 in Retrospect." *Robotics Today* (August 1986): pp. 31-33.

Schreiber, Rita R. "Whither Sensors?" *Manufacturing Engineering*, February 1988, pp. 54-58.

Schuster, Michael. "Gain Sharing: Do it Right the First Time." *Sloan Management Review*, Winter 1987, pp. 17-25.

Schwartz, Perry A. "When You're Asked to Cost-Justify Systems." *Computerworld*, 3 August 1987, pp. 46-52.

Schwendinger, James R. "Manufacturing Cost Management in a CIM World." *American Production and Inventory Control Society, 1987 Conference Proceedings*, St. Louis, MO, October 19-23, 1987, pp. 659-662.

Schwind, Gene. "AS/RS: New Needs, New Roles in Manufacturing." *Material Handling Engineering* 40 (November 1985): pp. 62-64.

Schwind, Gene. "Automatic Guided Vehicle Systems: More Choices Today Than Ever." *Material Handling Engineering* 41 (November 1986): pp. 76-82.

Sclack, Mark. "Flex Drives Quality Route to Automotive Success." *Plastics World*, July 1988, pp. 48-53.

Scott, Cynthia D. and Jaffe, Dennis T. "Survive and Thrive in Times of Change." *Training & Development Journal* 42 (April 1988): pp. 25-28.

Scrimgeour, Jack. "Don't Ignore CIM Any Longer." *Canadian Machinery and Metalworking*, January 1988, pp. 34-36.

Scully, John; Olson, Kenneth; Hanson, Robert; and Murrin, Thomas. NCMS Autofact 1988 Plenary, SME/CASA Autofact 1988 Conference, Chicago, IL., October 30-November 2, 1988.

Seither, Mike. "EDI Reaches Critical Mass, Creates Opportunities for Integrators." *Mini-Micro Systems*, July 1988, pp. 17-19.

Senia, Al M. "A Showplace FMS." *Production* 100 (January 1988): pp. 69-71.

Senia, Al M. "Air Force Launches Factory of the Future Program." *Production* 99 (November 1987): p. 10.

Senia, Al M. "Keeping Track of Parts: In and out of Process." *Production* 99 (August 1987): pp. 64-69.

Senia, Al M. "Making Moves for Improved Production Efficiencies." *Production* 99 (October 1987): pp. 48-53.

Sharplin, Arthur D. "Low-Cost Strategies to Improve Worker's Job Security." *Journal of Business Strategy*, pp. 90-93.

Shaum, Loren. "A Flexible Automation Material Handling Concept." *NMBTA 4th Biennial International Manufacturing Technology Conference Proceedings*, Chicago, IL, September 7-14, 1988, pp. 6-193 – 6-201.

Shenkar, Oded. "Robotics: A Challenge for Occupational Psychology." *Journal of Occupational Psychology* 61 (March 1988): pp. 103-112.

Sheridan, John H. "Calling in 'Hired Guns'." *Industry Week*, 4 July 1988, p. 61.

Sheridan, John H. "Toward the CIM Solution." *Industry Week*, 16 October 1989, pp. 35-80.

Shim, Jae K. and Rice, Jeffrey S. "Expert Systems Applications to Managerial Accounting." *Journal of Systems Management* 39 (June 1988): pp. 6-13.

Shingo, Shingeo. *A Revolution in Manufacturing: The SMED System*. Cambridge, MA: Productivity Press, 1985.

Shinohara, Isao. *NPS: New Production System –JIT Crossing Industry Boundaries*. Cambridge, MA: Productivity Press, 1985.

Shute, David C. "Standard Cost Variables: Help or Hindrance to Inventory Management." *American Production and Inventory Control Society, 1987 Conference Proceedings*, St. Louis, MO, October 19-23, 1987, pp. 636-638.

Sibbald, George W. "A Methodology for Industrial Revitalization and CIM/CIE Implementation." *SME/CASA Autofact, 1988 Conference Proceedings*, Chicago, IL, October 30-November 2, 1988, pp. 1-21.

Siegel, S.L. "Simulation of Scheduling Rules Helps Decision-Making on Various Objectives in Manufacturing Plant." *Industrial Engineering* 19 (October 1987): pp. 40-46.

Sierra, Enrique. "How Developing Countries Can Get Ahead." *Quality Progress*, December 1988, pp. 52-55.

Sinetar, Marsha. "The Informal Discussion Group – A Powerful Agent for Change." *Sloan Management Review*, Spring 1988, pp. 61-65.

Skinner, Wickham. "The Focused Factory." *Harvard Business Review*, May-June 1974, pp. 113-121.

Skinner, Wickham. *The Focused Factory: Survival Strategies for American Industry*. New York: John Wiley & Sons, 1983.

Skinner, Wickham. *Manufacturing: The Formidable Competitive Weapon*. New York: John Wiley & Sons, 1985.

Skinner, Wickham. "Manufacturing: Missing Link in Corporate Strategy." In *Survival Strategies for American Industry*, pp. 99-113. New York: John Wiley & Sons, 1983.

Skinner, Wickham. "The Productivity Paradox." *Harvard Business Review*, July-August 1986, pp. 55-59.

Skinner, Wickham. "What Matters to Manufacturing." *Harvard Business Review*, January-February 1988, pp. 10-16.

Slipkowsky John N. "The Volvo Way of Financial Reporting." *Management Accounting*, October 1988, pp. 22-26.

Slutaker, Gary. "They Thought I Was a Madman." *Forbes*, 19 May 1986, pp. 100-102.

Smeltzer, Larry R. and Kedia, Ben L. "Training Needs of Quality Circles." *Personnel* 64 (August 1987): pp. 51-56.

Smith, Donald N. and Heytla, Peter, Jr. *Industrial Robots Forecasts and Trends* Dearborn, MI: SME, 1985.

Smith, George David. "Reinventing the Company." *Business Month*, July-August 1988, pp. 60-64.

Smith, L. Murphy; Thompson, James H.; and Bassett, John M. "Management Stock Option Plans Maximize Corporate Performance." *The CPA Journal*, April 1987, pp. 79-80.

Smith, R.E. "Robotic Vehicles Will Perform Tasks Ranging From Product Retrieval to Sub-Assembly Work in Factory of Future." *Industrial Engineering* 15 (September 1983): pp. 62-72.

Smith, Richard. "CIM Case Study: Ingersoll Milling Machine Company." *APICS 1987 Total Manufacturing Performance Seminar Proceedings*, Washington, DC, July 13-15, 1987, pp. 1-14.

Snyder, Jan. "Counting the Ways CASA Helps." *Computerworld*, 6 June 1988, pp. S14-S15.

Snyder, Kenton R. "Barriers to Factory Automation: What are They, and How Can They be Surmounted?" *Industrial Engineering*, April 1988, pp. 44-51.

Society of Manufacturing Engineers. *Directory of Manufacturing Education Programs in Colleges, Universities and Technical Institutes, 1987-1988 Edition.* Dearborn, MI, 1987

Soroka, Daniel P. "Lasers and Flexible Automation." *NMBTA 3rd Biennial International Machine Tool Technical Conference Proceedings*, Chicago, IL, September 3-11, 1986, pp. 9-28—9-48.

Stalk, George, Jr. "Time—the Next Source of Competitive Advantage." *Harvard Business Review* 66 (July-August 1988): pp. 41-51.

Stalk, George, Jr. and Hout, T.M. *Competing Against Time.* New York: The Free Press, 1990.

Stanislawski, Michael. "CNCS for the Factories of the Future." *NMBTA 4th Biennial International Manufacturing Technology Conference Proceedings*, Chicago, IL, September 7-14, 1988, pp. 5-207—5-218.

Stankard, Martin F. "Paying for Productivity: A Concept Revisited." *Small Business Report*, April 1988, p. 6.

Starr, Stephanie. "Seventh Annual Survey of IIE Members' Views on Productivity Contrasts With JIIE Survey." *Industrial Engineering* 20 (April 1988): pp. 60-64.

Stauffer, J.L. "Product Design for Finish Quality." *Manufacturing Engineering* 101 (September 1988): pp. 75-77.

Stauffer, Robert N. "Converting Customers to Partners at Ingersoll." *Manufacturing Engineering* 101 (September 1988): pp. 41-44.

Stauffer, Robert N. "Robots and Vision: Tying the Technologies Together." *Manufacturing Engineering* 101 (August 1988): pp. 103-105.

Steedle, Lamont F. "Has Productivity Measurement Outgrown Infancy?" *Management Accounting*, August 1988, p. 15.

Stephan, Eric; Mills, Gordon E.; Pace, R. Wayne; and Ralphs, Lenny. "HRD in the Fortune 500." *Training & Development Journal* 42 (January 1988): pp. 26-33.

Stevens, Al. "How to Get Top Management Aboard on the Journey Toward World-Class Manufacturing." *American Production and Inventory Control Society, 1987 Conference Proceedings*, St. Louis, MO, October 19-23, 1987, pp. 196-198.

Stevens, Lawrence. "Users Say EDI Cuts Inventory Costs, Speeds Product Delivery." *Computerworld*, 2 May 1988, pp. 53-57.

Stickler, Michael J. and Buker, David W. "How to Implement SMED." *American Production and Inventory Control Society, 1987 Conference Proceedings*, St. Louis, MO, October 19-23, 1987, pp. 81-84.

Stiles, Edward M. "Engineering the 1990's Quality Function." *SME/CASA Autofact, 1987 Conference Proceedings,* Dearborn, MI, November 9-12, 1987, pp. 7-49—7-59. Detroit, MI, 1987.

Stimson, Richard. "Air Force Takes Initiative in Industrial Modernization Incentive Program for DOD." *Industrial Engineering*, November 1986, pp. 32-38.

Stoll, Henry W. "The Four Cs." From speeches given by Henry Stoll based on his article "4 Cs," *Production Engineering*, 1988.

Stratton, Brad. "Salaries for Quality Personnel Creep Forward." *Quality Progress* 21 (September 1988): pp. 16-29.

Strycula, James. "Creating a Tool Management Network." *NMBTA 3rd Biennial International Machine Tool Technical Conference Proceedings*, Chicago, IL, September 3-11, 1986, pp. 6-60-6-70.

Suby, Carol. "Test Shops: Your Chips or Your Money Please." *Electronic Business*, 15 February 1988, pp. 30, 32-33.

Sullivan, John F. "The Future of Merit Pay Programs." *Compensation and Benefits Review* 20 (May-June 1988): pp. 22-31.

Sullivan, L.P. "Seven Stages in Company-Wide Quality Control." *Quality Progress* 19 (May 1986): pp. 77-83.

Suzaki, Kiyoshi. "Corporate Culture for JIT." *APICS 1984 Zero Inventory Philosophy and Practices Seminar Proceedings*, St. Louis, MO, October 29-31, 1984, pp. 246-255.

Suzaki, Kiyoshi. "Japanese Manufacturing Techniques: Their Importance to U.S. Manufacturers." *Journal of Business Strategy*, Winter 1985, pp. 10-20.

Suzaki, Kiyoshi. *The New Manufacturing Challenge.* New York: The Free Press, 1987.

Swinehart, Kerry; Boulton, William R.; and Blackstone, John H. "Current State of Robotics in Japan: Some Implications." *Production and Inventory Management* 28 (3rd Quarter 1987): pp. 44-49.

Swoyer, H. Samuel. "Just-In-Time Product Costing Employing the OPT Principles." *APICS 1984 Zero Inventory Philosophy and Practices Seminar Proceedings*, St. Louis, MO, October 29-31, 1984, pp. 162-173.

Szekenyl, Robert. "Tips for Improving the Management of Manufacturing Technology." *Industrial Engineering*, November 1987, pp. 18-20.

Taguchi, G. and Clausing, D. "Robust Quality." *Harvard Business Review*, January-February 1990, pp. 65-75.

Tang, Thomas Li-Ping; Tollison, Peggy Smith; and Whiteside, Harold D. "The Effect of Quality Circle Initiation on Motivation to Attend Quality Circle Meetings and on Task Performance." *Personnel Psychology* 40 (Winter 1987): pp. 799-815.

Tappan, David S., Jr. "Project Management of the Future." *Columbia Journal of World Business* 20 (1986): pp. 27-29.

Tatikonda, Lakshimi U. "Production Managers Need a Course in Cost Accounting." *Management Accounting* 68 (June 1987): pp. 26-30.

Taylor, Alex, III. "Why Fords Sell Like Big Macs." *Fortune*, 21 November 1988, pp. 122-129.

Teresko, John. "CIM: Much More Than Adding Computers." *Industry Week* 232 (9 February 1987): pp. 47-52.

Teresko, John. "Making it Simpler." *Industry Week*, 18 April 1988, pp. 67-68.

Teresko, John. "Managing in the 90s: Remaking the Way We Make Things." *Industry Week* 236 (18 April 1988): pp. 59-60.

Teresko, John. "Modernizing Maintenance." *Industry Week*, 29 June 1987, pp. 41-44.

Teresko, John. "Remaking the Way We Make Things." *Industry Week*, 18 April 1988, pp. 59-60.

Teresko, John "Speeding the Product Development Cycle." *Industry Week* 237 (18 July 1988): pp. 40-42.

Teresko, John. "Strategic Manufacturing: Managing the Process Plan." *Industry Week* 236 (7 March 1988): pp. 35-40.

Teresko, John; Goldstein, Mark; and Patterson, William Pat. "The Promise of CAD/CAM/CAE." *Industry Week* 232 (23 March 1987): pp. 48-53.

Teresko, John and Rohan, Thomas M. "Automation and the Bottom Line." *Industry Week* 229 (26 May 1986): pp. 41-94.

Teschler, Leland. "Who is Designing the Factory of the Future?" *Machine Design*, 7 May 1987, p. 4.

Thackray, John. "Restructuring is the Name of the Hurricane." *Euromoney*, February 1987, pp. 106-117.

Therrien, Lois. "The Rival Japan Respects." *Business Week*, 13 November 1989, pp. 108-118.

Therrien, Lois and Finch, Peter. "Mr. Rust Belt." *Business Week*, 17 October 1988, pp. 72-83.

Thomas, Barry and Olson, Madeline Hess. "Gain Sharing: The Design Guarantees Success." *Personnel Journal*, May 1988, pp. 73-79.

Thompson, Donald B. "A Missing Link?" *Industry Week*, 20 April 1987, pp. 20-21.

Thompson, Donald B. "World-Class Ambitions." *Executive Reports* 6 (May 1988): pp. 119-128.

Thorne, Paul. "Pay As An Incentive: The Road to Ruin?" *International Management*, February 1987, p. 58.

Tiersten, Sylvia. "Can MAP Make it on the Factory Floor in 1988?" *Electronic Business* 14 (1 May 1988): pp. 114-118.

Timmerman, Barry. "AGVs for FMSs." *NMBTA 3rd Biennial International Machine Tool Technical Conference Proceedings*, Chicago, IL, September 3-11, 1986, pp. 1-210—1-219.

Tincher, Michael. "Small Lot Production: Fundamental to Higher Quality Levels." *American Production and Inventory Control Society, 1987 Conference Proceedings*, St. Louis, MO, October 19-23, 1987, pp. 10-11.

Titone, Richard C. "Productivity Through People." *American Production & Inventory Control Society, 1987 Conference Proceedings*, St. Louis, MO, October 19-23, 1987, pp. 643-644.

Tombak, Mihkel and De Meyer, Arnoud. "Flexibility and FMS: An Empirical Analysis." *IEEE Transactions on Engineering Management* 35 (May 1988): pp. 101-107.

Tompkins, J.A. and Smith, J.D. "Keys to Developing Material Handling System for Automated Factory are Listed." *Industrial Engineering* 15 (September 1983): pp. 48-50, 52-54.

Tonkin, Lea. "John Deere Harvests the Fruits of VA." *Purchasing*, 16 June 1988, pp. 82-87.

Tonkin, Lea. "Value Analysis: Diversity in Application and Benefits." *Datapro* 2 (October 1988): pp. 6-8.

Touche Ross Logistics Consulting Services. "Implementing Just In Time Logistics." National Survey on Progress, Obstacles and Results, 1988.

Touche Ross & Company. "Manufacturing Technology Challenges: North American Manufacturing Survey." 1987.

Touche Ross & Company. "Manufacturing Technology Trends: North American Manufacturing Survey." 1988.

Touche Ross & Company. *Issues in Competitive Manufacturing: Evaluating Just-in-Time as a Manufacturing Strategy.* Cleveland, OH, 1988.

Treadwell, James. "CIM Case Study: Martin Marietta Energy Systems." *APICS 1987 Total Manufacturing Performance Seminar Proceedings*, Washington, DC, July 13-15, 1987.

Trowbridge, Alexander B. "Avoiding Labor-Management Conflict." *Management Review* 77 (February 1988): pp. 46-49.

Troxler, Joel W. and Blank, Leland T. "Value Analysis for Manufacturing Technology Investments." *NMBTA 4th Biennial International Manufacturing Technology Conference Proceedings*, Chicago, IL, September 7-14, 1988, pp. 24-29.

Tyson, Thomas N. "Quality & Profitability: Have Controllers Made the Connection?" *Management Accounting*, November 1987, pp. 38-43.

Tyson, Thomas N. and Sadhwani, Arjan T. "Bar Codes Speed Factory Floor Reporting." *Management Accounting* 69 (April 1988): pp. 41-46.

U.S. Bureau of Standards. *Survey of Flexible Manufacturing Systems Implementation.* Gaithersberg, MD, 1986.

U.S. Department of Commerce. *A Comprehensive Assessment of the U.S. Manufacturing Automation Equipment Industries.* Washington, DC, July 1984.

U.S. Department of Commerce. *Current Industrial Reports: Manufacturing Technologies: 1988.* 1989.

U.S. Department of Defense. *Bolstering Defense Industrial Competitiveness.* Report to the Secretary of Defense by the Under-Secretary of Defense, July 1988.

U.S. Department of Defense. *Industrial Modernization Incentives Program (IMIP).* Washington, DC, August 1986, Dod 5000 44-G.

U.S. Department of Navy. "Report of Survey Conducted at Litton Systems Inc. Guidance and Control Systems Division." *Best Manufacturing Practices Series,* October 1985.

U.S. Department of Navy. "Report of Survey Conducted at Honeywell Inc. Underseas Systems Division." *Best Manufacturing Practices Series*, January 1986.

U.S. Department of Navy. "Report of Survey Conducted at Texas Instruments Defense Systems and Electronics Group." *Best Manufacturing Practices Series,* May 1986.

U.S. Department of Navy. "Report of Survey Conducted at General Dynamics Pomona Division." *Best Manufacturing Practices Series,* August 1986.

U.S. Department of Navy. "Report of Survey Conducted at Harris Corporation Government Support Systems Division." *Best Manufacturing Series,* September 1986.

U.S. Department of Navy. "Report of Survey Conducted at IBM Corporation Federal Systems Division." *Best Manufacturing Practices Series,* October 1986.

U.S. Department of Navy. "Report of Survey Conducted at Control Data Corporation Government Systems Division." *Best Manufacturing Practices Series,* December 1986.

U.S. Department of Navy. "Report of Survey Conducted at Hughes Aircraft Company Radar Systems Group." *Best Manufacturing Practices Series,* January 1987.

U.S. Department of Navy. "Report of Survey Conducted at ITT Defense Technology Corporation Avionics Division." *Best Manufacturing Practices Series,* September 1987.

U.S. Department of Navy. "Report of Survey Conducted at Rockwell International Collins Defense Communications." *Best Manufacturing Practices Series,* October 1987.

U.S. Department of Navy. "Report of Survey Conducted at Unisys Corporation Computer Systems Division."*Best Manufacturing Practices Series,* November 1987.

U.S. International Trade Administration. *Competitive Assessment of the U.S. International Construction Industry.* Washington, DC: U.S. Government Printing Office, 1984.

U.S. Office of Technology Assessment. "Making Things Better—Competing in Manufacturing." 1989.

Underwood, Michael L. "Productivity in the 1990 Factory: Financial Leadership Put to the Test." *Financial Executive,* March-April 1988, pp. 40-42.

Urbanski, Al. "Incentives Get Specific." *S&MM,* April 1986, pp. 98-102.

Utecht, Robert. "Need Help Justifying Automation?" *Production,* June 1986, pp. 87-88.

Uttal, Bro. "Speeding New Ideas to Market." *Fortune* 115 (2 March 1987): pp. 62-66.

Van Der Hooning, Robert G. "Manufacturing Strategy and the Implementation Problem." *SME/CASA Autofact, 1987 Conference Proceedings,* Dearborn, MI, November 9-12, 1987, pp. 7-85—7-97.

Varney, Glenn. "The Future of American Organizations: An Interview With Marshall Sashkin." *Group & Organization Studies* 12 (June 1987): pp. 125-136.

Vasconcellos e Sa, Jorge A. "The Influence of Technology on Key Success Factors in the Manufacture of Mature Industrial Products." *International Studies of Management & Organization* 17 (Winter 1987): pp. 105-124.

Vasilash, Gary S. "Buried Treasure and Other Benefits of Quality." *Production* 100 (March 1988): pp. 33-39.

Vasilash, Gary S. "Capital Spending Survey: the Reaction." *Production* 100 (January 1988): pp. 50-55.

Vasilash, Gary S. "Hertel Can't Stand Still With CIM." *Production* 100 (May 1988): pp. 76-80.

Vasilash, Gary S. "Honda is World-Class in Ohio." *Production* 100 (July 1988): pp. 64-67.

Vasilash, Gary S. "Playing for High Stakes in High Technology." *Production* 99 (October 1987): pp. 54-59.

Vasilash, Gary S. "Robotics: Appropriate Technology Coming of Age." *Production* 99 (October 1987): pp. 40-47.

Vasilash, Gary S. "Robotics: No One Said it Would Be Easy Part I." *Production* 100 (April 1988): pp. 38-49.

Vasilash, Gary S. "Simultaneous Engineering." *Production* 99 (July 1987): pp. 36-41.

Vasilash, Gary S. "The Common Wisdom and Other Facets of Forging." *Production* 100 (June 1988): pp. 52-61.

Vasilash, Gary S. "The Elements of Modern Machine Tools." *Production* 100 (August 1988): pp. 38-42.

Vasilash, Gary S. "When Do You Measure? and Do You?" *Production* 100 (February 1988): pp. 38-50.

Vasilash, Gary S. "Whither Technology?" *Production* 100 (January 1988): pp. 46-48.

Vassallo, Helen and Oden, Howard. "How Tomorrow's Manufacturing Management Must Change." *NMBTA 4th Biennial International Manufacturing Technology Conference Proceedings*, Chicago, IL, September 7-14, 1988, pp. 3-175 – 3-181.

Veltrop, Bill and Harrington, Karin. "Roadmap to New Organizational Territory." *Training & Development Journal*, June 1988, pp. 22-33.

Venjara, Yusef. "Reducing Set-Up Time on CNC Machines," *NMBTA 4th Biennial International Manufacturing Technology Conference Proceedings*, Chicago, IL, September 7-14, 1988, pp. 4-215 – 4-240.

Verespej, Michael A. "Bluecollar Incentives." *Industry Week*, 4 July 1988, pp. 41-46.

Verespej, Michael A. "Bye-Bye Piecework." *Industry Week*, 20 June 1988, pp. 21-22.

Vester, John and Venckus, John. "Systems Integration and the Material Spine." *Industrial Engineering*, 19 (January 1987): pp. 48-53.

Viola, Thomas J. and Johnson, Terry. "JIT/TQC: Pathways to Competitive Manufacturing." *APICS 1987 Total Manufacturing Performance Seminar Proceedings*, Washington, DC, July 13-15, 1987.

Voelcker, John. "Global Automobile-Lesson Learned: The Experts Speak Out." *IEEE Spectrum* 24 (October 1987): pp. 57-60.

Vogel, Todd. "There's No Word for Chapter 11 in Dutch." *Business Week*, 30 November 1987, pp. 62-66.

Vogeler, Walter J. "JIT: Why it is so Urgent." *American Production and Inventory Control Society, 1987 Conference Proceedings*, St. Louis, MO, October 19-23, 1987, pp. 352-354.

Vogt, Judith F. and Hunt, Bradley D. "What Really Goes Wrong With Participative Work Groups?" *Training & Development Journal*, 42 (May 1988): pp. 96-100.

Vogus, Clinton E. "Critical Success Factors in Factory Automation." *NMBTA 3rd Biennial International Machine Tool Technical Conference Proceedings*, Chicago, IL, September 3-11, 1986, pp. 5-118 – 5-127.

Vonderembase, Mark A. and Wobser, Gregory S. "Steps for Implementing a Flexible Manufacturing System." *Industrial Engineering*, 19 (April 1987): pp. 38-44.

Voss, C.A. and Clutterbuck, D. *Just-in-Time: A Global Report*, Springer-Verlag, 1989, p. 8.

Voss, C.A. and Harrison, A. "JIT in the Corporate Strategy." *SME/CASA Autofact, 1987 Conference Proceedings*, Dearborn, MI, November 9-12, 1987, pp. 58-73.

Wagel, William H. "A Software Link Between Performance Appraisals and Merit Increases." *Personnel*, March 1988, pp. 10-14.

Wagel, William H. "Working (and Managing) Without Supervisors." *Personnel*, 64 (September 1987): pp. 8-12.

Wallace, Thomas F. *MRPII – Making it Happen: The Implementors' Guide to Success With Manufacturing Resource Planning.* Boston: Oliver Wright Limited Productions, 1985.

Wantuck, Kenneth "Changing to JIT Means Changing the Measurements." *American Production and Inventory Control Society, 1987 Conference Proceedings*, St. Louis, MO, October 19-23, 1987, pp. 412-414.

Warndorf, Paul R. and Merchant, Eugene M. "Trends in CIM in the USA." *Production*, February 1986, pp. 64-65.

Warnecke, Hans-Juergen and Gzik, Herbert. "Trends in Robot Development With Special Reference to Welding Technology." *Welding and Cutting*, April 1987, pp. E52-E55.

Wartzman, R. "McDonnell's Shakeup Still Reverberates." *Wall Street Journal*, 7 February 1990, Sec. A, p. 8.

Weaver, Ray L. "Manufacturing Excellence: A Strategy for Competitive Excellence." *American Production and Inventory Control Society, 1987 Conference Proceedings*, St. Louis, MO, October 19-23, 1987, pp. 225-228.

Webb, J.C. "Programmable Flow: Pathways to Glory." *Production Engineering*, 31 (September 1984): pp. 94-98.

Weber, David. "Bigger Bag of Technology Sharpens Machine Vision." *Electronics Week*, 58 (2 September 1985): pp. 41-45.

Wechsler, Dana. "A Comeback in Cubicles." *Forbes*, 21 March 1988, pp. 54-56.

Wehrenberg, Stephen B. "Supervisors as Trainers: The Long-Term Gains of OJT." *Personnel Journal*, 66 (April 1987): pp. 48-51.

Weimer, George. "The Yellow Brick Road?" *Production Engineering*, February 1987, p. 7.

Weimer, George; Knill, Bernie; Teresko, John; and Potter, C. "Integrated Manufacturing IV." *Production Engineering*, 34 (August 1987): pp. IM1-IM15.

Weimer, George; Knill, Bernie; Welter, Therese; and Mills, Robert. "Quality: The Driving Force in Manufacturing." *Automation*, 35 (May 1988): pp. IM1-IM16.

Wells, George N. "Designing the Support for the Factory of Tomorrow." *American Production and Inventory Control Society, 1987 Conference Proceedings*, St. Louis, MO, October 19-23, 1987, pp. 495-498.

Wells, George N. "The Buyer/Planner Partnership." *APICS 1987 Total Manufacturing Performance Seminar Proceedings*, Washington, DC, July 13-15, 1987, pp. 282-283.

Wells, Mary Kay. "Now Technology Means Change, Change Requires Training." *SME/CASA Autofact, 1987 Conference Proceedings*, Dearborn, MI, November 9-12, 1987, pp. 8-41 – 8-46.

Welter, Therese R. "Compensation System Called Unfair." *Industry Week*, 26 January 1987, pp. 17-18.

Welter, Therese R. "The Goal is Working Smarter Not Longer." *Industry Week*, 18 April 1988, pp. 65-66.

Wenzel, Sarah. "Flexible Work Force: A Foundation For JIT." *American Production & Inventory Control Society, 1987 Conference Proceedings*, St. Louis, MO, October 19-23, 1987, pp. 510-513.

Werner, Kenneth I. "Design Automation." *IEEE Spectrum*, 25 (January 1988): pp. 35-37.

Werner, Kenneth I. "Technology '87: Design Automation." *IEEE Spectrum*, 24 (January 1987): pp. 39-41.

Wesberry, James P., Jr. "The United States Constitution and International Accounting Standards." *Government Accountants Journal*, 36 (Winter 1987): pp. 32-38.

Weston, Frederick C., Jr. "Computer Integrated Manufacturing Systems: Fact or Fantasy?" *Business Horizons*, July-August 1988, pp. 64-68.

Wheatman, Victor S. "Just Getting Started." *Software Magazine*, March 1988, pp. 52-58.

Wheeler, William A., III. "Social and Strategic Implications of JIT in the Future." *APICS 1987 Total Manufacturing Performance Seminar Proceedings*, Washington, DC, July 13-15, 1987.

Wheelwright, Steven C. "Manufacturing Strategy: Defining the Missing Link" *Strategic Management Journal*, 5 (1984): pp. 77-91.

Wheelwright, Steven C. "Restoring the Competitive Edge in U.S. Manufacturing." *California Management Review*, 27 (Spring 1985): pp. 26-42.

Wheelwright, Steven C. and Hayes, Robert H. "Competing Through Manufacturing." *Harvard Business Review*, 63 (1985): pp. 99-108.

Whitney, Daniel E. "Manufacturing By Design." *Harvard Business Review*, July-August 1988, pp. 83-91.

Whitney, James L. "Pay Concepts for the 1990's." *Compensation and Benefits Review*, 20 (May-June 1988): pp. 45-50.

Whitney, W.A. "The Fabricator's 1988 Precision Cutting Chart." *Fabricator* 18 (July/August 1988): pp. 10-14.

Whybark, D. Clay. "Strategic Manufacturing Management." Indiana University Working Paper, IRMIS Working Paper #601, February 7, 1986.

Wicker, John. "Computer Integrated Manufacturing Management." *NMBTA 4th Biennial International Manufacturing Technology Conference Proceedings*, Chicago, IL, September 7-14, 1988, pp. 1-147 – 1-154.

Wicker, John. "Dynamic Scheduling." *NMBTA 4th Biennial International Manufacturing Technology Conference Proceedings*, Chicago, IL, September 7-14, 1988, pp. 3-129 – 3-135.

Wilke, David M. "Finite Forward Scheduling – An Absolute Necessity for Successful CIM Implementations." *SME/CASA Autofact, 1987 Conference Proceedings*, Dearborn, MI, November 9-12, 1987, pp. 9-35 – 9-52.

Willard, Gary E. and Savara, Arun M. "Patterns of Entry: Pathways to New Markets." *California Management Review*, 30 (Winter 1988): pp. 57-77.

Willax, Paul A. "Insight Must Precede Action in Expense Control." *American Banker*, 7, July 1988, pp. 4-9.

Williams, V.A. "Spotlight on Machining Centers." *Production*, 93 (May 1984): pp. 33-35, 39-68.

Wilmot, Robb. "Innovation: How to Harness it." *Accountancy*, 100 (August 1987): pp. 106-108.

Wilson, Michael J. "Computer Integrated Flexible Manufacturing." *Production*, December 1985, pp. 48-51.

Windau, Wolf. "New Safety Solutions for FMS and Robotics." *NMBTA 4th Biennial International Manufacturing Technology Conference Proceedings*, Chicago, IL, September 7-14, 1988, pp. 1-291 – 1-297.

Winter, Drew. "Technical Outsourcing – Reach Out and Touch Someone." *Ward's Autoworld*, March 1988, pp. 43-56.

Wireman, Terry. "Integrating Maintenance and CIM." *SME/CASA Autofact 1987 Conference Proceedings*, Dearborn, MI, November 9-12, 1987, pp. 15-1 – 15-5.

Wittet, G.B. "New Controls Can Extend Life of Automated Storage & Retrieval Systems." *Industrial Engineering*, 15 (April 1983): pp. 58-62.

Wolf, Barnet. "U.S. Manufacturers Lagging." *Columbus Dispatch*, 2 November 1988, sec. B, p. 1.

Wood, Chris G. *Achieving Strategic Change: Managing the Transition to a CIM Environment.* New York: Coopers & Lybrand, pp. 63-69.

World Competitiveness Forum, Autofact 1988 Conference, October 30-November 2, 1988, Chicago, IL.

Worley, Gary. "Improving Productivity Through People." *Training & Development Journal*, 42 (June 1988): pp. 39-42.

Worthy, Ford S. "Accounting Bores You? Wake Up." *Fortune*, 12 October 1987, pp. 43-50.

Xerox Corporate Quality Office. *Competitive Benchmarking: What It Is and What It Can Do For You.* Stamford, CT, 1984.

Young, John A. "The Quality Focus at Hewlett-Packard." *Journal of Business Strategy*, pp. 6-9.

Zedeblick, William. "Generative Process Planning Systems." *NMBTA 4th Biennial International Manufacturing Technology Conference Proceedings*, Chicago, IL, September 7-14, 1988, pp. 8-87 – 8-99.

Zemke, Ron and Schaff, Dick. *The Service Edge: Inside 101 Companies That Profit from Customer Care.* New York: NAL, 1989.

Zellner, Wendy. "GM's New Teams Aren't Hitting Any Homers." *Business Week*, 8 August 1988, pp. 46-47.

Zuech, Nello. "Machine Vision: Exploring the State of the Field of Machine Vision." *Robotics Today* 8 (April 1986): pp. 35-37.

Zygmont, Jeffrey. "Flexible Doesn't Always Fly." *Across the Board*, 23 (December 1986): pp. 52-58.

Zylstra, Kirk D. "The Human Aspects of CIM." *SME/CASA Autofact, 1987 Conference Proceedings*, Dearborn, MI, November 9-12, 1987, pp. 10-19 – 10-29.

Index

absenteeism, reduction in 141
accounting systems *See* cost accounting systems
action programs 79; European 105; flexibility 118-120; Japanese 104-106; linking with strategies 113; U.S. 105
activity analysis, use by GE 162-163
activity-based costing 161-167
advanced manufacturing technology 129, 152; implementing 316, 320-323
affinity diagram 55
affordable cost concept 167
after-sales service 65-68, 82, 104
Airbus Industries 21
Aisin Warner, use of Quality Function Deployment 53
Allen, Bob 245
Allen Bradley Company: certification of suppliers 217; use of CIM 308
alliances, strategic *See* strategic alliances
Allied-Signal Inc. 249
allowable cost 168
American Express Company 68
American Productivity and Quality Center 198, 296
American Society for Quality Control 59, 133, 198
American Society for Training and Development 274
American Standard Corporation 244
AMP Incorporated 216
antitrust laws 17
A.O. Smith Corporation, structure of work teams 279
Apple Computer, Inc. 75
Armstrong International 68
arrow diagram 55
Arthur D. Little Center for Product Development 209
artificial intelligence systems 237
asset turnover 175
AT&T Company 63, 101, 254; joint venture with Olivetti 245; pay-for-performance systems 296; reduction in product introduction time 191; reorganization into strategic business units 245
Auld, David 60
automated manufacturing: cells 306; cost accounting 153
automation: justifying 168-173; to reduce costs 117
automobile industry 2; Japanese 136-137; market shares 4, 302; obsolescence of knowledge base in 289; throughput-time reduction in 136-137
awards, quality 58; *See also* Malcolm Baldrige Award for Quality

Baby Boomers 265-266
balance of trade *See* trade balance
Bandit project 191, 323
Bank of America 296
Baxter International 22, 54, 59, 60, 66; cooperative education program 276;
quality programs 196; restructuring of 242
BCI Holdings Corporation 130
Beijing Automotive Works 249, 251
benchmarking 31, 149, 174-180
benefits, employee 15
best manufacturing practices 76, 204; equipment and facilities management 205; focused factories 83, 87; human asset management 204; just-in-time techniques 204; product/process simplification 205; production planning and control 204; quality management 204; Total Product Maintenance 205
bills of materials, strategic 98, 202, 228
Black and Decker Corporation, use of JIT purchasing 217
BMW A.G. 89
Boeing Company 21, 308
bonus programs 294-295
bottlenecks 185, 187
boundary management 82-86, 329
brainstorming 55
breakeven point, lowering the 333
British Petroleum Inc. 243
Brunswick Corporation 101
business unit, strategic *See* strategic business unit
business unit strategy 72

CAD/CAE 207, 208, 212
CAD/CAM 119-122, 170-171

CAM-I 153
Canon Business Machines Inc. 89, 212
capacity of manufacturer 80
capital expenditures, reduction through JIT 154, 215
career paths 15, 290
Caterpillar Inc. 71, 136
cause-and-effect diagrams 54
cellular production *See* manufacturing cells
Center of Advanced Television Studies 254
Chandler, Colby B. 63
channels of distribution, vertical control over 14
Chaparral Steel Company 43
Chase Manhattan Bank 60
Chrysler Corporation 54, 75, 89, 137; alliance with Mitsubishi 249; joint venture with Beijing Automotive Works 249-250; *Quality Is Excellence* program 248; speed in product introductions 191; supplier program 219
Clark, Kim 7, 139, 314, 332
Coca-Cola Company 254
Colgate-Palmolive Company, employment process 271
commitment to quality 9, 22; by European manufacturers 24; by top management 26
commitment to the customer 9
compensation 291, 297; bonus programs 294-295; fringe benefits 15; gains-sharing plans 294; incentives 208, 297-298; individual incentives 294-295; merit pay programs 291-293; pay-for-knowledge programs 295-297; pay-for-performance programs 295-297; planning 297; profit-sharing plans 293; stock purchase plans 293
competitive advantage: in customer service 95, 98; in delivery 115, 122-124; in flexibility 96, 104, 115, 118-120; in price 78, 96, 114, 115-118; in technology 96; in quality 95-96, 114, 120-122; in time-based capabilities 100-104, 136-138
competitive priorities: Japanese 93-94, 104; U.S. 94
computer-integrated manufacturing 236, 306, 308, 324
computerized numerical control 305
computers, on the shop floor 52
Concept to Customer program 191
concurrent engineering *See* simultaneous engineering
conformance quality 10, 3, 99, 117, 132
consortia, R&D *See* R&D consortia
consumer electronics market share 5, 302
continuous improvement 7, 8, 15, 22-23, 70, 160
cooperative education programs 276-277; Boston Compact 275; Chicago 276; Fannie Mae 276
Corning Glass Works 60, 66; employee involvement in quality at 61; quality programs at 60
corporate culture 13, 29; basing on quality 21; changing 22, 61, 259-260
corporate staff, reducing 243
corporate strategy 72
cost: activities that add 161; affordable concept *See* affordable cost concept; allowable 168; centers 151; direct labor 155; improvement 141; information 166; management 157; nonvalue-added 12, 165; of capital 154; of labor 147; of materials 147; of overhead 147, 159; of quality 34, 39, 46, 49; product 147, 160; reduction 147, 234; related to investment justification 168-173; target 12; total of product cycle 157; value-added 12; variable 151-152
cost accounting systems 12, 150; for automated manufacturing 153;
collecting/allocating costs 152; investment in technology 152; Japanese 154-155; using multiple systems 159-161
cost drivers 12, 161, 164, 165; cost of correlation 165; cost of measurement 165; nonfinancial 183; nonvalue-added 160
costing: activity-based *See* activity-based costing; operational 160; velocity 194
Crosby, Philip 14, 23, 34-35, 49, 117, 196
cross-functional teams 40, 43, 47, 77
cross training 277
customer: building loyalty 59, 65; closeness to 10, 141; cycle time 192; feedback 65, 68; satisfaction 4, 23, 54, 59
customer service 68, 96, 98; after-sales 65-68, 82, 104; and JIT 214; and quality 329

customer-company relationships 67-69, 81
customer-driven manufacturing strategy 76-78, 148, 332
cycle, customer 192

Daewoo, alliance with General Motors 249
Dahaitsu 168
Dai Nippon, use of TPM 239
Daimler-Benz AG, alliance with Mitsubishi Corporation 250
data management 16
Deal, Terrence E. 259
Deere & Co., use of technology at 310
defect rates, reduction through JIT 218
Delco Remy, value analysis workshop 234
delivery: competing on 100, 115, 122-124, 189; cycle-time reductions 136; flexibility as part of 120; JIT 217; measuring performance 190-194; on-time 11; trend toward faster 11, 103
delivery time: European action programs 124; Japanese action programs 124; U.S. action programs 122
Deming Application Prize 58
Deming Prize 58, 196, 199
Deming, W. Edwards 23, 33-34, 117
demographics 15, 266; declining birthrate 266; education 269; illiteracy 269; population growth 266, 267
design engineering: administrative process 208; engineering change orders 206, 207; problems of 206-208; use of CAD/CAE 207, 208, 212
design requirements, using QFD to capture 54
Design for Manufacture program 120, 208
Design of Experiments 36, 37-38, 52, 65
Diamond Star Motors, employment practices 270
Digital Equipment Corporation 54, 60, 254
direct labor: as fixed asset 155; cost 155, 161; efficiency 185, 186, 188; hours 152; reducing cost of 312-313; reduction through JIT 215; reporting 186
discrete-product plants 52
Dorn, Robert L. 213
downsizing, corporate 243, 254
downtime, machine 187
Drucker, Peter 152, 331
Duke University, partnership with Baxter 60
DuPont (E.I. DuPont deNemours & Co.) 2, 60, 66, 254

Eastman Kodak Company 22; accelerated learning program 275; Colby B. Chandler 63; entrepreneurial spirit in 63; restructuring of 241-242
Eaton Corporation 246, 248
economic outcomes, absolute 126-127
education 269; continuing 290; cooperative programs 275-277; decline of interest in science and math 269, 287; drop-out rate 269; improving basic 305; in technology 304, 320-321; of engineers 287; of workforce 337; remedial 274-275; required by future jobs 268
education system, overhauling 289-290
efficiency, direct labor 185, 186, 188
electronic data interchange 219, 307
Electronic Data Systems Corp. 109
electronics, consumer *See* consumer electronics
Emhart Corporation, use of simultaneous engineering at 213
employee: demographics 15, 266; involvement 8, 25, 40, 44, 45, 61, 277-284; motivation 278; organizational change and the 256-258; participation 22; reward systems *See* compensation and reward systems; teams 39; training 56, 281; turnover 62
employment process: at Japanese-owned US companies 270; at U.S. plants 271
engineering: change orders 139-140, 206, 207; design *See* design engineering; over-the-wall 205; simultaneous *See* simultaneous engineering
engineers: in Japan 287; in U.S. 286, 287, 290; manufacturing 287, 288
entrepreneurial spirit 63
environmental regulations 193
equipment decisions 80, 186
Ethernet networks 307
European companies: flexibility-related programs 120; growth by acquisition 91;

quality programs 24-25; use of manufacturing cells 310
experiment control 35
expert systems 237
external integration 81
Exxon Corporation, downsizing effort of 243, 254

facilities, decisions about 80
factories: focused *See* focused factories; off-shore 93; service 78; source 93; within factories 13, 246
factory automation *See* flexible manufacturing systems
Federal Express Corporation 43
Feigenbaum, Armand 23, 35
financial: performance 142; reporting systems 161
flexibility 12, 333; achieving in JIT environment 223; competing on 96, 115, 118-120; enhancing production 295; linkages to 118; product 104, 192; volume 104; worker 117
flexible manufacturing systems 25, 104, 306, 307, 323
Florida Power and Light Company, employment process 271-272
focused factories 83-84, 87, 245, 332
focused manufacturing 245-247
Ford, Henry 2, 78, 89
Ford Motor Company 22, 29, 35, 54, 77, 89, 203, 254, 278; alliance with IBM 249; alliance with Mazda 249; *Concept to Customer* program 191; *Quality Is Job 1* program 29, 62, 248; restructuring during 1980s 241; speed in product introductions 191; supplier program 62, 219; training hourly workers 281; Wixom plant 131; zero defects crusade 62
Forum Group, Inc. 62
Foxboro Company 246
fringe benefits 15
Fuji-Xerox 174, 212

gains-sharing plans 294
Gallup Quality Survey 133-136
Gantt, Henry 2
GE Motors 77
General Electric Company (GE) 68, 77, 100-101, 134, 149, 162-163, 166, 249, 309, 339; cooperative education program 275; creation of Total Productive Maintenance 238; quality programs 196; reduction in product cycle 191; use of activity analysis 162-163; Work-Out Program 61; originators of Value Analysis 233
General Foods Corporation, work teams at 279, 281
General Mills, Inc. 43
General Motors Corporation 54, 89, 134, 154, 177, 254, 314, 321, 339; advanced technology at 310; alliance with Daewoo 249; alliance with Isuzu 249; alliance with Suzuki 249; alliance with Toyota 249; Cadillac Motor Car Division 213; employee education program 274-275; employee involvement in quality 61; increasing vertical control by 248; JIT purchasing at 217; pay-for-performance systems by 296; simultaneous engineering at 213; speed in product introductions 191; supplier program 219; *Targets for Excellence* program 219, 248
Gilbreth, Frank 2
Gillette Company 249, 311-312
Globe Metallurgical 198
Goldratt, Eli 184
gross dollars paid per capita (GPD) 2
gross national product (GNP) growth 3
group technology 122, 323-324
growth: acquisition 91; GNP 3; productivity 3;

Hall, Robert 8, 28, 222
Harley-Davidson Inc. 78, 101, 183; elimination of direct labor reporting at 186; use of JIT by 215
Hayes, Robert 6, 7, 139, 314
headcount: reduction in 141; in measuring productivity 185
Hewlett-Packard Company 29, 31, 99-101; incentives for engineers 290; JIT and on-time delivery 220-221;

nontraditional recruiting of workers 272; reduction in product introduction time 191; supplier feedback program 218; *Total Quality Control* program 29; use of JIT purchasing 216
high-performance teams 280
Hitachi, savings through VA/VE 234
H.J. Heinz Company 249, 256
Honda Motor Company, Inc. 4, 89, 101, 148, 175, 212, 248; competition with Yamaha 104; repositioning as automobile manufacturer 89; speed in product introductions 191; threat to U.S. automakers 72
Horicon Works *See* John Deere & Company
horizontal management styles 208
Houghton, James R. 60
Hughes, Kent 1
human assets 14, 337-338
human resource policies 80
Hutchens, Spencer, Jr. 59

Iaccoca, Lee 75
IBM (International Business Machines Corporation) 22, 66, 138, 243; *affordable cost* concept 167; alliance with Ford Motor Co. 249; automated materials handling system 310; customer service 65; education of workers 320-321; outsourcing of components 248
illiteracy, in U.S. adults 269
immigrants, in workforce 268
improvement, continuous *See* continuous improvement
incentives *See* reward systems
indirect expenses, accounting for 150
indirect labor, reduction through JIT 215
Industrial Technology Institute 209
information, activity-based *See* activity-based information
information: and employee involvement 44; capturing real-time 159; cost 50, 166; lack of access to technical 302; management 16, 66; nonfinancial 166; systems 45, 129, 130, 303
Ingersoll International Inc., use of simultaneous engineering 213
Ingersoll-Rand Company, remedial education program 274
innovation 13, 63
integration: backward/forward 90; external/internal 81; vertical 2
International Harvester *See* Navistar International Corporation
International Manufacturing Futures Project (MFP) 89, 94, 99, 103
International Motor Vehicle Program 138
Intertek Services Corporation 59
inventory: low levels in JIT 214; pull system 137; reductions 215; turnover 137, 141; valuation of 150, 160
investment justification 168-173
Isuzu, alliance with General Motors 249
ITT Rayonier Inc., work teams at 280, 281

Jaguar Motor Inc., speed in product introductions 191
Japan: application of Design of Experiments 65; assault on foreign markets 91; competitive priorities 104; cost accounting systems 154-155; engineers 287; focus on motivation 155; growth by acquisition 91; importance of product flexibility to 104; integration strategies 90; inventory turnover rates 137; investment in industry 1; manufacturing strategy 75; market repositioning 89; Ministry of International Trade and Industry (MITI) 18; new product development in 58; plants 139; production methods 137; productivity during 1980s 3; promotion of quality in 58; quality circles 41, 58, 59; quality training programs 56, 58; Quality Control Medal 58, 196; Quality Control Society 196; suggestion programs 283-284; use of statistical methods 57; use of Statistical Process Control 64; use of target cost concept 167-168; use of VA/VE 234
J.D. Powers customer satisfaction ratings 4
JIT *See* Just-In-Time
job classifications, reduction in 13
job enlargement programs 115, 120
Jobs, Steven 75

John Deere & Company: Consumer Products Division 72; Horicon Works 87
Johnson & Johnson 63, 249, 272
Johnsonville Foods 43
joint ventures 249-252
journal vouchering 150
Juran, Joseph 23, 32, 49
Just-in-Time (JIT) 8, 52, 87, 136, 137, 138, 157, 185, 213-228, 334; adoption in U.S. 224; and MRP 230-233; benefits of 215, 221-222; capital expenditures reductions 215; changes in organization required 222, 227-228; deliveries 217; flexibility 222-223; implementation difficulties 216, 226-228; implementation in Japan 226; Kanban production control 223; purchasing 216-219; steady rate of production in 223; suppliers 215, 218, 221-222, 248; visibility in 222; waste elimination through 53, 214-215

KAIZEN tools 54-55; affinity diagram 55; arrow diagram 55; matrix data analysis diagram 55; matrix diagram 55; process decision program chart 55; relationship diagram 55; tree diagram 55
kanban production control 137, 223
Kelly, Donald 130
Kelsey Hayes Co., use of Quality Function Deployment 54
Kennedy, Allan A. 259
Kennedy, John F. 301
Komatsu Limited 71, 136
Komatsu MEC 53
Kraft Foods (Kraft Inc.) 254
KT-Swasey Co. 277

labor: costs 147; hours 152; utilization 185, 186
Lamb, Charles 213
lasers 306
Leadership through Quality program 29
Lee, W. David 209
legislation 17
linkages: between marketing and manufacturing 96; strategies and action programs 113; with flexibility 118
L.L. Bean 174
local area networks (LANs) 309
Lockheed Corporation 309, 320
Loucks, Vernon R., Jr., 242

machine downtime 187
machine tool market: Japanese share of 6, 302; U.S. share of 6, 302
machine utilization, measuring 186-188
macropolitical climates, contrasting 46
macropolitical process in small groups 46
maintenance management 236-239; preventive maintenance 237; Total Productive Maintenance 238-239
Malcolm Baldrige Award for Quality 18, 30, 60, 62, 196-199
management: approach 9, 10, 330-331; boundary *See* boundary management; cost *See* cost management; lack of commitment to quality 70; maintenance *See* maintenance management; manufacturing 315-316; participatory 13, 263, 279-283; performance 143; scientific 2; styles 208; support for JIT 228; technology 303, 316, 320-321, 330
management levels, reduction in 15, 21, 265
manufacturing: American *See* American manufacturing; automation protocol 307; capabilities 9, 11, 333-334; cells 13, 87, 306, 310, 323-324; closed system 82; computer-integrated 236, 306, 324-326; cost improvement 141; customer-driven 77, 148; European *See* European manufacturing; excellence 7; flexible 25, 306, 307, 323; focused *See* focused manufacturing; infrastructure 15-16; intelligence systems 160; investment in 17; jobs 269; management in 315-316; open system 82; performance criteria 141-144, 184-185; process-focused 246; product-focused 247; structure 15; success 126; tax incentives for 17; technology 303-307
Manufacturing Resources Planning (MRPII) 228; and JIT 230-233; implementation problems 229-230

manufacturing strategy 9, 10-11, 23, 73-78, 332-333; action programs 79, 104-106, 113 118-120; American cost reducing 117; as a pattern of choices 78-81; behaviorally defined 114; capacity 80; competing on delivery 115, 122-124; competing on flexibility 115, 118-120; competing on price 78, 114, 115-118; competing on quality 114, 120-122; customer-driven 73-74, 76-77, 83, 110, 332; customer interface 141; development of 74, 107-109; effect on cost improvement 141; European cost reducing 117; external integration 81; facilities 80; global 115-125; holistic 142, 143; human resource policies 80; infrastructure 15, 80-81, 142; integration choices 81, 142; internal environment 141; internal integration 81; inventory turnover and 141; Japanese cost reducing 117; manufacturing strength 129; materials systems 80, 141; organizational structure 81; performance measurement 81, 126, 142, 149; planning cycle of 108; problems in implementing 109-110; process technology and equipment 80; production planning and control 80; quality 80; reducing complexity 141; sales forecast 141; sharing technology 81; strategic intent 111, 332; structural choices 80, 142; supplier quality 141; technology 143; time-based 124; total factor productivity 141; vertical control 247, 248; vertical integration 80, 247, 248

manufacturing-marketing linkages 96

Manuflex Corporation 254

market direction, strategic 87-89

Massachusetts Institute of Technology 138; Commission on Industrial Productivity 322

material: cost 147; flow 141; handling 310; management 247; storage 80; transport 80

Material Requirements Planning (MRP) 204, 228-233

Matsushita Electric Corp. 101

Maytag Corporation 68

Mazda Motors 35, 249

McDonnell Douglas Corporation 21-22, 309

McKinsey & Co. 190, 206

Mead Corporation: focus around product lines 246; Stevenson Mill 263

measuring performance *See* performance measurement

Mercedes-Benz A.G. 90, 191

Merck, Sharp and Dohme Pharmaceuticals 113

merit pay programs 291-293

Microelectronic and Computer Technology Corporation (MCC) 254

middle managers 284

Miles, L.D. 233

Milliken & Co. 22, 60, 198

Ministry of International Trade and Industry (MITI) 18

minorities, in workforce 268

Mitsubishi Motor Corporation: alliance with Chrysler 249; alliance with Daimler-Benz AG 249

Morita, Akio 30

motivation: employee 278; in Japan 155

Motorola, Inc. 22, 29, 30, 31, 101, 138, 149; Bandit project 191, 322; business unit for Japanese market 245; corporate culture 31; daily delivery of orders 31; employee education program 30, 274; Malcolm Baldrige Award for Quality 30, 62, 198; participative management 31; Partnership Growth Advisory Board 61-62; pay-for-performance and knowledge programs 296-297; *Six Sigma* program 29, 30, 62, 196; team bonuses at 296; total cycle time reduction 30; worker training 320

MRP/MRP II *See* Manufacturing Resources Planning

multinational companies, U.S. 91

Nakane, Jinichiro 105

Nashua Corporation 22

National Aeronautics and Space Administration (NASA) 210; 301

National Center for Manufacturing Sciences, Inc. (NCMS) 252, 254, 334

National Cooperative Research Act of 1984 18, 252

National Institute of Standards and Technology (NIST) 196
Navistar International Corporation 101, 242
NEC Electronics, Inc. 212
NeXT Inc. 73, 305, 310
Nippon Electric Corporation, use of quality circles 59
Nippon Kokan Steelworks, savings from quality circles 59
Nippondenso, employment practices 270-271
Nissan Motor Corporation 136-137; single sourcing of parts 217; Smyrna, Tennessee plant 271
Nissan Tochigi auto plant, use of TPM 238
Nonaka, Ikujiro 212
nonfinancial performance measurement 166, 183, 184
nonvalue-added: activities 157, 165; costs 12
North American Manufacturing Futures Project 23, 73, 76, 107
North American Manufacturing Technology Survey 95, 99, 107, 140, 310
Northern Telecom Inc. 73
not-invented-here syndrome 322
numerical control, computerized 305

off-shore production 91-92
Ohno, Taiichi 104, 137, 214-215
Olivetti, joint venture with AT&T 245
on-time delivery 11, 190, 220-221
open systems 82
operating: framework 8-9; margin 174
operational costing 160
O'Reilly, J.F. 256
organization 13; around product lines 13; changes required to by JIT 227-228; changing culture in the 258-259; communication within 66; designing quality into the 69; downsizing the 15, 243; lean 13, 138-139, 332, 334; manufacturing 325; participatory management in 282; reducing complexity of 141; relationship between units of 81; repositioning 89; restructuring 81, 90-91, 130, 204, 243, 334, 336; software supporting the 303
organizational change 255-261; effect on employees 256-258; implementing new technology 258-259
Orlicky, Joseph 228
outsourcing 248
overhead 12; accounting for 150; allocating in Japan 155; cost 147, 159

parallel approach 212
Pareto charting 54, 164
participatory management 13, 40, 263, 279-284; suggestion programs 284-285; work teams 279-284
parts: interchangeable 1; managing 247; reduction in 141
pay, merit *See* merit pay programs
pay-for-knowledge programs 295-297
pay-for-performance programs 295-297
Pepsi Cola (Pepsico, Inc.) 254
perceived quality, superior 132
performance: awards for 81, 149; customer-driven 148; manufacturing 141-144,176, 186-187
performance measurement 9, 12, 81, 156, 160, 202, 335-336; business unit 143, 176; cost drivers 12, 147; nonfinancial 159, 185, 186, 201-202; systems 148-149; throughput 191-194; time-based 12, 189-190;
Perkin-Elmer Corporation, employee involvement in quality 61
Peters, Tom 126, 256-257
Peterson, Donald F. 278, 281
Phillips Industries, Inc., use of VA/VE 234 pilot projects 166
Pioneer International 68
planning process 110
plant within a plant 83
Plant with a Future program 71
plants, manufacturing: discrete-product 52; high-performance 138-139; leanness of Japanese 139
Plastics Recycling Foundation 254
population growth 266, 267
Pratt & Whitney Company Inc. 247, 309
President's Commission on Industrial Competitiveness 307

preventive maintenance 237
price 148; competition 96, 113, 115-118; strategies 167-168
problem solving tools: cause-and-effect diagrams 54; check sheets 55; control charting 55; fishbone techniques 54; graphs 55; histograms 54; KAIZEN 54-55; Pareto charting 54; scatter diagrams 55
problem-solving model 235-236
process communications, upgrading 16
Process Decision Program Chart (PDPC) 55
process focus 246
process simplification 205
Procter & Gamble Company 54, 249, 254, 272
product: closeness to 10; cost 147, 160; cycle 157; design 120, 205, 206; development 58; flexibility 104, 192; focus 13, 242, 246; introduction 118, 191; materials technology 303-304; quality 194-196; simplification 205; specification changes 207
product-market matrix 87-89
product/process design 303, 308
production: capacity 215; control 80, 228; floor 16; methods 137; off-shore 91;
physical processes 303; planning 80, 228; process 52, 141, 308; rate in JIT environment 215, 223
productivity 138, 296; as measure of excellence 185; correlation with quality 139; growth 3; total factor 141, 170-171; worker 115
profit margin 127
Profit Impact of Market Strategy (PIMS) 131, 132
profit-sharing plans 293
profitability, financial measurements of 177
profits, short-term 302
project management, phased approach 210, 211
pull system 137, 223
purchasing, JIT 216-219
push system 228

QFD *See* Quality Function Development
quality: achieving 10, 22, 23, 69, 80, 134, 141, 194; and JIT 214; awards 58, 196;competition 94, 95, 114, 120-122; conformance 10, 23, 117, 132; continuum 26, 27; control 52, 58; correlation with productivity 139; cost of 39, 46, 49, 50; culture 29; customer satisfaction 23, 59, 327; customer service 329; durability factors 134; employee involvement in 61; European action programs 120; flexibility as part of 120; gurus 23, 32; improvement in 117, 147; Japanese action programs 120; measurement systems 12, 50, 62, 185, 195; metrics 200; of kind 23; of process 329; performance factors 134; policy 80; problems 70; programs 29, 39, 129, 196, 197; promoting in Japan 58; reducing variation in 65; revolution 21, 23; strategy 31; toolkits 27; training courses in 57; U.S. action programs 120
quality circles 23, 25, 40, 47; in Japan 41, 58, 59; in U.S. 41, 277; white- vs. blue-collar 41
quality control tools *See* Kaizen tools
Quality Excellence program 248
Quality Function Deployment (QFD) 23, 40, 53, 54, 204, 333
Quality Is Job 1 program 29, 62, 248

R&D consortia 252-256, 260, 304, 336; benefits for members of 253; Center of Advanced Television Studies 254; legal concerns of 255; Microelelectronic and Computer Technology Corporation 254; National Center for Manufacturing Sciences 252, 253; National Cooperative Research Act 252; Plastics Recycling Foundation 254
Rawl, Lawrence G. 243
Ray, Jodie 322
recruitment practices 272-273
regulation, government 193
relational database management systems 309
relationship diagram 55
relative perceived quality 132
reporting, financial *See* financial reporting systems
research, government funding of 304
restructuring 90-91, 129, 130, 209, 243
retraining employees 15

return on assets 127, 177
reward systems 15, 160, 208, 297-298; bonus programs 294-295; compensation 291, 297; gains-sharing plans 294; individual incentives 294-295; individual vs group 297; merit pay programs 291-293; nonfinancial 291; nontraditional 296; pay-for-knowledge programs 295-297; pay-for-performance programs 295-297; profit-sharing plans 293; stock purchase plans 293
robotics industry 6, 306, 307
robust design 35
Rohm and Haas Company, Bayport Plant 271, 272, 279-280, 282
Royal Dutch/Shell 243
Rubbermaid Incorporated 44

Saab-Scania, Inc., speed in product introductions 191
scatter diagrams 55
Schneider, Herb 77, 166-167
Schonberger, Richard 8
school year 269
Sears, Roebuck and Co., cooperative education program 276
semi-conductor market 3, 4, 302
service, after-sales *See* after-sales service
service factory 78
Shainin, Dorian 23, 37, 52
Sheffield Machine Tool Company 254
Sherman Antitrust Act of 1890 252
Shigeru, Mizuno 55
shop floor, computers on the 52
Simmonds Precision Products 170-171
simultaneous engineering 12, 82, 204, 209-213, 333; Japanese approach 212; use of CAD/CAE with 212; use of work teams with 212
Single European Act 18
Six Sigma program 29, 30, 62, 196
Skinner, Wickham 73, 83, 87, 107-109
small-group activities 46
Society of Japanese Value Engineers 234
Society of Manufacturing Engineers 287
Sony Corporation 30
source factories 93
SPC *See* Statistical Process Control
staff, downsizing corporate 243
Stanadyne Inc. 27, 28
statistical methods, Japanese use of 57
Statistical Process Control (SPC) 23, 25, 36, 40, 51-53, 64, 117
Stimpson Industries 217
stock purchase plans 293
strategic: alliances 249; bills of materials 95-98; business unit 114, 143, 174, 244; intent 111, 332; manufacturing planning 10, 11; market direction 87-89
Strategic Planning Institute 132
strategy: corporate *See* corporate strategy; formulation *See* manufacturing strategy formulation
suggestion programs 283-284
superior perceived quality 132
superteams 43-44, 279
supervisors 284-286, 299
suppliers: kanban system for 137; networks 14; quality 141, 218; reduction in 61, 141, 216-219; relationships with 13, 14, 62, 80, 81; vertical control of 82; vertical integration of 14; zero defects program for 62
Suzuki, alliance with General Motors 249

Taguchi, Genichi 23, 35-36, 52
Takeuchi, Hirotaka 212
target cost concept 12, 168
Targets for Excellence program 219, 248
tax incentives 17
tax laws 17
Taylor, Frederick 2
team building 281
team members 282-283
teams 13, 15, 39, 61, 279, 283, 337; career paths of members 282; cross-functional 40, 43, 47, 77; decision making authority of 279; embedding into hierarchical structures 282; facilitators 286; high-performance 280; macropolitical climate of 46; multi-disciplinary 319; performance of 281; problems of 44; self-managed 43, 279; structures for

279-280; super *See* superteams; training members of 281
Technical Assistance Research Programs (TARP) 63
technology 143, 303; advanced manufacturing 129, 316, 320-323; automated manufacturing cells 306; base 302; competing in 16, 96; creating industry wide standards for 304; education in 304; electronic data interchange 307; Ethernet networks 309; impact on organization 258-259, 318, 320; implementation of 15, 89, 258-259, 289, 312, 316, 318, 338-338; justification process 152, 168-173, 318; lack of U.S. investment in 312; lasers 306; leadership in 96; local area networks 309; manufacturing automation protocol 309; manufacturing cells 310; manufacturing *See* manufacturing technology; policy 305-307, 321; product/process design 308; relational database management systems 309; robots 307; transfer 81, 302
television industry 5
Tennant Company 22
Texas Instruments Inc. 195, 254, 334
Textron Inc., use of VA/VE 234
Thomasville Furniture Industries, Inc. 98
3M 43, 63
throughput 184, 192; performance 136, 191-193; time reductions 136, 138
time: as measure of excellence 183; delivery 189; reduction, effects on product cycle 190; study 2
time-based: capabilities 11, 12, 99-103; competition 99-103, 135-138; performance measurement 189-190
Tokyo Juki Kogyo Co., use of Quality Function Deployment 53
top management, leadership of 330
Toshiba International Corporation 84-85
total factor productivity 141, 188-189
Total Productive Maintenance 205, 238-239
Total Quality Control program, Hewlett-Packard 29
Total Quality Control (TQC) system 35
Total Quality Management System (TQMS) 21, 22, 87, 204
Toyota Motors Company 89, 136-137, 226, 339; alliance with General Motors 249; Georgetown Kentucky plant 270, 271; JIT and the 214-215; production system 104; quality circles 58; Quality Function Deployment 53; Total Productive Maintenance (TPM) 238
TQMS *See* Total Quality Management System
trade balance 2
trade policies 17
training: employee *See* employee training; in quality 57; programs 281, 337;
technology 320-321
tree diagram 55
turnover: asset 175; employee *See* employee turnover; inventory 141; work in process 141

U.S. Council on Competitiveness 1
U.S. Department of Commerce 196, 307
U.S. Department of Defense 17
U.S. Department of Education 269
U.S. Navy, Bureau of Ships, Value Engineering program 233
United Airlines Inc., cooperative education program 276
United Technologies Corporation 254
utilization, machine see machine utilization

VA/VE 234-236
valuation, inventory 160
value-added costs 12
value analysis 120, 265, 233
value engineering 205, 233
variable costs 151-152
variance reporting 150
variation research 37
VCR marketplace 5
velocity costing 194
vendors *See* suppliers
vertical control 14, 247

vertical integration 2, 14, 247-248
vertical management styles 208
Volkswagen A.G. 90
volume flexibility 104
Volvo Corporation, speed in product introductions 191

wages 265
Waseda University 105
Welch, Jack 61
Westinghouse Electric Corporation 138; implementation of VA/VE process 235-236; winner of Malcolm Baldrige award 198
Wheelwright, Steven 6, 7, 314
Wight, Oliver 228
work assignments, flexibility in 22
work in process 141
work teams *See* teams
worker: flexibility 117, 227; productivity 115; reward systems *See* reward systems; skills of 295; throughput per, in Japan 137
workforce: decline of entry-level workers 266; demographic changes to U.S. 15, 264, 265-270; displacement by technology 273; education needs of 265, 269, 273; flexibility of in JIT environments 222; improving basic education of 268, 303, 305; pool of obsolete workers 265; recruiting 272; reduction in management ranks 265; training requirements 265, 273; world-class manufacturers 6, 7, 8, 328-340

Xerox Corporation 22, 29, 31, 49, 63, 77, 149, 217, 249; benchmarking at 31, 173, 178, 179; JIT purchasing 216; *Leadership through Quality* program 29; nontraditional recruiting of workers 272; quality programs 196; reduced defect rates 32, 218; reducing number of suppliers 61; winner of Malcolm Baldrige Award 198; worker training 320

Yamaha Motor Corporation, competition with Honda 104
yen, world valuation of 94
Young, John 99

Zenith Electronics Corporation 5
zero defects 25, 29, 62, 120